REVISE CHEMISTRY

Letts Study Aids

A COMPLETE REVISION COURSE FOR GCSE

Bob McDuell BSc (Hons)
Deputy Headmaster, Berry Hill High School, Stoke-on-Trent

Charles Letts & Co Ltd
London, Edinburgh & New York

First published 1979
by Charles Letts & Co Ltd
Diary House, Borough Road, London SE1 1DW

Revised 1981, 1983, 1986, 1987
Reprinted 1988

Illustrations: Ian Foulis and Associates

British Library Cataloguing in Publication Data
McDuell, G. R.
Revise chemistry: a complete revision course for GCSE. – 5th ed. – (Letts study aids)
1. Chemistry
I. Title
540 QD33

ISBN 0 85097 776 2

Printed and bound in Great Britain by
Charles Letts (Scotland) Ltd

PREFACE

Revise Chemistry has been prepared as a complete guide for all students preparing for GCSE examinations. It has been prepared using the latest syllabuses, the information contained in the *National Criteria* for chemistry and *Draft Grade Criteria*. The author's practical experience of being an examiner and assessor enables the help given in this book to be directly useful to the student.

The earlier editions of *Revise Chemistry* have been the leading chemistry study aid for 16+ examinations since 1979. They have been successful with students and accepted by teachers because they are not 'crammers'. They are a complete revision programme which, used methodically and in sufficient time, can only benefit the student.

This present edition has been thoroughly rewritten to cater for the new examination system. In it you will find analysis of all GCSE syllabuses in England and Wales, plus syllabuses in Scotland. New GCSE courses include less factual content but this is still important. The factual core has therefore been reduced. Even in 1979 *Revise Chemistry* was putting emphasis on social, economic and environmental chemistry. These are essential in all GCSE Chemistry courses and are fully reflected in the book. Every unit has its own self test to ensure that the factual core is understood.

There is a section explaining what GCSE is about in simple terms and giving the student some advice about choosing papers to be taken, etc. This section may also be useful to parents wanting to know something about GCSE Chemistry.

All possible types of examination question are analysed with advice on how to get a good mark in your answers. There are a large number of questions including questions from sample papers of examination groups. The answers given to these questions are the author's and are given as sample answers.

Revise Chemistry tackles the important task of ensuring that you get the best mark from your coursework assessment – whether it is practical, project or oral assessment. The section that deals with this is based largely on the author's experiences in training assessors and carrying out the assessment of work from a large number of schools.

The *Foundation Skills Chemistry* books volumes I, II and III, this revision chemistry guide and the chemistry A level course companion ensure that a student is helped from his first days in chemistry right through to entry to university!

In writing this book I am grateful for the help of my panel of advisors, all experienced chemistry teachers well aware of modern trends in teaching and examining. They are Ron Frost, Dennis Garvie in Scotland, Michael James, Michael Martin and Roy Williams. Also I am grateful for the work of the staff of Charles Letts & Co Ltd.

I must take this opportunity also to thank all of the staff in examination boards for the permission to use questions and for the help and advice that they give me.

Finally I must thank my wife Judy and my sons Robin and Timothy for their patience and encouragement and my own pupils for their helpful suggestions.

Bob McDuell 1987

CONTENTS

INTRODUCTION

Knowing how best to prepare for an examination is most important. *Revise Chemistry* will help you by providing a complete revision course. The book contains all of the necessary factual material, self test units, hints on taking examinations and on coursework, sample questions of all types and sample answers. To get the best from this book you should start the following procedure.

Guide to Using this Book

1 Using the table of analysis of examination syllabuses:
Find the column on pages xvi–xvii containing information about the syllabus you are studying. (Remember several examination groups have more than one syllabus and there are often alternatives within a syllabus.) If you are in any doubt which syllabus you are following you should ask your teacher. In the column you will find:

a the number of written examination papers and their duration

b details of teacher assessment (this is the mark awarded by your teacher for the work you have done in school)

c details of other papers and the percentage of marks allocated to them

d the units of knowledge that are required

The core of the book has been divided into 32 units but not all of these units are required for each syllabus. The key to the symbols in the column is:

● unit required for your syllabus (blanks indicate that units are not required).

A, B, C, D only the part of the unit labelled with the appropriate letter is required.

2 You are advised to work through as many of the required units as possible as well as some of the optional units. Your teacher will advise you which options you should study. You may find it useful to consult the syllabus from the examination group as it may give you additional information. You can write to one of the examination boards on page xviii and buy copies of the syllabus and samples of examination questions.

3 After you have worked through a unit you should complete the corresponding self test (pages 122–134) and check your answers. If you find you are getting less than half marks in a self test you should go back and reread the unit. Having worked through the units and self tests you should have a good knowledge and understanding of the chemistry syllabus. This is not enough for success at GCSE. You are now ready to work through the sections which deal with types of examination question and sample examination questions. At the end of this you should be better prepared to take the examination.

You will find the section on coursework (pages 177–196) useful during the course in ensuring that you produce your best work.

You will also find useful sections following this introduction giving hints on learning, taking examinations, and a background to the new GCSE examination, etc. Remember, GCSE is a new examination for all – students, teachers, examiners, employers and parents.

Learning and Remembering

Revision for examinations is most important and effective revision should start as early as possible. Certainly regular revision should start early in the fifth year.

Revision is more than just sitting down and reading through a book. It is important to produce a plan. List all of the topics in the syllabus and make an assessment of the topics that you think you know well and the ones you know less well. Concentrate on mastering the topics that you do not fully understand. Do not leave out big sections of the syllabus because examinations are designed to cover as much of the syllabus as possible.

Learning is best done in fairly small chunks. Half an hour of intensive study followed by a short break and a change of subject is probably best but everybody is different and you have to learn what is best for you. Set yourself definite revision targets – to complete certain sections by a certain date – and make sure you achieve them!

As you work through the sections write down the important points. This is helpful for triggering your memory and, if you keep these notes, for last minute revision.

Learn basic definitions as many questions revolve around knowing these.

Hints on Taking Examinations

In this section there is information that you should find useful both in revising and preparing for your examination and also in taking examinations. As a chief examiner responsible for setting and marking GCSE papers and as an assessor in assessing coursework, all of the time I see candidates not achieving what they should because of basic flaws in the preparation or examination technique. This section is intended to be very practical and to ensure that you do your best both in preparation and in taking the examinations.

1 UNDERSTANDING THE SYLLABUS

Syllabuses are very clear about what should be studied. However, you should understand some of the language of the syllabus. The following terms frequently appear on syllabuses.

'The characteristics of' You are required to describe the main properties/reactions of the substance concerned.

'The identification of' Recognizing the substance by means of simple test tube reactions is required.

'The formation of' Knowledge of the reaction or reactions leading to the making of the substance is required without details of apparatus and/or collection methods.

'The manufacture of' A knowledge of the essential reactions involved in the commercial production of the material is required. Details of the industrial plant are not required.

'The preparation of' Knowledge of the reaction(s) and apparatus necessary for the obtaining, purifying and collection of the material is required.

'The use of' A brief statement of both commercial and domestic applications of the material is required with emphasis being placed on the property of the material that makes it suitable.

Often syllabuses provide useful amplifying notes that tell you, and your teachers, the detail which is required.

2 THE IMPORTANCE OF READING THE QUESTIONS THOROUGHLY AND CAREFULLY

It is frequently stated in reports by examiners that candidates misread questions or fail to use the information given in the question. If the question concerns the industrial preparation of nitric acid, there could be no credit for a candidate whose answer refers to the laboratory preparation of nitric acid. If information (e.g. a table comparing two allotropes, an equation, relative atomic masses, a graph etc.) is given in a question, it must be required to answer the question fully.

The question 'Explain what you would *see* if excess iron filings are added to copper(II) sulphate solution' requires more than a correct equation. You would be expected to mention that copper(II) sulphate solution is blue and when excess iron filings are added the blue solution goes colourless and a brown solid (copper) is deposited. It is obvious that most of the marks are awarded for these observations.

Where there is a choice of questions to be made, read all the questions through carefully before making any choice.

There are a number of instructions that appear on examination papers and candidates are not clear what they mean. Here are some of the common instructions:

'Define' Just a literal statement or definition is required.

'State' Just a short answer is required with little or no supporting argument.

'State and explain' Again a short answer is required but also a little more reasoning is required.

'Describe' This is often used with reference to a particular experiment. The answer should include reference to what is seen during the experiment.

'Outline' The answer must be brief and the main points must be picked out.

'Predict' You are not expected to recall the answer but logically link other pieces of information together. No supporting material is required in your answer.

'Suggest' Either there is no unique answer or you are applying your knowledge to a situation outside the syllabus.

'Find' This is a general term which may mean calculate, measure, determine, etc.

'Calculate' A numerical answer is required. You should show your working in order to get an answer. Do not forget to put in the correct units.

3 SPENDING TOO MUCH TIME ON ONE QUESTION OR ON PART OF THE EXAMINATION

A candidate who completes only half of the questions required can only achieve a maximum of 50 per cent and can have little chance of being successful. It is important to divide your time equally between the questions. It is a good idea to prepare a timetable for the examination before entering the examination room and then to stick to it.

For example, if you are preparing to take a $2\frac{1}{2}$ hour examination paper consisting of 40 fixed-response questions and four longer questions to be chosen from eight, your plan might be:

13.30. Start the fixed response questions. There is no point in reading all the fixed response questions before you start as they are compulsory.

13.55. You should have reached item 20.

14.30. Complete fixed response questions. Note that more time is allowed for later items in the fixed response section. Read the eight longer questions and select the four to be attempted.

14.40. Attempt the question you feel you can do best. This is important as it will increase your confidence and will benefit you if, for any reason, you cannot complete the paper. If you have not finished at 15.00, stop writing and move on to the next question.

15.00. Start the second long question of your choice.

15.20. Start the third long question of your choice.

15.40. Start the fourth long question of your choice.

16.00. Examination finishes.

If you have any time left go back and complete any unfinished question.

It is unwise to abandon a question when you have spent some time on it in order to start another question. You will be wasting time with no certainty that you can do better with the other question.

4 DOING THE WRONG NUMBER OF QUESTIONS

Despite clearly stating the number of questions to be attempted, an examiner sees many papers where candidates have attempted the wrong number of questions.

It is believed by some candidates that if too many questions are attempted the examiner will mark all the questions and credit the candidate with marks from the best answered questions. This is not so and the examiner will award marks for the first questions attempted. It can never benefit the candidate to do more questions than required. If you find you have time to spare that enables you to attempt other questions, your answers may not be sufficiently detailed.

If you have done too many questions, make sure you have crossed out thoroughly any question that you do not want to be marked.

5 LACK OF PLANNING AND POOR PRESENTATION

It is important to plan your answers carefully. This becomes more important as the questions get longer and more involved. A plan takes only a few minutes and it will help you to ensure that you have covered the full extent of the answer required.

Some candidates believe that there is some credit given for long answers. When marking a longer question an examiner is looking for the inclusion of certain facts or statements. Marks are awarded when these are included. Unless these long answers contain the required facts or statements, no marks can be awarded.

You can lose marks for bad presentation and untidy work. Examiners cannot award marks for answers they cannot read.

6 CHEMICAL EQUATIONS

Your answers to questions in chemistry should include equations whenever relevant. If you are in doubt about the relevance of an equation, include it in your answer.

An equation is a useful summary of a chemical reaction. If three marks are awarded for an equation, one mark will be awarded for the correct word equation, one mark for the correct formulae throughout the equation and one mark for correctly balancing the equation. You are advised to include, therefore, both word and symbol equations.

If the question asks for a test for carbon dioxide, an equation (see 18.3) should be included for the reaction between limewater and carbon dioxide.

Check that the symbol equation is balanced before you move on. If it is impossible to balance the equation completely, it suggests that you may have missed one or more of the reactants or products or that you may have written one or more of the formulae incorrectly.

State symbols, e.g. (s), (l), (g) or (aq), are used in this book. They are a useful addition to your equations.

7 CHEMICAL CALCULATIONS

All chemistry papers contain chemical calculations. They are usually not very well attempted by candidates. The figures are chosen to minimize arithmetic and it should not be necessary to resort to mathematical tables, slide rules or calculators.

On many candidates' papers only answers are given and the working is not shown. If the answer is incorrect the examiner cannot award any marks if no working is shown. If the working is given, despite the wrong answer, it may still be possible to award a good mark. Remember most of the marks will be awarded for the essential *chemistry* rather than incidental arithmetic.

Remember to give units, where appropriate, to your answer. Before moving on to the next question, check that your answer is reasonable. I have seen a candidate, after making an arithmetical mistake, write 'because the ratio 41977:28493 is a simple ratio, the law of multiple proportions is verified'. If he had realized that this ratio is not simple he might have looked back and found the mistake.

8 ATTENTION TO DETAIL

One of the distinguishing features between a good candidate and an average candidate is the ability to incorporate details into the answer. As a guide, in any answer, you should include:

a Names of chemicals used and produced. Include states, colours and concentrations of the chemicals.

b Give the conditions of any reaction, i.e. temperature, catalysts, etc.

c Explain why the reaction takes place. Relate your answer, if possible, to the Periodic Table or the reactivity series.

d Write the equations in words and symbols.

9 DRAWING DIAGRAMS

Diagrams should be included whenever they improve your answer. They should help you avoid having to write a long descriptive account of your experiment.

Do not spend a long time doing an artistic diagram. The important feature of your diagrams must be clarity. Draw the diagrams about half as big again as the diagrams in this book. Draw your diagrams freehand with a pencil and have an eraser available in case you make any mistakes. You may use stencils but they restrict your diagrams in size. When the diagram is finished, look carefully to make sure you have not made any obvious mistakes, e.g. in a gas preparation, a thistle funnel must enter the solution in the flask or the gas will escape. Finally, label every piece of apparatus and all chemicals in ink.

The following points are worth remembering:

a Do not waste time drawing stands and clamps. The examiner assumes you will support the apparatus correctly. They also detract from the important features of the diagram.

b Do not waste time drawing Bunsen burners for heating part of your apparatus. Just draw an arrow and label it HEAT.

c Try to draw each piece of apparatus to the correct size in relation to the other pieces of apparatus.

d Draw a round-bottomed flask if heating is required and a flat-bottomed or conical flask if it is not. If a round-bottomed flask is used, it must be drawn above the level of other apparatus to enable heat to be applied (see Fig. 2.3).

e A common mistake in diagrams is to omit corks and bungs.

f If a gas is to be bubbled through a liquid, ensure that the tube goes below the level of the liquid in the wash bottle, and the outlet for the gas is above the level of the liquid (see figure below).

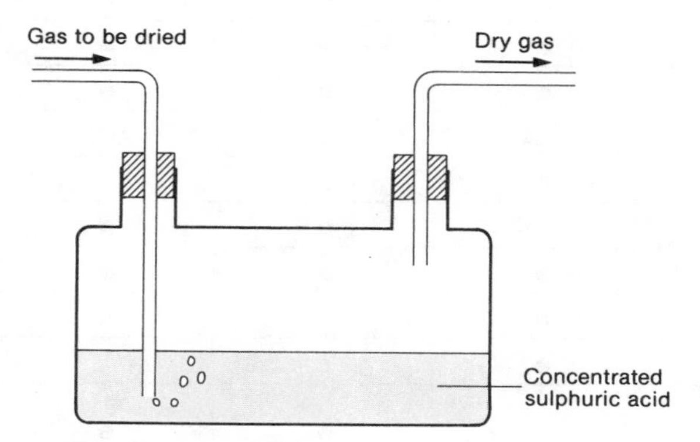

g When collecting a gas in a gas jar using a beehive shelf (e.g. Fig. 5.1), ensure that the level of the water in the trough is above the top of the beehive shelf, that there is water inside the beehive shelf and that there is a hole in the top of the beehive shelf through which the gas can pass.

h When collecting a gas by upward or downward delivery (e.g. Fig. 18.2 or Fig. 25.1), ensure that the delivery tube reaches to the end of the gas jar.

i Having drawn an arrangement to dry a gas, do not collect it over water.

Table of Analysis of Examination Syllabuses

		LONDON AND EAST ANGLIAN GROUP		MIDLAND EXAMINING GROUP		
	Syllabus	A	B (Nuffield)	—	Nuffield	Salters
	Number of papers (Written) [Duration of papers (hours)]	Grades A–D 2 (1+2); Grades C–G 2 (1+2)	Grades A–D 2 (1+2); Grades C–G 2 (1+2)	Grades A–G 3 (1+1+$1\frac{1}{4}$); Grades C–G 2 (1+1)	Grades A–G 2 ($1\frac{1}{2}+1\frac{1}{2}$); Grades C–G 1 ($1\frac{1}{2}$)	Grades A–G 2 ($2+2\frac{1}{4}$); Grades C–G 1 (2)
	Teacher assessment	20%	25%	20%	25% (Practical 15%) (Option Topics 10%)	40% (Practical 25%) (Project 15%)
	Objective questions	● (40%)	● (40%)	● (40%)	●	●
	Free response questions	● (40%)	● (35%)	● (40%)	●	●
1	States of matter	●	●	●	●	●
2	Separation techniques in chemistry	●	●	●	●	●
3	Elements, mixtures and compounds	●	●	●	●	●
4	Chemical equations	●	●	●	●	●
5	Hydrogen	●		●	●	●
6	Metals and nonmetals	●	●	●	●	●
7	Atomic structure and bonding	●	●	●	●	●
8	The reactivity series of metals	●	●	●	●	●
9	Chemical families and the Periodic Table	●	●	●	●	●
10	Oxidation and reduction	A	A	●	●	●
11	Acids, bases and salts	●	●	●	●	●
12	The effect of electricity on chemicals	A	A	A	●	A
13	Rates of reaction	●	●	●	●	●
14	Reversible reactions and equilibrium	A	A	A	●	A
15	Energy changes in chemistry	ABC	AB	A	ABC	ABC
16	Social, economic and environmental considerations	●	●	●	●	●
17	Oxygen and the air	●	●	●	●	●
18	Chalk, limestone and marble	●	●	●	●	●
19	Water	●	●	●	●	●
20	Chemicals from petroleum	●	●	●	●	●
21	Ethanol and ethanoic acid	A	AD	A	ACD	ACD
22	Fuels	●	●	●		●
23	Extraction of metals	●	●	●	●	●
24	Corrosion of metals	●	●	●	●	●
25	Ammonia and nitric acid	A	●	●	A	A
26	Feeding the world	●	●	●		●
27	Sulphur and sulphuric acid	●		●	●	●
28	Salt and chemicals from salt	●	●	●	●	●
29	The mole and chemical calculations	● †	●	●	●	●
30	Quantitative volumetric chemistry					
31	Radioactivity					
32	Qualitative analysis	●		●	●	

† Elementary treatment for Paper 2 (aimed at grades C–G)
More detailed treatment for Paper 3.

	NORTHERN EXAMINATION ASSOCIATION		SOUTHERN EXAMINING GROUP		WELSH JOINT EDUCATION COMMITTEE	SCOTTISH CERTIFICATE OF EDUCATION	NORTHERN IRELAND SCHOOLS EXAMIN-ATIONS COUNCIL
	A	B (Nuffield)	—	Alternative (Nuffield)	—	—	—
	Grades A–G 2 $(2+1\frac{1}{2})$ / Grades C–G 1 (2)	Grades A–G 2 $(2+1\frac{1}{2})$ / Grades C–G 1 (2)	Grades A–G 2 $(2+1\frac{1}{2})$ / Grades C–G 1 (2)	Grades A–G 3 $(1+1+1\frac{1}{4})$ / Grades C–G 2 (1+1)	Grades A–G 2 (2+2) / Grades C–G 1 (2)	Credit Level (Grades 1–3) (2) / General Level (Grades 3–5) (2)	Grades A–G 3 $(1+1\frac{1}{2}+1\frac{1}{2})$ / Grades C–G 2 $(1+1\frac{1}{2})$
	20%	25% (Practical 20%) (Optional Topic 5%)	20%	20%	10%	●	20%
	●	●		●	●		●
	●	●	● (80%)	●	●	●	●
1	●	●	●	●	●	●	●
2	●	●	●	●	●	●	●
3	●	●	●	●	●	●	●
4	●	●	●	●	●	●	●
5	●						●
6	●	●	●	●	●	●	●
7	●	●	●	●	●	●	●
8	●	●	●	●	●	●	●
9	●	●	●	●	●	●	●
10	●	●	●	●	A	●	●
11	●	●	●	●	●	●	●
12	A	A	A	●	A	A	A
13	●	●	●	●	●	●	●
14	A	●	A	●	A	A	A
15	ABC	ABC	AB	●	●	ABC	A
16	●	●	●	●	●	●	●
17	●	●	●	●	●	●	●
18	●	●	●	●	●	●	●
19	●	●	●	●	●	●	●
20	●	●	●	●	●	●	●
21	A	AD	AB	AD	AB	●	AB
22	●	●	●	●	●	●	●
23	●	●	●	●	●	●	●
24	●	●	●	●	●	●	●
25	A	A	●	A	●	●	●
26	●	●	●	●	●	●	●
27	●	●	●	●	●		●
28	●	●	●	●	●		●
29	●	●	●	●	●	●	●
30	●					●	●
31	●	●		●	●		
32					●	●	

Examination Boards Addresses

NORTHERN EXAMINING ASSOCIATION

JMB	Joint Matriculation Board Devas Street, Manchester M15 6EU
ALSEB	Associated Lancashire Schools Examining Board 12 Harter Street, Manchester M1 6HL
NREB	North Regional Examinations Board Wheatfield Road, Westerhope, Newcastle upon Tyne NE5 5JZ
NWREB	North-West Regional Examinations Board Orbit House, Albert Street, Eccles, Manchester M30 0WL
YHREB	Yorkshire and Humberside Regional Examinations Board Harrogate Office – 31–3 Springfield Avenue, Harrogate HG1 2HW Sheffield Office – Scarsdale House, 136 Derbyshire Lane, Sheffield S8 8SE

MIDLANDS EXAMINING GROUP

Cambridge	University of Cambridge Local Examinations Syndicate Syndicate Buildings, 1 Hills Road, Cambridge CB1 2EU
O & C	Oxford and Cambridge Schools Examinations Board 10 Trumpington Street, Cambridge CB2 1QB and Elsfield Way, Oxford OX2 8EP
SUJB	Southern Universities' Joint Board for School Examinations Cotham Road, Bristol BS6 6DD
WMEB	West Midlands Examinations Board Norfolk House, Smallbrook Queensway, Birmingham B5 4NJ
EMREB	East Midlands Regional Examinations Board Robins Wood House, Robins Wood Road, Apsley, Nottingham NG8 3NR

LONDON AND EAST ANGLIAN GROUP

London	University of London Schools Examinations Board Stewart House, 32 Russell Square, London WC1B 5DN
LREB	London Regional Examinations Board Lyon House, 104 Wandsworth High Street, London SW18 4LF
EAEB	East Anglian Examinations Board The Lindens, Lexden Road, Colchester, Essex CO3 3RL

SOUTHERN EXAMINING GROUP

AEB	The Associated Examining Board Stag Hill House, Guildford, Surrey, GU12 5XJ
Oxford	Oxford Delegacy of Local Examinations Ewert Place, Summertown, Oxford OX2 7BZ
SREB	Southern Regional Examinations Board Avondale House, 33 Carlton Crescent, Southampton, SO9 4YL
SEREB	South-East Regional Examinations Board Beloe House, 2–10 Mount Ephraim Road, Tunbridge Wells TN1 1EU
SWEB	South-Western Examinations Board 23–9 Marsh Street, Bristol, BS1 4BP

WALES

WJEC	Welsh Joint Education Committee 245 Western Avenue, Cardiff CF5 2YX

NORTHERN IRELAND

NISEC	Northern Ireland Schools Examinations Council Beechill House, 42 Beechill Road, Belfast BT8 4RS

SCOTLAND

SEB	Scottish Examinations Board Ironmills Road, Dalkeith, Midlothian EH22 1BR

THE GCSE

A great deal has been said and written about the General Certificate of Secondary Education (GCSE) introduced in 1986 for first examinations in England and Wales in June 1988. Similar changes are taking place in Scotland where the Scottish Examinations Board (SEB) are introducing new Standard Grade certificates at three levels – foundation, general and credit. In this section of *Revise Chemistry* there is information about GCSE, and GCSE Chemistry in particular, which is important to you as a student and also to parents and employers.

GCSE is a single examination for all students at 16+. It caters for students of all abilities and it is intended to enable all students to show what they can do.

The examinations are conducted by four examination groups in England, one in Wales and one in Northern Ireland. Each group is made up from a number of independent examination boards (see page xviii). Each group has produced chemistry syllabuses, often more than one, and these have been approved by the Secondary Examinations Council (SEC) who are monitoring GCSE for the Department of Education and Science (DES).

In 1985 the SEC produced *National Criteria* in major subjects including chemistry. Each chemistry syllabus proposed by an examination group must contain the basic chemistry outlined in the *National Criteria* and this must make up 60 per cent of the content of the syllabus. Additional material, either extending the depth of treatment or introducing new topics, makes up the other 40 per cent.

The results of conforming to the *National Criteria* include:

1 the reduction in the factual content of syllabuses;

2 making syllabuses throughout the country more similar;

3 ensuring that all chemistry students have a basic core of knowledge.

Students of all abilities are catered for by differentiation. In chemistry this usually means that alternative written papers are available and a candidate is entered for the combination of papers which enables him/her to show positively what they can achieve in chemistry.

In addition to written papers, candidates have to complete coursework which must include an assessment of practical skills and possibly project work. Candidates have to show an appreciation of social, economic, industrial and environmental aspects of chemistry.

In GCSE grades A, B, C, D, E, F and G are awarded. Grades A, B and C roughly coincide with GCE grades A, B and C. Grades D, E, F and G will still show positive achievement and will roughly coincide with CSE grades 2–5. Allocation of grades to candidates will not purely be a statistical exercise but will depend on what the candidate can do.

The new GCSE resembles in some ways the car driving test of which we have been familiar for many years. It is a *criterion-referenced* system. In order to pass a driving test a candidate must possess knowledge and understanding to answer correctly the examiner's questions about driving and the highway code. In addition he/she must satisfy the examiner that they possess certain basic skills, e.g. reversing, turning in the road, etc. They are not competing with other candidates directly but trying to achieve certain basic standards which mean they are fit to drive on the roads alone.

The GCSE is very similar. The candidate has to show the examiner that they possess certain clearly-defined skills before an award can be made. GCSE differs from a driving test, however, because a driving test is just a pass/fail test but GCSE recognizes different levels of achievement and rewards them by the awarding of different grades.

It is hoped that this new system will be more appropriate and will give students a better introduction to Chemistry in the modern world.

Changes in Chemistry with GCSE

All GCSE Chemistry courses contain less factual material to be learnt. The content is considered to be relevant to the needs of young people entering a technological age. Emphasis in all syllabuses must be given to social, economic, environmental and industrial aspects of chemistry. Topics such as pollution and acid rain, use of fertilizers, recycling of water products and renewable energy sources become vital parts of a GCSE course.

Chemistry for the purposes of assessment has been divided into three areas or 'domains'.

These are:

knowledge and understanding
handling information and solving problems
experimental skills and investigations

The 'knowledge and understanding' domain is the domain that is probably easiest to understand. This has been the basis of most chemistry syllabuses up to now. As you work through the Units 1–32 (pages 1–121) and the self tests on these units (pages 122–134) you will learn sufficient chemistry to be able to satisfy the examiners in the knowledge and understanding domain.

The 'handling information and solving problems' domain is a most important part of GCSE Chemistry courses. It is important that you are able to find information that you do not know and are able to order and display data, detect patterns, interpret chemical behaviour and solve scientific problems. This domain includes chemical calculations. Your performance in this domain will improve as you work through *Revise Chemistry* especially in the sections on sample examination questions (pages 147–176) and practical coursework questions (pages 182–193).

The 'experimental skills and investigations' domain emphasizes the importance of practical enquiry in chemistry. It covers planning, conduct and interpretation of investigations and requires competence in a range of practical skills and techniques. The section of *Revise Chemistry* on coursework (pages 177–182) will encourage the development of your work in this domain.

All three are equally important. The levels of achievement expected of grade A, C and F candidates in each of the domains has been established. As a candidate it is up to you to reach the highest standards you can and you will be given the grade that your performance deserves, irrespective of the performance of others.

All GCSE Chemistry syllabuses require coursework. This may consist of practical work, project work or oral work or, more likely, a combination of two of them or all three. Your coursework will be assessed in the first instance by your teacher. The overall standards will be established by an assessor working for the examination group. The section on pages 177–182 is intended to help you with this vital area of coursework.

Choosing the Right Combination of Papers

One of the basic ideas behind GCSE is differentiation. This means that each candidate should sit papers that they can do and can show 'positive achievement'.

In practice this means either:

1 All candidates sit common papers which can give grades up to grade C. An additional extra and harder paper is then available which candidates can choose to do. Success on this paper can lead to grades A and B. Or:

2 All candidates sit a common paper and then a choice is made either to sit a paper aimed at grades C–F or a paper aimed at grades A–C.

First evidence that comes to us from trial examinations in 1986 is that too many candidates opt to do the additional paper or the harder alternative. The additional paper or the harder alternative should only be taken when *success at Grade C level is pretty certain*. Otherwise the easier alternative may be better and an award of a grade C is acceptable for all future needs. As an examiner it is far more pleasing to see candidates achieving 90 per cent + on the easier alternative than less than 10 per cent on the difficult alternative.

The introduction of GCSE is an exciting change in assessment at 16+. The chemistry course that you are following leads to a qualification which will be valuable to you in later life.

1 STATES OF MATTER

1.1 States of Matter

There are three states of matter – **solid**, **liquid** and **gas**. Any substance can exist in each of these three states depending on the conditions of temperature and pressure. Figure 1.1 shows the relationship between these states of matter.

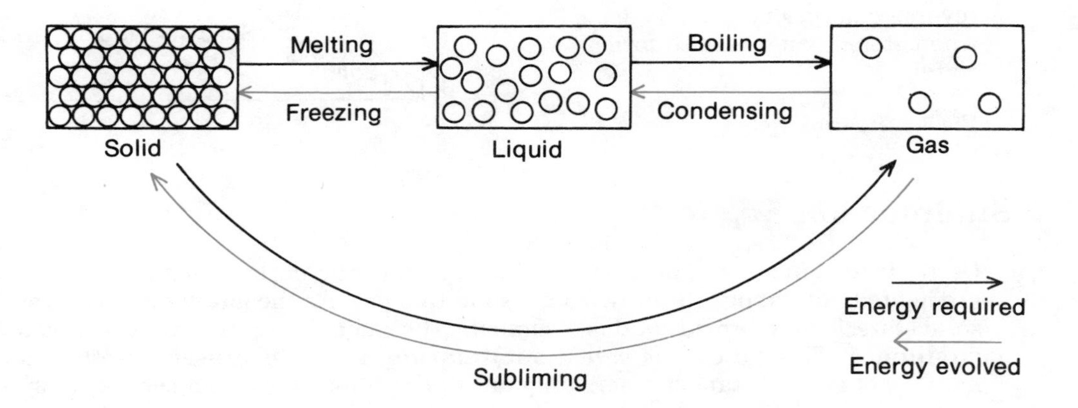

Fig 1.1 States of matter and their interconversion

When a solid is heated it **melts** and forms a liquid. The temperature at which both solid and liquid can exist is called the **melting point** (or freezing point). When a liquid is heated to its **boiling point**, it **boils** and forms a gas (or vapour).

A simple two-dimensional representation of the particles in a solid, liquid and gas is shown in Fig. 1.1.

In the solid state the particles are usually **regularly** arranged and **rigidly** held in position. The particles can only vibrate. The particles vibrate more as the temperature is increased.

In the liquid state the particles are able to move much more than in the solid. Liquids are usually less dense than their corresponding solids because the particles are more widely spaced. Ice is, however, exceptional because it is less dense than water at 0 °C and therefore floats on water. When a solute is dissolved in a liquid, the solute particles fill the spaces between the particles of the liquid. In a liquid there are still forces holding the particles together.

Table 1.1 Comparison of solids, liquids and gases

Property	*Solid*	*Liquid*	*Gas*
Volume	Definite	Definite	Variable – expands or contracts to fill container
Shape	Definite	Takes up shape of bottom of container	Takes up the shape of the whole container
Density	High	Medium	Low
Expansion when heated	Low	Medium	High
Effect of applied pressure	Very slight	Slight decrease in volume	Large decrease in volume
Movement of particles	Very slow	Medium	Fast

In the ideal gas state the particles are completely independent and are moving randomly in all directions. As temperature increases the particles move faster and therefore collide more often. Gases are very compressible because of the large spaces between particles and this also causes the density to be low.

The properties of solids, liquids and gases are summarized in Table 1.1.

Water can exist in these three states – ice (solid), water (liquid) and steam (gas). Figure 1.2 shows the graph obtained when a sample of ice is heated with a steady source of energy. When a change of state is taking place, e.g. solid → liquid or liquid → gas, the temperature remains constant despite a continuing supply of energy. This energy, which is not being used to raise the temperature, is called **latent heat**. Latent heat is used to supply the particles with the extra energy they require as the state changes. It is evolved when the reverse changes take place, e.g. when steam condenses to form water.

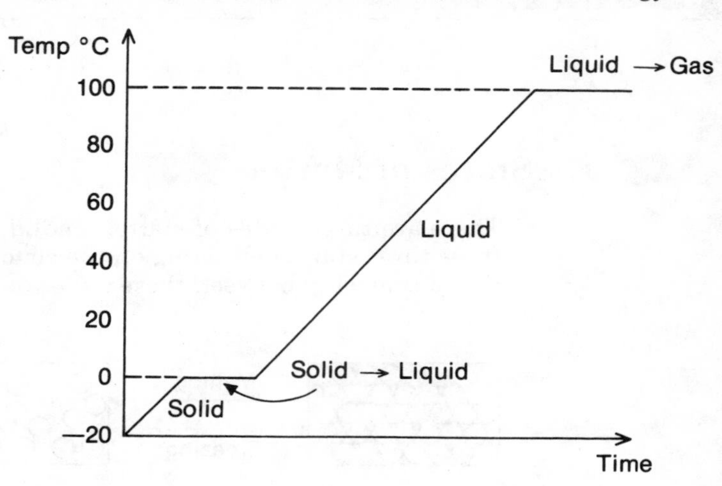

Fig 1.2 Heating water

1.2 Sublimation

Certain substances (e.g. ammonium chloride and solid carbon dioxide) do not melt when they are heated but change directly from a solid to a gas. As the gas cools it returns directly to the solid state. The reformed solid is chemically the same as the substance heated and is called the **sublimate**. This process is called **sublimation**, and it is a useful method for separating a mixture of two substances where only one of the substances sublimes, e.g. ammonium chloride and sodium chloride. Iodine is frequently quoted as an example of a substance that sublimes. However, when iodine crystals are heated they are usually seen to melt.

1.3 Kinetic Theory

The fact that all particles in a solid, liquid or gas are in a state of constant motion is called the kinetic theory. It explains two fundamental concepts – **diffusion** and **Brownian motion**.

1.4 Diffusion

If a drop of liquid bromine is dropped into a gas jar containing air and the gas jar is covered, the liquid bromine vaporizes and, after a while, the bromine vapour has spread evenly throughout the gas jar. The bromine particles and the constituent particles of the air have mixed thoroughly together. This movement of particles to spread out to fill the whole container is called **diffusion**.

Diffusion also takes place in liquids and solids but it is much slower because the particles are moving more slowly in liquids and solids. If a crystal of purple potassium permanganate is dropped into water, the colour spreads throughout the water after about a week.

If a piece of cotton wool soaked in concentrated hydrochloric acid (giving off hydrogen chloride fumes) and a piece of cotton wool soaked in concentrated ammonia solution (giving off ammonia fumes) are put in the opposite ends of a dry 100 cm long glass tube, a ring forms after about five minutes as shown in Fig. 1.3.

$$NH_3(g) + HCl(g) \rightleftharpoons NH_4Cl(s)$$

ammonia + hydrogen chloride ⇌ ammonium chloride

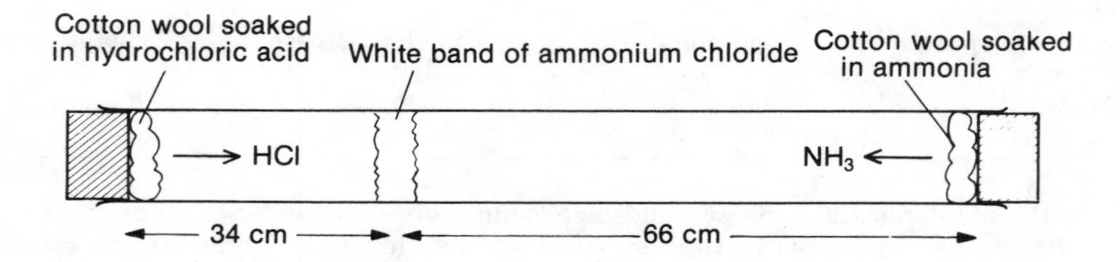

Fig 1.3 Diffusion of ammonia and hydrogen chloride

The ring does not form immediately because:

1 the particles are not moving just in one direction.
2 the tube is filled with air.

The ammonia particles are moving about twice as fast as the hydrogen chloride particles and so the ring of ammonium chloride formed when the gases meet is nearer the piece of cotton wool soaked in concentrated hydrochloric acid.
Generally, smaller (or lighter) particles move faster than larger (or heavier) particles.

1.5 Brownian Motion

In 1827, **Robert Brown** observed that fine pollen grains on the surface of water were not stationary but were in a state of constant random motion. The movement of these pollen grains is caused by collisions between water particles (these are too small to be seen) and pollen grains. Three facts are worth noting:

1 The water particles must be moving rapidly in order to move the larger pollen grains.
2 The direction of movement of a pollen grain is determined by the frequency and direction of these collisions.
3 There is no pattern in the movement of pollen grains or water particles.

1.6 Summary

All substances are made up of tiny particles which are constantly moving.
All substances can exist in three states of matter depending upon temperature and pressure. These states of matter are solid, liquid and gas. In the solid state the particles are close together and can only vibrate. When energy is given to melt the solid and turn it to a liquid, the particles move more. Further energy will make the liquid boil and form a gas. In the gas the particles are much more widely spaced and are moving much faster.
Diffusion is evidence for the movement of particles: a few drops of perfume can soon be smelt throughout a room.

2 SEPARATION TECHNIQUES IN CHEMISTRY

A chemist is frequently involved in separating mixtures or purifying substances. There are a number of methods available as outlined below.

2.1 Separating Insoluble Impurities from a Soluble Substance

E.g. removing sand and other impurities from salt solution

When crushed rock salt is added to water, the salt **dissolves** and forms a salt **solution**. The sand and other impurities remain undissolved, and this material can be removed by **filtration** (Fig. 2.1).

The salt solution passing through the filter paper contains no solid impurities. In order to obtain a sample of pure salt, the salt solution is **evaporated** (Fig. 2.2).

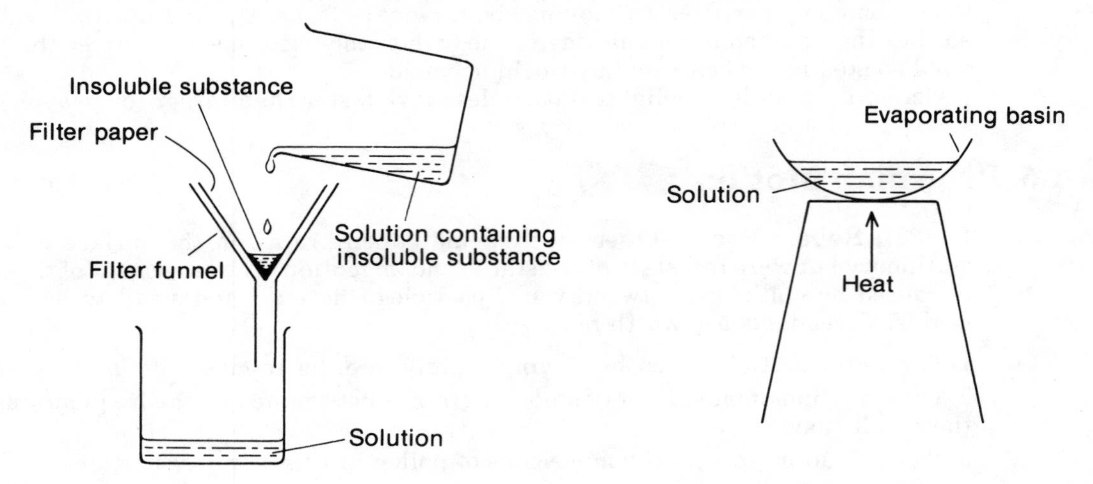

Fig 2.1 Filtration

Fig 2.2 Evaporation

Undissolved material can also be removed by **centrifuging**. This is used when small quantities of material are used, e.g. blood samples.

2.2 Separating a Liquid from a Solution of a Solid in a Liquid

E.g. producing pure water (distilled water) from sea water

This process is called **distillation.** When the flask is heated the solution boils and steam passes into the condenser. In the condenser, the steam is cooled by cold water passing through the outer condenser tube. The steam **condenses** and the **distillate** (distilled water) collects in the receiver. The impurities are left in the flask (Fig. 2.3).

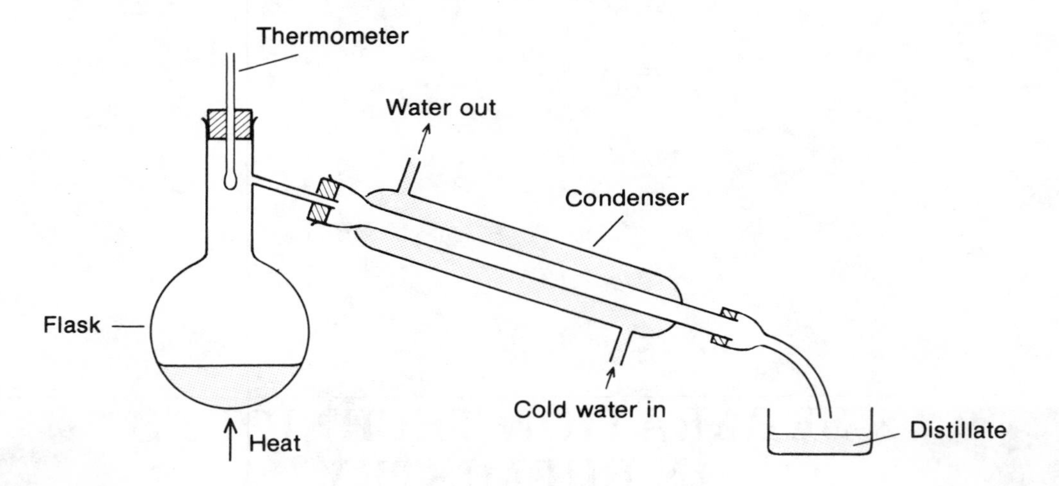

Fig 2.3 Distillation

2.3 Separating a Liquid from a Mixture of Miscible Liquids

E.g. removing ethanol from a mixture of ethanol and water

(Miscible liquids are liquids that mix together completely to form a single layer.)

This process is called **fractional distillation** (Fig. 2.4). It relies on the difference in boiling points of the two liquids (e.g. water 100 °C, ethanol 78 °C).

When the flask is heated the ethanol boils more readily than the water. The water (with the higher boiling point) condenses in the fractionating column and drips back into the flask. The ethanol distils over first. When all the ethanol has distilled over, the temperature (recorded on the thermometer) rises, and water distils over and is collected in a different receiver.

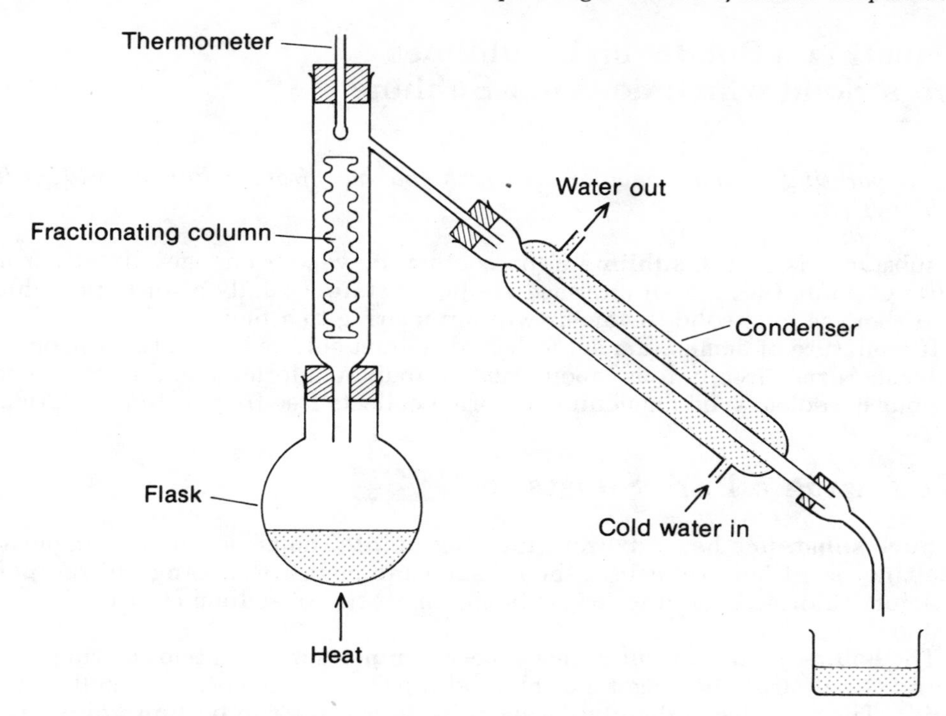

Fig 2.4 Fractional distillation

2.4 Separating a Mixture of Immiscible Liquids

E.g. separating a mixture of water and hexane

Water and hexane are **immiscible** and form two separate layers. The hexane layer forms above the water (or aqueous) layer because water is denser than hexane. These two liquids could be separated using a tap funnel.

2.5 Separating a Mixture of Similar Compounds in Solution

E.g. separating the dyes present in a sample of ink

This process is called **chromatography**. It is a very sensitive method that can be used to separate similar compounds in solution.

If a spot of dye solution is put on to a filter paper and the spot enlarged by slowly dropping solvent on to the centre of the spot, the different components of the dye spread out at different rates. Each component forms a definite ring on the filter paper.

Chromatography experiments are often carried out using square sheets of filter paper. Spots of dye solutions are put along the base line of a sheet of filter paper. The filter paper is coiled into a cylinder and the cylinder is put into a tank containing a small volume of solvent. The lid is put on the tank and the solvent slowly rises up the filter paper. When the solvent has nearly reached the top of the filter paper, the cylinder is removed and the position that the solvent has reached is marked.

In Fig. 2.5, dyes A and B are either pure substances or a mixture of dyes not separated with the solvent used. Dye C is composed of a mixture of A and B, because the original spot has separated into two spots corresponding to A and B.

Chromatography was originally devised to separate coloured substances in solution. It can, however, be used to separate colourless substances in solution, which can then be seen by spraying or dipping the filter paper into a suitable chemical (called a locating agent), which colours the spots produced.

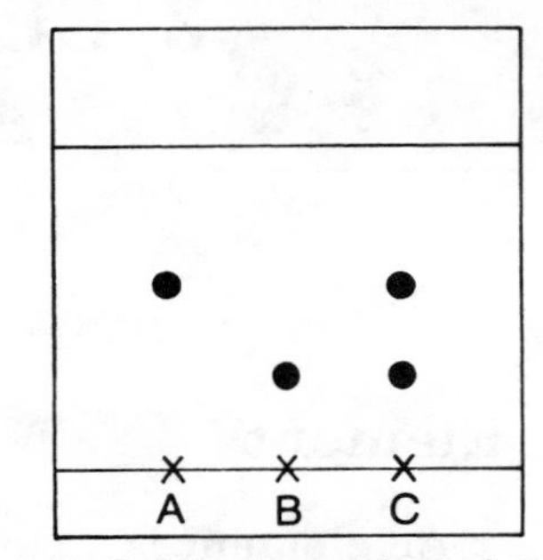

Fig 2.5 A chromatogram

2.6 Separating a Solid which Sublimes from a Solid which does not Sublime

E.g. separating ammonium chloride (which sublimes) from sodium chloride (which does not sublime)

A substance is said to **sublime** if, on cooling, its vapour changes directly from gas to solid without going through an intermediate liquid state. Usually a substance which sublimes also changes from solid to vapour without melting to a liquid.

If a mixture of ammonium chloride and sodium chloride is heated, the ammonium chloride turns directly to a vapour but the sodium chloride remains unchanged. When the vapour is cooled, solid ammonium chloride collects free from sodium chloride.

2.7 Recognizing a Pure Substance

A **pure substance** has a **definite melting point**. The presence of an impurity **lowers the melting point** but also causes the substance to melt over a **range of temperature**. Calcium chloride is used to lower the melting point of sodium chloride in the extraction of sodium.

The boiling point of a substance depends on pressure. At atmospheric pressure, the boiling of a pure substance takes place at a particular temperature called the **boiling point**. The presence of dissolved impurities **increases the boiling point** slightly.

2.8 Summary

A pure substance contains no impurities and melts at a definite melting point. An impure substance melts at a lower temperature and over a range of temperature.

An insoluble substance, e.g. sand, can be removed from a solution by filtering. Centrifuging is an alternative method when working on a small scale.

Evaporation can be used to recover dissolved salt from a salt solution. The salt solution would pass through a filter paper without change.

Distillation can be used to recover a solvent from a solution, e.g. getting water from a salt solution. Distillation involves boiling followed by condensation. Distilled water is produced when a solution is distilled where water is the solvent.

Fractional distillation can be used to separate mixtures of liquids with different boiling points. Fractional distillation is used when the liquids mix completely, i.e. they are miscible. If the liquids are immiscible, they can be separated using a tap funnel.

Chromatography is used to separate a mixture of compounds which is dissolved in a solvent. It is often used to separate mixtures of coloured compounds.

Sublimation is used to separate a mixture where one of the substances sublimes. Ammonium chloride, which sublimes, is often one of the substances in the mixture.

3 ELEMENTS, MIXTURES AND COMPOUNDS

3.1 Elements

An **element** is a pure substance which cannot be split up by chemical reaction.

There are 105 chemical elements; most of these occur naturally but a few are man-made. Most of the elements are *solid* and *metallic*. However, bromine and mercury are *liquids* at room temperature and pressure, while hydrogen, helium, nitrogen, oxygen, fluorine, neon, chlorine, argon, krypton, xenon and radon are *gases* at room temperature and pressure.

Elements that have been mixed together can still be separated. For example, a mixture of iron and sulphur can be separated with a magnet.

3.2 Symbols

Each element is represented by a symbol. This is either one or two letters and is a shorthand that all chemists understand:

e.g.	Element	Symbol	
	hydrogen	H	the first letter of the name
	carbon	C	
	calcium	Ca	the first two letters of the name
	helium	He	
	chlorine	Cl	the first letter and one other letter in the name
	magnesium	Mg	
	iron	Fe	two letters not coming from the English name
	sodium	Na	

(N.B. The first letter is a capital letter and the second letter is a small letter.) A full list of the elements and their symbols will be found in Fig. 9.2 on page 28.

3.3 Compounds

Certain mixtures of elements react together or combine (usually when heated) to form **compounds**. The formation of a compound from its constituent elements is called **synthesis**, and energy is usually liberated or released during this process. The compounds thus formed have very different properties from the elements of which they are composed. Splitting up a compound into its constituent elements is not an easy process. The proportions of the different elements in a compound are fixed.

3.4 Examples of Compound Formation

1 When a mixture of powdered iron and sulphur is heated, a reaction takes place with the evolution of energy forming iron(II) sulphide.

$$Fe(s) + S(s) \rightarrow FeS(s)$$

iron + sulphur → iron(II) sulphide

2 When a mixture of the gaseous elements hydrogen and oxygen is exploded, a reaction takes place forming water (a liquid). The properties of water are different from the properties of hydrogen and oxygen.

$$2H_2(g) + O_2(g) \rightarrow 2H_2O(l)$$

hydrogen + oxygen → water

3.5 Naming compounds

1 Compounds ending in **-ide** contain two elements.

E.g. copper(II) oxide is a compound of copper and oxygen;
calcium chloride is a compound of calcium and chlorine.

(Exceptions are compounds such as sodium hydroxide, which contains the elements sodium, hydrogen and oxygen.)

2 Compounds ending in **-ate** or **-ite** contain oxygen. There is a greater proportion of oxygen in the compound ending in -ate.

E.g. sodium sulphate Na_2SO_4
sodium sulphite Na_2SO_3

3 Compounds with a prefix **per-** contain extra oxygen.

E.g. sodium oxide Na_2O
sodium peroxide Na_2O_2

4 Compounds with a prefix **thio-** contain a sulphur atom in place of an oxygen atom.

E.g. sodium sulphate Na_2SO_4
sodium thiosulphate $Na_2S_2O_3$

3.6 Formulae

Each compound is represented by a formula which gives the proportions of the different elements in the compound by mass. The formula of any compound could be found by carrying out a suitable experiment. The formulae of many compounds can be found by use of the list of ions in Table 3.1.

Table 3.1 List of common ions

Positive ions		*Negative ions*	
Sodium	Na^+	Chloride	Cl^-
Potassium	K^+	Bromide	Br^-
Silver	Ag^+	Iodide	I^-
Copper(II)	Cu^{2+}	Hydroxide	OH^-
Lead	Pb^{2+}	Nitrate	NO_3^-
Magnesium	Mg^{2+}	Nitrite	NO_2^-
Calcium	Ca^{2+}	Hydrogencarbonate	HCO_3^-
Zinc	Zn^{2+}	Sulphate	SO_4^{2-}
Barium	Ba^{2+}	Sulphite	SO_3^{2-}
Iron(II)	Fe^{2+}	Carbonate	CO_3^{2-}
Iron(III)	Fe^{3+}	Oxide	O^{2-}
Aluminium	Al^{3+}	Sulphide	S^{2-}
Ammonium	NH_4^+	Phosphate	PO_4^{3-}
Hydrogen	H^+		

In forming the compound the number of ions used is such that the number of positive charges equals the number of negative charges:
e.g. sodium chloride is made up from Na^+ and Cl^- ions. Since a sodium ion has a single positive charge and a chloride ion has a single negative charge, the formula of sodium chloride is NaCl.

Sodium sulphate is made up from Na^+ and SO_4^{2-} ions. Twice as many sodium ions as sulphate ions are necessary in order to have equal numbers of positive and negative charges. The formula of sodium sulphate is Na_2SO_4. Table 3.2 contains further examples.

Table 3.2 Further examples of formulae

Compound	*Ions present*	*Formula*
Copper(II) oxide	$Cu^{2+}O^{2-}$	CuO
Ammonium chloride	$NH_4^+Cl^-$	NH_4Cl
Silver nitrate	$Ag^+NO_3^-$	$AgNO_3$
Magnesium chloride	$Mg^{2+}Cl^-$	$MgCl_2$
Magnesium hydroxide	$Mg^{2+}OH^-$	$Mg(OH)_2$
Aluminium nitrate	$Al^{3+}NO_3^-$	$Al(NO_3)_3$
Aluminium oxide	$Al^{3+}O^{2-}$	Al_2O_3
Hydrochloric acid	H^+Cl^-	HCl
Sulphuric acid	$H^+SO_4^{2-}$	H_2SO_4
Nitric acid	$H^+NO_3^-$	HNO_3

N.B. **1** Acids contain H^+ ions.

2 A small number after a bracket multiplies everything inside the bracket.
E.g. $Mg(OH)_2$ is composed of one magnesium, two oxygen and two hydrogen atoms. These are formed into three ions – one Mg^{2+} ion and two OH^- ions.

All of the compounds above are composed of ions. However, many compounds are not ionized. The formulae of some of these compounds are shown in Table 3.3.

Table 3.3 Formulae of some common compounds

Compound	*Formula*	*Compound*	*Formula*
Water	H_2O	Sulphur dioxide	SO_2
Carbon dioxide	CO_2	Sulphur trioxide	SO_3
Carbon monoxide	CO	Ammonia	NH_3
Nitrogen monoxide	NO	Hydrogen chloride	HCl
Nitrogen dioxide	NO_2	Methane	CH_4

3.7 Summary

An element is a pure substance which cannot be split up by chemical reaction. Elements can be represented by symbols composed of one or two letters.

Most elements are solid at room temperature and are metallic.

Elements can be combined together in fixed proportions to form compounds. The properties of a compound are different from the properties of the elements which make it up. Each compound can be represented in shorthand by writing a formula. It is most important to be able to write formulae correctly.

4 CHEMICAL EQUATIONS

This unit is written in two parts:

Part A	4.1–4.2	Chemical equations
Part B	4.3	Ionic equations

4.1 Chemical Equations

Chemical equations are widely used in textbooks and examination papers. An equation is a useful summary of a chemical reaction, and it is always theoretically possible to obtain an equation from the results of an experiment. It is advisable, however, to be able to write important equations in an examination.

THE STEPS IN WRITING A CHEMICAL EQUATION ARE AS FOLLOWS:

1 Write down the equation as a word equation using either the information given or your memory. Include all the reacting substances and products.

E.g. calcium hydroxide + hydrochloric acid → calcium chloride + water

Often the information you require is given to you in a jumbled form in the question; e.g. iron(II) chloride is produced when dry hydrogen chloride gas is passed over heated iron. The other product is hydrogen. The word equation for this reaction is:

iron + hydrogen chloride → iron(II) chloride + hydrogen

2 Fill in the correct formulae for all the reacting substances and products.

$$Ca(OH)_2 + HCl \rightarrow CaCl_2 + H_2O$$

3 The equation then needs to be balanced. During any chemical reaction, atoms cannot be created or destroyed. There must be the same number of atoms before and after the reaction. Only the proportions of the reacting substances and products can be altered to balance the equation – not the formulae.

$$Ca(OH)_2 + 2HCl \rightarrow CaCl_2 + 2H_2O$$

4 Finally, the states of reacting substances and products can be included in small brackets after the formulae. Thus:

(s) for solid – though sometimes (c) is seen for representing a crystalline solid
(l) for liquid
(g) for gas
(aq) for a solution with water as solvent

These state symbols are not given or expected by all examination boards.

$$Ca(OH)_2(aq) + 2HCl(aq) \rightarrow CaCl_2(aq) + 2H_2O(l)$$

Writing equations requires a great deal of practice. As you work through this book you will find many equations.

4.2 Information Provided by an Equation

$$2Mg(s) + O_2(g) \rightarrow 2MgO(s)$$

magnesium + oxygen → magnesium oxide

This equation gives the following information:

2 moles of magnesium atoms (48 g) combine with 1 mole of oxygen molecules (32 g) to produce 2 moles of magnesium oxide (80 g). It also tells us that magnesium and magnesium oxide are solids and oxygen is a gas.

A chemical equation gives:

1 the chemicals reacting together (called reactants) and the chemicals produced (called products).
2 the physical states of reactants and products.
3 the quantities of chemicals reacting together and produced.

The equation does not, however, give information about energy changes in the reaction, the rate of the reaction and the conditions necessary for the reaction to take place.

Calculating quantities of chemicals reacting together and produced in a reaction is considered in Unit 29.

4.3 Ionic equations

Ionic equations are useful because they emphasize the important changes taking place in a chemical reaction. For example, the (unbalanced) equation for the neutralization reaction between sodium hydroxide and hydrochloric acid is:

$$NaOH(aq) + HCl(aq) \rightarrow NaCl(aq) + H_2O(l)$$

sodium hydroxide + hydrochloric acid → sodium chloride + water

Since all of the reactants and products (except water) are composed of ions, this equation could be written:

$$Na^+(aq)OH^-(aq) + H^+(aq)Cl^-(aq) \rightarrow Na^+(aq)Cl^-(aq) + H_2O(l)$$

An equation should show change and therefore anything present before and after the reaction can be deleted. The simplest ionic equation, deleting $Na^+(aq)$, and $Cl^-(aq)$, is therefore:

$$OH^-(aq) + H^+(aq) \rightarrow H_2O(l)$$

The same ionic equation also applies to similar reactions, e.g. calcium hydroxide and nitric acid.

In addition to balancing in the usual way, the sum of the charges on the left-hand side must equal the sum of the charges on the right-hand side.

Other common ionic equations include:

$$2Fe^{2+}(aq) + Cl_2(g) \rightarrow 2Fe^{3+}(aq) + 2Cl^-(aq)$$
$$2Br^-(aq) + Cl_2(g) \rightarrow Br_2(aq) + 2Cl^-(aq)$$
$$Zn(s) + 2H^+(aq) \rightarrow Zn^{2+}(aq) + H_2(g)$$

4.4 Summary

A chemical equation is a useful way of summarizing a chemical reaction which takes place. A word equation is simpler to write but is not as useful as a symbol equation.

Word equation:

Copper(II) oxide + hydrochloric acid → copper(II) chloride + water

symbol equation:

$$CuO(s) + 2HCl(aq) \rightarrow CuCl_2(aq) + H_2O(l)$$

A symbol equation, apart from summarizing the reaction, can provide information about the reaction.

5 HYDROGEN

5.1 Laboratory Preparation of Hydrogen

Hydrogen is prepared by the action of dilute sulphuric acid on zinc.

$$Zn(s) + H_2SO_4(aq) \rightarrow ZnSO_4(aq) + H_2(g)$$

zinc + sulphuric acid → zinc sulphate + hydrogen

The apparatus is shown in Fig. 5.1.

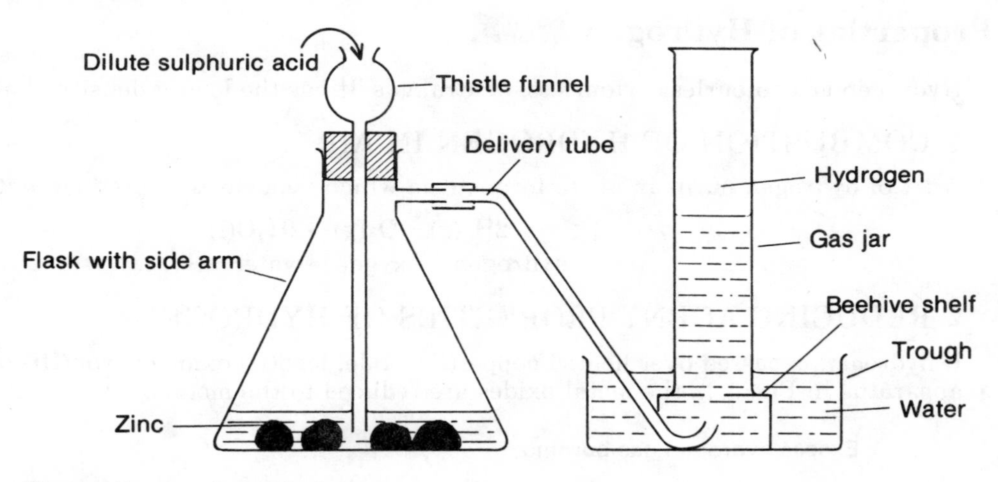

Fig 5.1 Preparation of hydrogen

Pure zinc reacts very slowly. The reaction can be speeded up by adding copper as a catalyst. (In practice, copper(II) sulphate solution is added.)

If the gas is required dry, it can be dried by passing it through concentrated sulphuric acid. It can then be collected by upward delivery.

5.2 Test for Hydrogen

When a lighted splint is put into hydrogen (preferably mixed with some air), it burns with a squeaky pop and the splint is extinguished.

$$2H_2(g) + O_2(g) \rightarrow 2H_2O(g)$$

hydrogen + oxygen → water

5.3 Other Laboratory Reactions producing Hydrogen

1 Reactions of metals with water (see Unit 19.9)

2 Reactions of certain metals with alkalis.
Zinc and aluminium react with sodium hydroxide solution producing hydrogen.

$$Zn(s) + 2NaOH(aq) \rightarrow Na_2ZnO_2(aq) + H_2(g)$$
zinc + sodium hydroxide → sodium zincate + hydrogen

$$2Al(s) + 2NaOH(aq) + 2H_2O(l) \rightarrow 2NaAlO_2(aq) + 3H_2(g)$$
aluminium + sodium hydroxide + water → sodium aluminate + hydrogen

5.4 Manufacture of Hydrogen

1 Hydrogen is manufactured from coke. Steam is passed over heated coke producing **water gas** (a mixture of carbon monoxide and hydrogen). The water gas is mixed with more steam and passed over an iron–chromium catalyst at 500 °C.

$$H_2(g) + CO(g) + H_2O(g) \rightarrow 2H_2(g) + CO_2(g)$$
water gas + water → hydrogen + carbon dioxide

The carbon dioxide is removed by dissolving it in water under pressure.

2 Hydrogen is manufactured from natural gas. Natural gas is mainly methane (CH_4). Methane is converted to hydrogen by partial oxidation at a high temperature (1300 °C). This is the usual method in industry today.

$$2CH_4(g) + O_2(g) \rightarrow 2CO(g) + 4H_2(g)$$
methane + oxygen → carbon monoxide + hydrogen

or by oxidation with steam in the presence of a nickel catalyst at 600 °C.

$$CH_4(g) + H_2O(g) \rightarrow CO(g) + 3H_2(g)$$
methane + water → carbon monoxide + hydrogen

The mixture of carbon monoxide produced by either method is then treated as in 1.

3 Hydrogen is also produced as a byproduct in the electrolysis of brine (sodium chloride solution) (see 28.3).

5.5 Properties of Hydrogen

Hydrogen is a colourless, odourless, neutral gas. It has the lowest density of all the elements.

1 COMBUSTION OF HYDROGEN IN AIR

A jet of hydrogen burns in air to form steam which condenses to produce water.

$$2H_2(g) + O_2(g) \rightarrow 2H_2O(l)$$
hydrogen + oxygen → water

2 REDUCING-AGENT PROPERTIES OF HYDROGEN

If hydrogen is passed over heated copper(II) oxide, lead(II) oxide or iron(III) oxide (using the apparatus in Fig. 5.2), the metal oxides are reduced to the metals.

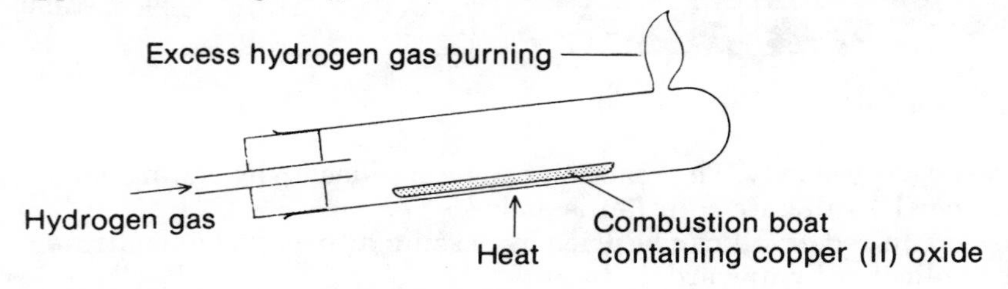

Fig 5.2 Reduction of copper(II) oxide

$$CuO(s) + H_2(g) \rightarrow Cu(s) + H_2O(g)$$
copper(II) oxide + hydrogen → copper + water

$$PbO(s) + H_2(g) \rightarrow Pb(s) + H_2O(g)$$
lead(II) oxide + hydrogen → lead + water

$$Fe_2O_3(s) + 3H_2(g) \rightarrow 2Fe(s) + 3H_2O(g)$$
iron(III) oxide + hydrogen → iron + water

5.6 Uses of Hydrogen

1 Hydrogen is used in the manufacture of ammonia (see 25) and hydrochloric acid.
2 Hydrogen is used in the manufacture of margarine (see 20.6).
3 Hydrogen is used for filling balloons. It is, however, inflammable.

5.7 Summary

Hydrogen can be prepared by the action of dilute sulphuric acid on zinc.
It is also produced when certain metals react with water and when zinc and aluminium react with alkalis.
Most hydrogen used in industry is produced from natural gas.

6 METALS AND NONMETALS

6.1 Introduction

We talk quite commonly about metallic properties and differences between metals and nonmetals. In this unit we are going to consider the physical and chemical properties of metals and how they differ from nonmetals. There are some elements such as silicon which are difficult to classify as metals or nonmetals because they have properties between the two. These elements are called **metalloids**.

Also, we are going to consider the advantages of mixtures of metals called **alloys** compared with pure metals.

6.2 Physical Properties of Metals

Table 6.1 compares the physical properties of metals and nonmetals.

Table 6.1 Comparison of properties of typical metals and nonmetals

Metals	*Nonmetals*
Solid at room temperature	Solid, liquid or gas at room temperature
Shiny	Dull
High density	Low density
Good conductor of heat and electricity	Poor conductor of heat and electricity
Can be beaten into thin sheets (malleable) and drawn into wire (ductile)	Brittle

There are many exceptions to the generalizations in Table 6.1. For example, mercury is a metal but is a liquid at room temperature. Carbon (in the form of graphite) is a nonmetal but conducts electricity. Silicon is a grey solid with a metallic appearance and could be mistaken for a metal if only physical properties are considered.

6.3 Chemical Properties of Metals

Many metals react with dilute hydrochloric or sulphuric acid to produce hydrogen gas.

E.g.

$$Mg(s) + H_2SO_4(aq) \rightarrow MgSO_4(aq) + H_2(g)$$

magnesium + sulphuric acid → magnesium sulphate + hydrogen

$$Mg(s) + 2HCl(aq) \rightarrow MgCl_2(aq) + H_2(g)$$

magnesium + hydrochloric acid → magnesium chloride + hydrogen

However, not all metals will produce hydrogen in this way.

A better method of distinguishing metals from nonmetals is to burn the element in air or oxygen and form the oxide. The oxide is then tested with Universal Indicator to find the pH of the oxide. If the oxide is pH 7 or above (neutral or alkaline) the element was a metal. If the oxide has a pH less than 7 (acid) the element was a nonmetal. This is a very reliable way of distinguishing between a metal and a nonmetal.

6.4 Alloys

Pure metals have a wide range of uses. Table 6.2 lists some of the uses of pure metals and the reasons for their use. For many purposes, however, mixtures of metals called alloys are preferred. An alloy is usually less malleable and ductile than a pure metal. A copper alloy is much stronger than pure copper and more suitable for coinage. Alloys also have lower melting points than pure metals.

Table 6.2 Uses of pure metals

Metal	*Use*	*A reason for use*
Copper	Electricity cables	Excellent conductor of electricity/v. ductile
Tin	Coating tin cans	Not poisonous
Aluminium	Kitchen foil	Very malleable
Iron	Wrought iron gates	Easy to forge and resists corrosion
Lead	Flashing on roofs	Soft, easy to shape, does not corrode

An alloy is made by weighing out correctly the different constituent metals and melting them together to form the alloy.

Steel is undoubtedly the most important alloy. It is an alloy of iron containing between 0.15 and 1.5 per cent of carbon with other metals possibly present.

Table 6.3 includes some of the common alloys and their uses.

Table 6.3 Examples of common alloys

Alloy	*Constituent elements*	*Uses*
Steel	Iron + between 0.15% and 1.5% carbon. The properties of steel depend on the percentage of carbon. Other metals may be present, e.g. chromium in stainless steel	Wide variety of uses including cars, ships, tools, reinforced concrete, tinplate (coated with tin)
Brass	Copper and zinc	Ornaments, buttons, screws
Duralumin	Aluminium, magnesium, copper and manganese	Lightweight uses e.g. aircraft, bicycles
Solder	Tin and lead	Joining metals (N.B. importance of low melting point)
Coinage bronze	Copper, zinc and tin	1p and 2p coins
Bronze	Copper and tin	Ornaments

6.5 Summary

Metals are usually solids with a shiny appearance and a silver or gold colour. They usually have a high density and conduct heat and electricity well. However, physical properties can be misleading. An element can sometimes look like a metal but, in fact, be a nonmetal.

A very reliable method of distinguishing a metal from a nonmetal involves forming the oxide and testing the oxide to find its pH with Universal Indicator. If the oxide is neutral or alkaline, the element is a metal. If the oxide is acid, the element is a nonmetal.

An alloy is a mixture of metals which is more useful, for some reason, than a pure metal. Common alloys include brass, bronze, solder, duralumin and steel.

7 ATOMIC STRUCTURE AND BONDING

7.1 Particles in Atoms

All elements are made up from **atoms**. An atom is the smallest part of an element that can exist.

It has been found that the atoms of all elements are made up from three basic particles and that the atoms of different elements contain different numbers of these three particles. These particles are:

proton	p	mass 1 u	charge + 1	
electron	e	mass $\frac{1}{1800}$ u (negligible)	charge − 1	(u = atomic mass unit)
neutron	n	mass 1 u	neutral	

Because an atom has no overall charge, the number of protons in any atom is equal to the number of electrons.

Atomic Number

The atomic number is the number of protons in an atom.

Mass Number

The mass number is the total number of protons and neutrons in an atom.

E.g. The mass number of carbon-12 is 12, and the atomic number is 6. Therefore a carbon-12 atom contains 6 protons (i.e. atomic number = 6), 6 electrons and 6 neutrons. This is sometimes written as $^{12}_{6}C$ (the atomic number is written under the mass number).

For sodium-23:

mass number = 23   atomic number = 11 (i.e. $^{23}_{11}Na$)

p = 11, e = 11, n = 23 − 11 = 12

It is possible, with many elements, to get more than one type of atom. For example, there are three types of oxygen atom:

oxygen-16	8p, 8e, 8n
oxygen-17	8p, 8e, 9n
oxygen-18	8p, 8e, 10n

These different types of atom of the same element are called **isotopes**. They are different because they contain different numbers of neutrons. (If they did not contain the same number of protons and the same number of electrons they would not be isotopes of oxygen.) Isotopes of the same element have similar chemical properties but slightly different physical properties.

There are two isotopes of chlorine – chlorine-35 and chlorine-37. An ordinary sample of chlorine contains approximately 75 per cent chlorine-35 and 25 per cent chlorine-37. This explains the fact that the relative atomic mass of chlorine is approximately 35.5. (The relative atomic mass of an element is the mass of an 'average atom' compared with the mass of a $^{12}_{6}C$ carbon atom – see 29.1.)

7.2 Arrangement of Particles in an Atom

The protons and neutrons are tightly packed together in the **nucleus** of an atom. The electrons move rapidly around the nucleus in distinct energy levels. Each energy level is capable of accommodating only a certain number of electrons. This is represented in a simplified form in Fig. 7.1

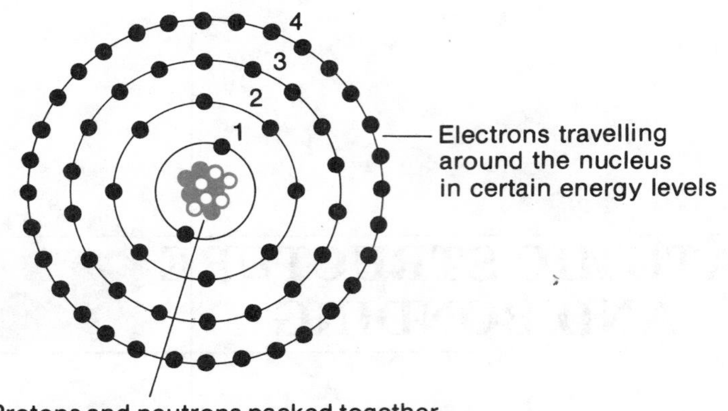

Fig 7.1 Arrangement of particles in an atom

1 The first energy level (sometimes called the K shell and labelled 1 in Fig. 7.1) can hold only two electrons. This energy level is filled first.

2 The second energy level (sometimes called the L shell and labelled 2 in Fig. 7.1) can hold only eight electrons. This energy level is filled after the first energy level and before the third energy level.

3 The third energy level (sometimes called the M shell and labelled 3 in Fig. 7.1) can hold a maximum of 18 electrons. However, when eight electrons are in the third energy level there is a degree of stability and the next two electrons added go into the fourth energy level (labelled 4 in Fig. 7.1). Then extra electrons enter the third energy level until it contains the maximum of 18 electrons.

4 There are further energy levels, each containing a larger number of electrons than the preceding energy level.

Table 7.1 gives the number of protons, neutrons and electrons in the principal isotopes of the first 20 elements. The electronic structure 2,8,1 denotes 2 electrons in the first energy level, 8 in the second, and 1 in the third. This is sometimes called the **electron configuration** of an atom.

Table 7.1 Numbers of protons, neutrons and electrons in the principal isotopes of the first 20 elements

Element	*Atomic number*	*Mass number*	*Number of* p	n	e	*Arrangement of electrons*
Hydrogen	1	1	1	0	1	1
Helium	2	4	2	2	2	2
Lithium	3	7	3	4	3	2,1
Beryllium	4	9	4	5	4	2,2
Boron	5	11	5	6	5	2,3
Carbon	6	12	6	6	6	2,4
Nitrogen	7	14	7	7	7	2,5
Oxygen	8	16	8	8	8	2,6
Fluorine	9	19	9	10	9	2,7
Neon	10	20	10	10	10	2,8
Sodium	11	23	11	12	11	2,8,1
Magnesium	12	24	12	12	12	2,8,2
Aluminium	13	27	13	14	13	2,8,3
Silicon	14	28	14	14	14	2,8,4
Phosphorus	15	31	15	16	15	2,8,5
Sulphur	16	32	16	16	16	2,8,6
Chlorine	17	35	17	18	17	2,8,7
Argon	18	40	18	22	18	2,8,8
Potassium	19	39	19	20	19	2,8,8,1
Calcium	20	40	20	20	20	2,8,8,2

7.3 Bonding

The joining of atoms together is called **bonding**. There are several types of bonding found in common chemicals. An arrangement of particles bonded together is called a **structure**.
Three methods of bonding will be discussed below.

7.4 Ionic (or Electrovalent) Bonding

This involves a **complete transfer of electrons** from one atom to another. Two examples are given below:

1 Sodium chloride
A sodium atom has an electronic structure of 2,8,1 (i.e. one more electron than the stable inert gas electronic arrangement of 2,8). A chlorine atom has an electronic arrangement of 2,8,7 (i.e. one electron less than the stable electronic arrangement 2,8,8).

If each sodium atom loses one electron (and forms a sodium ion Na^+) and each chlorine atom gains one electron (and forms a chloride ion Cl^-), both the ions formed have **stable electronic arrangements**. The ions are held together by strong electrostatic forces.

It is incorrect to speak of a 'sodium chloride *molecule*'. A sodium chloride *crystal* consists of a regular arrangement of equal numbers of sodium and chloride ions. This is called a **lattice** (Fig. 7.2).

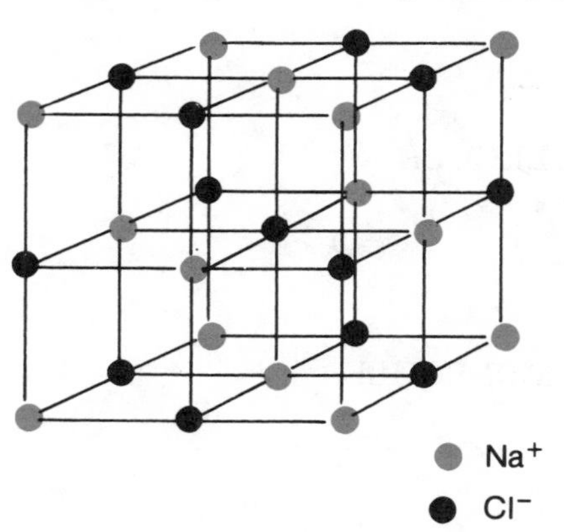

Fig 7.2 Structure of sodium chloride

2 Magnesium oxide

Electronic arrangement in magnesium atom	2,8,2
Electronic arrangement in oxygen atom	2,6

Two electrons are lost by each magnesium atom to form Mg^{2+} ions. Two electrons are gained by each oxygen atom to form O^{2-} ions.

Loss of one or two electrons by a metal during ionic bonding is common, e.g. NaCl or MgO. If three electrons are lost by a metal the resulting compound shows some covalent character, e.g. $AlCl_3$.

7.5 Covalent Bonding

Covalent bonding involves the **sharing of electrons** rather than complete transfer. Two examples are given below:

1 chlorine molecule (Cl_2)
A chlorine atom has an electronic arrangement of 2,8,7. When two chlorine atoms bond together they form a chlorine molecule. If one electron was transferred from one chlorine atom to the other, only one atom could achieve a stable electronic arrangement.

Instead, one electron from each atom is donated to form a pair of electrons which is shared between both atoms holding them together. This is called a **single covalent bond**. Figure 7.3 shows a simple representation of a chlorine molecule. This is often shown as Cl—Cl.

2 oxygen molecule (O_2)
An oxygen atom has an electronic arrangement of 2,6. In this case each oxygen atom donates two electrons and the four electrons (two pairs) are shared between both atoms. This is called a **double covalent bond**. Figure 7.4 shows a simplified representation of an oxygen molecule. This is usually shown as O=O.

Shared pair of electrons

Cl Cl

O O

X Electrons

Fig 7.3 Chlorine molecule

Fig 7.4 Oxygen molecule

Other common examples of covalent bonding and the different molecular shapes are shown in Fig. 7.5.

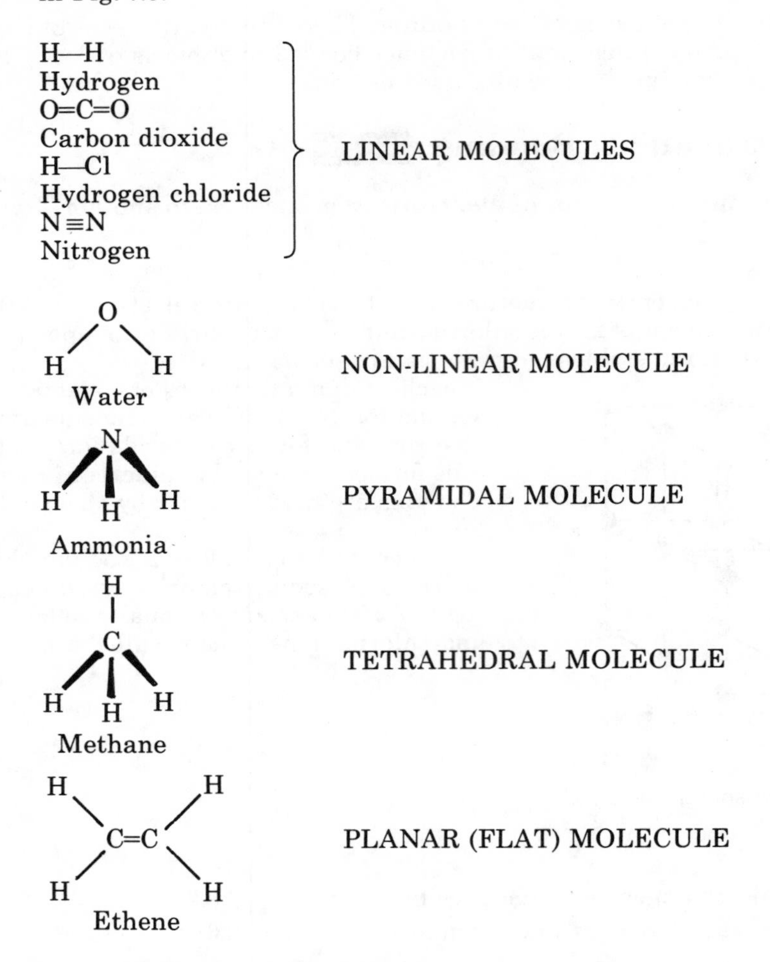

Fig 7.5 Shapes of simple molecules

7.6 Metallic Bonding

Metallic bonding is found only in metals. A metal consists of a close-packed regular arrangement of positive ions, which are surrounded by a 'sea' of electrons that bind the ions together. Figure 7.6 shows the arrangement of ions in a single layer. There are two alternative ways of stacking these layers. The arrows in Fig. 7.6 indicate that the layer shown continues in all directions. Around any one ion in a layer there are six ions arranged hexagonally.

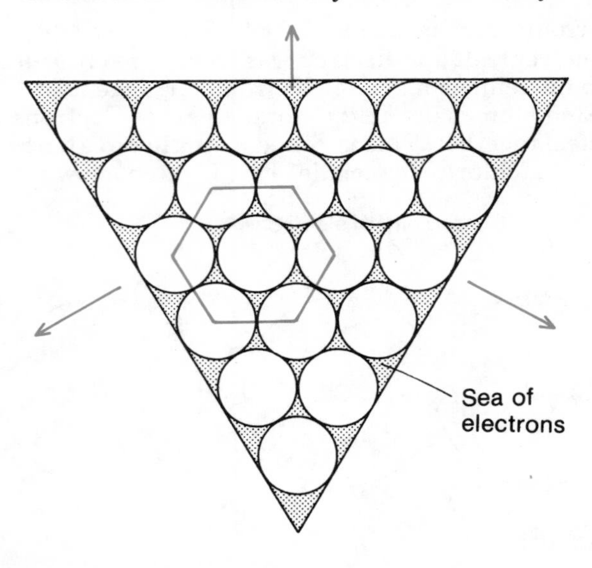

Fig 7.6 Arrangement of ions in a metal layer

7.7 Effects of Bonding on Properties of Substances

Compounds containing ionic bonds have certain properties in common – they have high melting and boiling points, and as a result are solids at room temperature. The ions are tightly held together in a regular lattice and energy (called **lattice energy**) is required to break up the lattice and melt the substance. The melting point of magnesium oxide is very high making it suitable for use as a **refractory** (i.e. for lining furnaces).

Substances containing ionic bonding usually dissolve in water (a **polar solvent**) but not in a nonpolar solvent. If they do not dissolve in water it is often because they have very high lattice energy.

Electricity passes through substances containing ionic bonds when the substances are molten or in solution in water but not when solid. These substances are called **electrolytes.**

Substances containing covalent bonding may be solid, liquid or gas at room temperature. They are usually insoluble in polar solvents but more soluble in nonpolar solvents. Generally they do not conduct electricity in any state.

Metals generally have high densities because the ions are close packed in the lattice. Because of the strong bonds between the ions caused by the **free electrons**, the melting points of most metals are high. The free electrons explain why metals are good conductors of heat and electricity.

7.8 Structures of Substances

The different types of structure found in pure materials are summarized in Fig. 7.7.

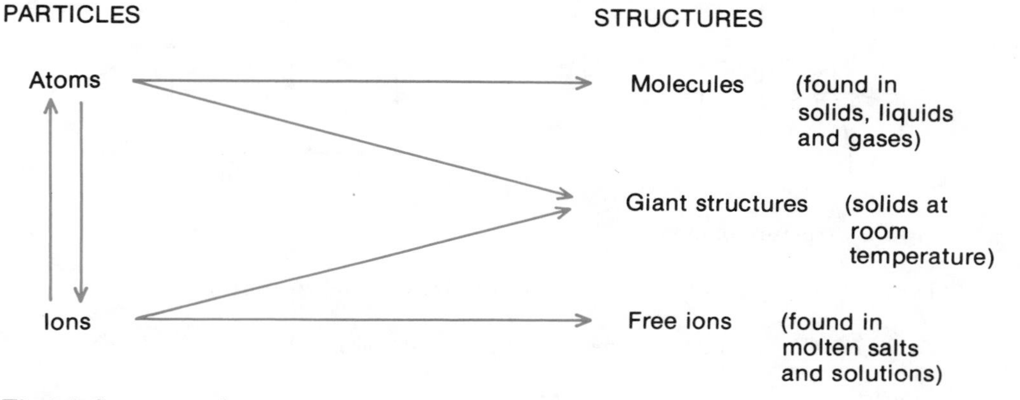

Fig 7.7 Summary of structures

A substance with a **molecular structure** contains separate groups of atoms called **molecules.** The groups of atoms are tightly held together within the molecules by covalent bonds but the forces between the molecules are much weaker. If the molecules are small, the substance is usually a liquid or a gas. A few solids do contain small molecules, but these solids have low melting points, e.g. sulphur and iodine. If the molecules are large the substance will be a solid with a high melting point. Large molecules are sometimes called **macromolecules** (e.g. polythene) or **giant molecular structures** (e.g. silicon dioxide). Substances with a molecular structure do not conduct electricity in any state because there are no free ions or electrons. They are usually not very soluble in water but dissolve in nonpolar solvents. Examples include sulphur, iodine, chlorine and carbon dioxide.

A substance with an ionic structure contains many ions bonded together into a **giant structure** of ions. Thus these substances always have high melting points as a large amount of energy is needed to break up the structure. Ionic substances conduct electricity when molten or dissolved in water because the ions are free to move. An example is sodium chloride.

Metals consist of giant structures (high melting points) except mercury, which is a liquid. Carbon is a nonmetal that has a giant atomic structure, and in the form of graphite it is also the only nonmetal that conducts electricity. Most other nonmetallic elements exist as small molecules, which is why many of them are gases under normal conditions. The noble gases exist as single atoms in a gaseous state.

7.9 Allotropy

Allotropy is the existence of two or more forms of an element in the same physical state. These different forms are called **allotropes**. Allotropy is caused by the possibility of more than one arrangement of atoms. For example carbon can exist in two allotropes – diamond and graphite. Sulphur can exist in two allotropes – α-sulphur and β-sulphur (Unit 27).

7.10 Allotropy of Carbon

Two crystalline allotropes exist – diamond and graphite.

1 DIAMOND

In the diamond structure each carbon atom is strongly bound (covalent bonding) to four other carbon atoms tetrahedrally. A large giant structure (three dimensional) is built up. All bonds between carbon atoms are the same length (0.154 nm). It is the strength and uniformity of the bonding which make diamond very hard, nonvolatile and resistant to chemical attack. Figure 7.8 shows the arrangement of particles in diamond.

2 GRAPHITE

Graphite has a layer structure. In each layer the carbon atoms are bound covalently. The bonds within the layers are very strong. The bonds between the layers, however, are very weak, which enables layers to slide over one another. This makes the graphite soft and flaky. Figure 7.9 shows the arrangement of particles in graphite.

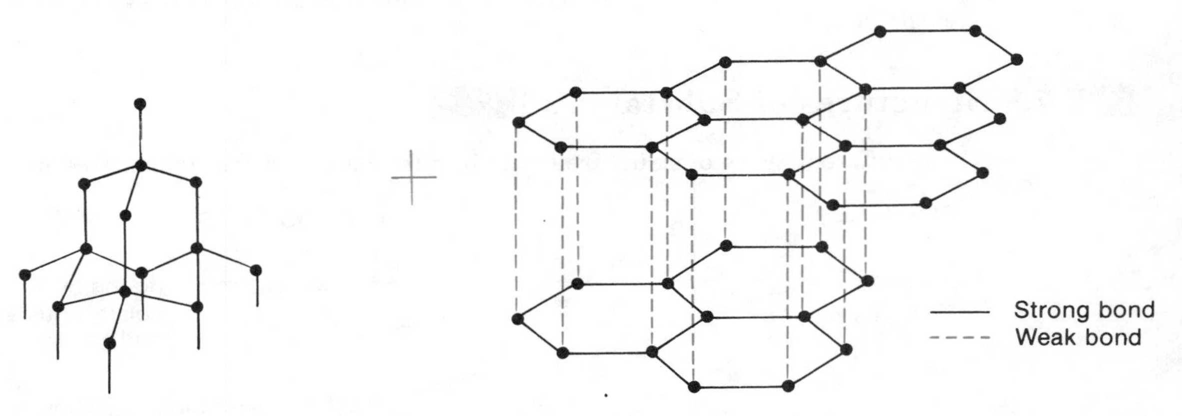

Fig 7.8 Structure of diamond

Fig 7.9 Structure of graphite

Table 7.2 compares the properties of diamond and graphite.

Table 7.2 Comparing the properties of diamond and graphite

Property	*Diamond*	*Graphite*
Appearance	Transparent, colourless crystals	Black, opaque, shiny solid
Density (g/cm³)	3.5	2.2
Hardness	Very hard	Very soft
Electrical conductivity	Nonconductor	Good electrical conductor
Burning in oxygen	Burns only with difficulty when heated to high temperature. Carbon dioxide	Burns readily to produce carbon dioxide. No residue produced. No residue

7.11 Summary

All elements are made up of atoms. Atoms are made up of three types of particle – protons (positive charge), neutrons (neutral) and electrons (negative charge).

In any atom the number of protons and the number of electrons are the same. Atoms of the same element may, however, contain different numbers of neutrons. Atoms of the same element containing different numbers of neutrons are called isotopes.

If an atom loses one or more electrons it forms a positive ion. If an atom gains one or more electrons it forms a negative ion.

In an atom the protons and neutrons are packed together in a positively charged nucleus. The negatively charged electrons move around the nucleus in certain energy levels.

There are different methods of joining or bonding atoms together.

Ionic bonding involves complete transfer of electrons from a metal to a nonmetal. Positive and negative ions are formed which are held together by strong electrostatic forces. Sodium chloride is an example of a substance with ionic bonding. Substances with ionic bonding have high melting and boiling points, are usually soluble in water (a polar solvent) and insoluble in organic (nonpolar) solvents such as hexane.

Covalent bonding involves a sharing of electrons. A single covalent bond involves one pair of shared electrons. Methane, CH_4, is a compound with covalent bonding.

Compounds with covalent bonding often have low melting and boiling points, are often insoluble in water and soluble in hexane.

Metallic bonding is found in metals. A metal consists of a regular arrangement of positive ions which are surrounded by a 'sea' of electrons that bind the ions together.

The existence of two forms of the same element in the same physical state is called allotropy. The different forms are called allotropes.

Two allotropes of carbon are diamond and graphite. The difference in the physical properties of diamond and graphite can be explained by the different arrangements of carbon atoms in the two allotropes.

8 THE REACTIVITY SERIES OF METALS

8.1 The Reactivity Series

In this unit the comparative reactivity of different pure metals is considered. Table 8.1 summarizes the reactivity of common metals.

Table 8.1 Reactions of some metals

Metal	*Metal heated in air*	*Metal with water or steam*	*Metal with acids*
Potassium Sodium Calcium	Burn to form oxide	React with cold water	Violent reaction with dilute acids
Magnesium Aluminium Zinc	Burn to form oxide	React with steam	React with dilute acids with decreasing ease
Iron	React slowly	Reacts reversibly with steam	React with dilute acids with decreasing ease
Lead Copper	React slowly	Not attacked by water or steam	Attacked only by oxidizing agents e.g. concentrated nitric acid
Mercury Silver	Reacts reversibly	Not attacked by water or steam	Attacked only by oxidizing agents e.g. concentrated nitric acid
Platinum	Do not react with air	Not attacked by water or steam	No reaction

Following the detailed consideration of these reactions and other similar reactions, it is possible to arrange metals in a **reactivity** (or activity) **series**. This is a list of metals in order of reactivity with the most reactive metal at the top of the list and steadily decreasing reactivity down the list.

The list is:

potassium	K
sodium	Na
calcium	Ca
magnesium	Mg
aluminium	Al
zinc	Zn
iron	Fe
lead	Pb
copper	Cu
mercury	Hg
silver	Ag
platinum	Pt

N.B. This is the same order as in Table 8.1. This list could be shorter or longer depending upon the number of metals you wish to consider. Do not try to learn this list. It will be given to you when you need it.

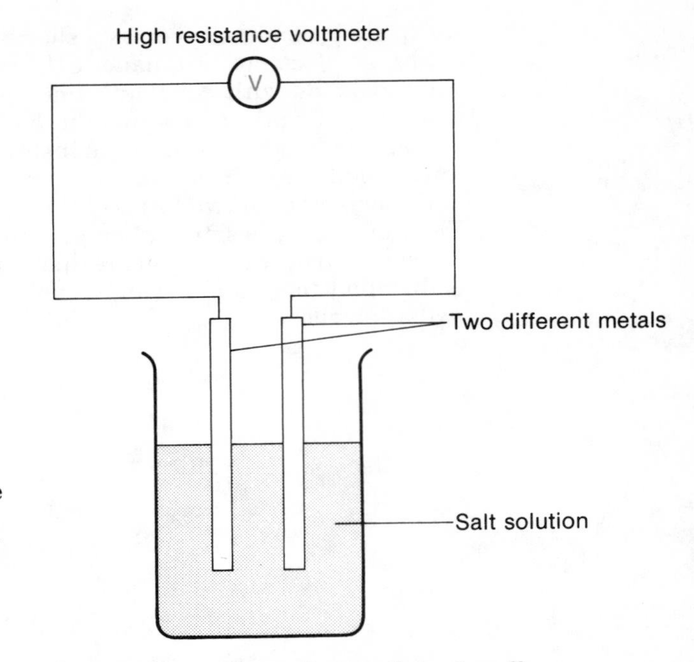

Fig 8.1 Measuring voltages of simple cells

A similar list could be obtained by measuring voltages. If two rods of different metals are dipped into salt solution, a small voltage is produced which can be measured by means of a high resistance voltmeter (Fig. 8.1).

By comparing voltages of these cells it is possible to come up with the same list but, strictly speaking, this is called the **electrochemical series** and the elements are in order of decreasing standard electrode potentials. In this book, the term 'reactivity series' will be used.

8.2 Use of the Reactivity Series

The reactivity series is very useful to the chemist in making predictions. It can be used to:

1 Explain the stability of compounds when heated.
2 Predict possible reactions involving metals.
3 Explain the corrosion of certain metals (Unit 24).
4 Explain the method used to extract a metal from its ore (Unit 23).

8.3 Stability of Compounds Containing Metals

Compounds of metals high in the reactivity series are not easily split up by heating. Compounds of metals low in the reactivity series are split up by heating.

Examples of this can be seen by comparing the action of heat on carbonates (Table 8.2). Sodium and potassium carbonates are not split up on heating. Also Table 8.3 compares the action of heat on metal nitrates. The lower the metal is in the reactivity series, the more readily and completely the nitrate of the metal splits up.

Table 8.2 Important properties of carbonates

Carbonate	*Formula*	*Colour*	*Solubility in water*	*Action of heat*
Potassium	K_2CO_3	White	Soluble	Not decomposed
Sodium	Na_2CO_3	White	Soluble	↓
Calcium	$CaCO_3$	White	Insoluble	↑
Magnesium	$MgCO_3$	White	Insoluble	
Zinc	$ZnCO_3$	White	Insoluble	Decomposed to the oxide of
Iron	$FeCO_3$	Light brown	Insoluble	the metal and carbon dioxide
Lead	$PbCO_3$	White	Insoluble	$PbCO_3(s) \rightarrow PbO(s) + CO_2(g)$
Copper	$CuCO_3$	Bluish-green	Insoluble	↓

Table 8.3 Action of heat on metal nitrates

Nitrate	*Formula*	*Colour*	*Equation*	*Colour of the solid residue*
Potassium nitrate	KNO_3	White	$2KNO_3(s) \rightarrow 2KNO_2(s) + O_2(g)$ potassium nitrate → potassium nitrite + oxygen	White
Sodium nitrate	$NaNO_3$	White	$2NaNO_3(s) \rightarrow 2NaNO_2(s) + O_2(g)$ sodium nitrate → sodium nitrite + oxygen	White
Calcium nitrate	$Ca(NO_3)_2$	White	$2Ca(NO_3)_2(s) \rightarrow 2CaO(s) + 4NO_2(g) + O_2(g)$ calcium nitrate → calcium oxide + nitrogen dioxide + oxygen	White
Magnesium nitrate	$Mg(NO_3)_2$	White	$2Mg(NO_3)_2(s) \rightarrow 2MgO(s) + 4NO_2(g) + O_2(g)$ magnesium nitrate → magnesium oxide + nitrogen dioxide + oxygen	White
Zinc nitrate	$Zn(NO_3)_2$	White	$2Zn(NO_3)_2(s) \rightarrow 2ZnO(s) + 4NO_2(g) + O_2(g)$ zinc nitrate → zinc oxide + nitrogen dioxide + oxygen	Yellow when hot, white when cold
Iron(III) nitrate	$Fe(NO_3)_3$	Brown	$4Fe(NO_3)_3(s) \rightarrow 2Fe_2O_3(s) + 12NO_2(g) + 3O_2(g)$ iron(III) nitrate → iron(III) oxide + nitrogen dioxide + oxygen	Reddish-brown
Lead nitrate	$Pb(NO_3)_2$	White	$2Pb(NO_3)_2(s) \rightarrow 2PbO(s) + 4NO_2(g) + O_2(g)$ lead nitrate → lead oxide + nitrogen dioxide + oxygen	Yellow
Copper(II) nitrate	$Cu(NO_3)_2$	Blue	$2Cu(NO_3)_2(s) \rightarrow 2CuO(s) + 4NO_2(g) + O_2(g)$ copper(II) nitrate → copper(II) oxide + nitrogen dioxide + oxygen	Black
Silver nitrate	$AgNO_3$	White	$2AgNO_3(s) \rightarrow 2Ag(s) + 2NO_2(g) + O_2(g)$ silver nitrate → silver + nitrogen dioxide + oxygen	Silvery

8.4 Predicting Chemical Reactions involving Metals

A knowledge of the order of the metals in the reactivity series can be used to predict and explain chemical reactions involving metals.

The reactivity series can also be written:

potassium
sodium
calcium
magnesium
aluminium
(carbon)
zinc
iron
lead
(hydrogen)
copper
mercury
silver
platinum

The inclusion of carbon and hydrogen in their correct places, although they are not metals, extends the usefulness of the reactivity series.

Metals above hydrogen will displace hydrogen from acids. The metals below will not displace hydrogen from dilute acids. Therefore, copper does not react with dilute hydrochloric acid.

Metals above carbon in the reactivity series are not produced by reduction of metal oxides with carbon. Metals below carbon can be produced by reduction of metal oxides with carbon.

If iron(III) oxide and aluminium powder are heated together, a reaction takes place because aluminium is more reactive than iron.

$$Fe_2O_3(s) + 2Al(s) \rightarrow 2Fe(s) + Al_2O_3(s)$$

iron(III) oxide + aluminium → iron + aluminium oxide

No reaction would take place if zinc oxide and copper were heated together because copper is less reactive than zinc.

If an iron nail is put into copper(II) sulphate solution, a reaction takes place. The blue solution turns almost colourless and brown copper is deposited.

$$Fe(s) + CuSO_4(aq) \rightarrow FeSO_4(aq) + Cu(s)$$

iron + copper(II) sulphate → iron(II) sulphate + copper

The reaction takes place because iron is more reactive than copper.

No reaction would take place if a piece of lead were put into a solution of magnesium sulphate.

Reactions of this type are called **displacement reactions** and it is important that the metal being added is more reactive than the metal already present in the compound if a reaction is to take place.

8.5 Summary

Metals can be arranged in order of decreasing reactivity (or activity) in a reactivity (or activity) series. For many purposes a short form of the reactivity series is useful:

sodium
calcium
magnesium
aluminium
zinc
iron
lead
copper

The reactivity series can be used to explain stability of compounds, i.e. how easily they are split up. It can also be used to understand and predict displacement reactions, where a reactive metal replaces a less reactive metal in a compound.

9 CHEMICAL FAMILIES AND THE PERIODIC TABLE

9.1 Chemical Families

The 105 known elements can be divided into different groups in several ways. All elements can be classified as metals or nonmetals and, from the appearance, it is easy to classify elements as solids, liquids or gases at room temperature. From a chemical point of view, it is useful to group elements together because they have similar chemical behaviour. These groups of elements are sometimes called **families** of elements. Elements in the same family are not identical but usually show marked similarities with other members of the same family.

9.2 The Alkali Metal Family

This is a group of very reactive metals. The most common members of the family are lithium, sodium and potassium and some of their properties are shown in Table 9.1.

Table 9.1 Alkali metals

Element	*Symbol*	*Appearance*	*Melting point* (°C)	*Density* (g/cm^3)
Lithium	Li	Soft grey metal	181	0.54
Sodium	Na	Soft light grey metal	98	0.97
Potassium	K	Very soft blue/grey metal	63	0.86

These metals have to be stored in oil to exclude air and water. They do not look much like metals, at first sight, but when freshly cut they all have a typical shiny metallic surface.

They are also very good conductors of electricity. Note, however, that they have melting points and densities that are low compared with other metals.

REACTION OF ALKALI METALS WITH WATER

When a small piece of the alkali metal is put into a trough of water, the metal reacts immediately, floating on the surface of the water and evolving hydrogen.

With sodium and potassium, the heat evolved from the reaction is sufficient to melt the metal.

The hydrogen evolved by the reaction of potassium with cold water is usually ignited and burns with a pink flame.

Sodium reacts more quickly than lithium and potassium reacts more quickly than sodium.

In each case the solution remaining at the end of the reaction is an alkali.

$$2Li(s) + 2H_2O(l) \rightarrow 2LiOH(aq) + H_2(g)$$
lithium + water → lithium hydroxide + hydrogen

$$2Na(s) + 2H_2O(l) \rightarrow 2NaOH(aq) + H_2(g)$$
sodium + water → sodium hydroxide + hydrogen

$$2K(s) + 2H_2O(l) \rightarrow 2KOH(aq) + H_2(g)$$
potassium + water → potassium hydroxide + hydrogen

N.B. These three equations are basically the same and, if the alkali metal is represented by M, these equations can be represented by:

$$2M(s) + 2H_2O(l) \rightarrow 2MOH(aq) + H_2(g)$$

REACTION OF ALKALI METALS WITH OXYGEN

When heated in air or oxygen, the alkali metals burn to form white solid oxides. The colour of the flame is characteristic of the metal:

lithium – red
sodium – orange
potassium – lilac

E.g.
$$4Li(s) + O_2(g) \rightarrow 2Li_2O(s)$$
lithium + oxygen → lithium oxide

or
$$4M(s) + O_2(g) \rightarrow 2M_2O(s)$$

The alkali metal oxides all dissolve in water to form alkali solutions.

E.g.
$$Li_2O(s) + H_2O(l) \rightarrow 2LiOH(aq)$$
lithium oxide + water → lithium hydroxide

or
$$M_2O(s) + H_2O(l) \rightarrow 2MOH(aq)$$

REACTION OF ALKALI METALS WITH CHLORINE

When a piece of burning alkali metal is lowered into a gas jar of chlorine, the metal continues to burn forming a white smoke of the metal chloride.

E.g.
$$2K(s) + Cl_2(g) \rightarrow 2KCl(s)$$
potassium + chlorine → potassium chloride

or
$$2M(s) + Cl_2(g) \rightarrow 2MCl(s)$$

It is because of these similar reactions that these metals are put in the same family. In each reaction the order of reactivity is the same, i.e. lithium is least reactive and potassium is the most reactive.

There are three more members of this family – rubidium (Rb), caesium (Cs) and francium (Fr). They are all more reactive than potassium.

9.3 The Halogen Family

This is a family of nonmetals. In the alkali metal family, the members of the family all have similar appearances. In the halogen family, the different members have different appearances but they are put in the same family on the basis of their similar chemical reactions. Their appearances are compared in Table 9.2.

Table 9.2 Halogens

Element	*Symbol*	*Appearance at room temperature*
Fluorine	F	Pale yellow gas
Chlorine	Cl	Yellow/green gas
Bromine	Br	Red/brown volatile liquid
Iodine	I	Dark grey crystalline solid

There is another member of the family called astatine (At). It is radioactive and a very rare element.

Fluorine is a very reactive gas and is too reactive to handle in normal laboratory conditions.

SOLUBILITY OF HALOGENS IN WATER

None of the halogens is very soluble in water. Chlorine is the most soluble. Iodine does not dissolve much in cold water and only dissolves slightly in hot water.

Chlorine solution (sometimes called chlorine water) is very pale green. It turns Universal Indicator red showing the solution is acidic. The colour of the indicator is quickly bleached.

Bromine solution (bromine water) is orange. It is very weakly acidic and also acts as a bleach.

Iodine solution is very weakly acidic and is also a slight bleach. The low solubility of halogens in water (a polar solvent) is expected because halogens are composed of molecules.

SOLUBILITY OF HALOGENS IN HEXANE (A NONPOLAR SOLVENT)

The halogens dissolve readily in hexane to give solutions of characteristic colour:

chlorine – colourless
bromine – orange
iodine – purple

REACTIONS OF HALOGENS WITH IRON

The halogens react with metals by direct combination to form salts. The name 'halogen' means salt producer. Chlorine forms chlorides, bromine forms bromides and iodine forms iodides.

If chlorine gas is passed over heated iron wire, an exothermic reaction takes place forming iron(III) chloride, which forms as a brown solid on cooling. Figure 9.1 shows a suitable apparatus for preparing anhydrous iron(III) chloride crystals.

$$2Fe(s) + 3Cl_2(g) \rightarrow 2FeCl_3(s)$$
iron + chlorine → iron(III) chloride

The anhydrous calcium chloride tube is to prevent water vapour entering the apparatus.

Bromine vapour also reacts with hot iron wire to form iron(III) bromide. When iodine crystals are heated, they turn to a purple vapour. This vapour reacts with hot iron wire to produce iron(II) iodide.

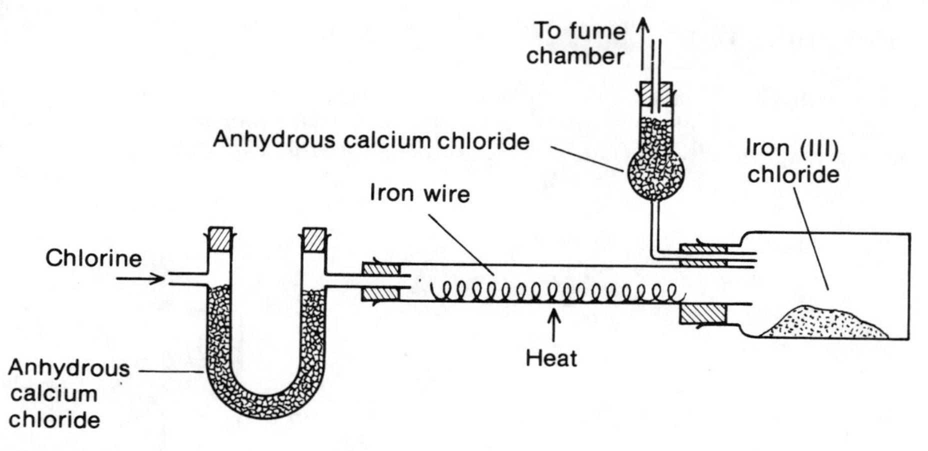

Fig 9.1 Preparation of iron(III) chloride

Order of reactivity of the halogens
From their chemical reactions the relative reactivities of the halogens are:

fluorine – most reactive
chlorine
bromine
iodine – least reactive

DISPLACEMENT REACTIONS OF THE HALOGENS

A more reactive halogen will displace a less reactive halogen from one of its compounds. For example when chlorine is bubbled into a solution of potassium bromide, the chlorine displaces the less reactive bromine. This means the colourless solution turns orange as the free bromine is formed.

$$2KBr(aq) + Cl_2(g) \rightarrow 2KCl(aq) + Br_2(aq)$$

potassium bromide + chlorine → potassium chloride + bromine

No reaction would take place if iodine solution were added to potassium bromide solution because iodine is less reactive than bromine.

9.4 The Noble or Inert Gas Family

This is a family of gases. They are put in the same family because they are all very unreactive. Until about 20 years ago these gases were believed to be completely without chemical reactions. Since then a number of compounds, including xenon tetrafluoride XeF_4, have been produced. Table 9.3 gives some information about the noble gases.

Table 9.3 Noble gases

Element	*Symbol*	*Boiling point* (°C)	*Density* (g/dm³)
Helium	He	−269	0.17
Neon	Ne	−246	0.84
Argon	Ar	−185	1.66
Krypton	Kr	−153	3.46
Xenon	Xe	−109	5.45
Radon	Rn	−62	8.9

The noble gases are important because of their lack of reactions and they form the basis of theories of bonding.

USES OF THE NOBLE GASES

Helium is used in balloons and airships. Although it is denser than hydrogen it has the advantage of not being inflammable.
Neon is used in advertising signs and **argon** is used to fill electric light bulbs. **Krypton** and **xenon** are used in lighthouse and projector bulbs.
Radon is radioactive and can be used to detect leaks in pipes.

9.5 The Periodic Table

The Periodic Table is an arrangement of all the chemical elements in order of increasing atomic number with elements having similar properties (i.e. of the same chemical family) in the same vertical column. The Periodic Table is shown in Fig. 9.2 in a modern form.

	I	II	Transition elements										III	IV	V	VI	VII	0
1							1 H Hydrogen 1											4 He Helium 2
2	7 Li Lithium 3	9 Be Beryllium 4											11 B Boron 5	12 C Carbon 6	14 N Nitrogen 7	16 O Oxygen 8	19 F Fluorine 9	20 Ne Neon 10
3	23 Na Sodium 11	24 Mg Magnesium 12											27 Al Aluminium 13	28 Si Silicon 14	31 P Phosphorus 15	32 S Sulphur 16	35.5 Cl Chlorine 17	40 Ar Argon 18
4	39 K Potassium 19	40 Ca Calcium 20	45 Sc Scandium 21	48 Ti Titanium 22	51 V Vanadium 23	52 Cr Chromium 24	55 Mn Manganese 25	56 Fe Iron 26	59 Co Cobalt 27	59 Ni Nickel 28	64 Cu Copper 29	65 Zn Zinc 30	70 Ga Gallium 31	73 Ge Germanium 32	75 As Arsenic 33	79 Se Selenium 34	80 Br Bromine 35	84 Kr Krypton 36
	85.5 Rb Rubidium 37	88 Sr Strontium 38	89 Y Yttrium 39	91 Zr Zirconium 40	93 Nb Niobium 41	96 Mo Molybdenum 42	98 Tc Technetium 43	101 Ru Ruthenium 44	103 Rh Rhodium 45	106 Pd Palladium 46	108 Ag Silver 47	112 Cd Cadmium 48	115 In Indium 49	119 Sn Tin 50	122 Sb Antimony 51	128 Te Tellurium 52	127 I Iodine 53	131 Xe Xenon 54
	133 Cs Caesium 55	137 Ba Barium 56	139 La Lanthanum 57	178.5 Hf Hafnium 72	181 Ta Tantalum 73	184 W Tungsten 74	186 Re Rhenium 75	190 Os Osmium 76	192 Ir Iridium 77	195 Pt Platinum 78	197 Au Gold 79	201 Hg Mercury 80	204 Tl Thallium 81	207 Pb Lead 82	209 Bi Bismuth 83	210 Po Polonium 84	210 At Astatine 85	222 Rn Radon 86
	223 Fr Francium 87	226 Ra Radium 88	227 Ac Actinium 89															

139 La Lanthanum 57	140 Ce Cerium 58	141 Pr Praseodymium 59	144 Nd Neodymium 60	147 Pm Promethium 61	150 Sm Samarium 62	152 Eu Europium 63	157 Gd Gadolinium 64	159 Tb Terbium 65	162.5 Dy Dysprosium 66	165 Ho Holmium 67	167 Er Erbium 68	169 Tm Thulium 69	173 Yb Ytterbium 70	175 Lu Lutetium 71
227 Ac Actinium 89	232 Th Thorium 90	231 Pa Protactinium 91	238 U Uranium 92	237 Np Neptunium 93	242 Pu Plutonium 94	243 Am Americium 95	247 Cm Curium 96	247 Bk Berkelium 97	251 Cf Californium 98	254 Es Einsteinium 99	253 Fm Fermium 100	256 Md Mendeleevium 101	254 No Nobelium 102	257 Lw Lawrencium 103

KEY:

Atomic mass
Symbol
Name
Atomic number

Fig 9.2 The Periodic Table of elements

9.6 Brief History of the Development of the Periodic Table

In the early 19th century many new elements were being discovered and chemists were looking for similarities between these new elements and existing elements.

Döbereiner (1829) suggested that elements could be grouped in three (triads). Each member of the triad has similar properties.

E.g. lithium, sodium, potassium
chlorine, bromine, iodine

Newlands (1863) arranged the elements in order of increasing relative atomic mass. He noticed that there was some similarity between each eighth element.

Li	Be	B	C	N	O	F	
Na	Mg	Al	Si	P	S	Cl	etc.

These were called **Newlands' Octaves**. Unfortunately the pattern broke down with the heavier elements and because he left no gaps for undiscovered elements. His work did not receive much support at the time.

Meyer (1869) looked at the relationship between relative atomic mass and the density of an element. He then plotted a graph of atomic volume (mass of 1 mole of atoms divided by density) against the relative atomic mass for each element. The curve he obtained showed periodic variations.

Mendeleef arranged the elements in order of increasing relative atomic mass but took into account the patterns of behaviour of the elements. He found it was necessary to leave gaps in the table and said that these were for elements not known at that time. His table enabled him to predict the properties of the undiscovered elements. His work was proved correct by the accurate prediction of the properties of gallium and germanium. The Periodic Table we use today closely resembles the table drawn up by Mendeleef.

A modification of the Periodic Table was made following the work of **Rutherford** and **Moseley**. It was realized that the elements should be arranged in order of atomic number, i.e. the number of protons in the nucleus. In the modern Periodic Table the elements are arranged in order of increasing atomic number with elements with similar properties in the same vertical column.

9.7 Structure of the Periodic Table

The vertical columns in the table are called **groups**. A group will contain elements with similar properties. The groups are given roman numbers as shown in Fig. 9.2.

The horizontal rows of elements are called **periods**.

The 'main block' elements are shaded in Fig. 9.2 and between the two parts of the main block are the heavy or transition metals.

9.8 Electron Arrangement and Reactivity in a Group

In Unit 7 the arrangement of electrons within an atom was explained. The chemical properties of an element are controlled by the number of electrons in the outer energy level.

As elements in the same group have similar properties, we should expect some similarity in their electronic arrangement.

Table 9.4 shows the arrangement of electrons in the alkali metal family (group I of the Periodic Table).

Table 9.4 Electron arrangement in group I

Element	*Atomic number*	*Arrangement of electrons*
Li	3	2,1
Na	11	2,8,1
K	19	2,8,8,1
Rb	37	2,8,18,8,1
Cs	55	2,8,18,18,8,1

Note that, in each case, the outer energy level contains just one electron. When an element reacts it attempts to obtain a full outer energy level.

Group I elements will lose one electron when they react and form a positive ion.

$$Na \rightarrow Na^{+} + e^{-}$$

We can explain the order of reactivity within the group. The electrons are held in position by the electrostatic attraction of the positive nucleus. This means that the closer the electron is to the nucleus, the harder it will be to remove it.

As we go down the group, the outer electron gets further away from the nucleus and so becomes easier to take away. This means as we go down the group, the reactivity should increase.

Table 9.5 shows the arrangement of electrons in the alkaline earth metal family (group II of the Periodic Table).

Table 9.5 Electron arrangement in group II

Element		*Atomic number*	*Arrangement of electrons*
Beryllium	Be	4	2,2
Magnesium	Mg	12	2,8,2
Calcium	Ca	20	2,8,8,2
Strontium	Sr	38	2,8,18,8,2
Barium	Ba	56	2,8,18,18,8,2
Radium	Ra	88	2,8,18,32,18,8,2

As the atoms all have two electrons in their outer energy level, they will lose two electrons to form positive ions.

$$Mg \rightarrow Mg^{2+} + 2e^{-}$$

More energy will be required to remove two electrons and so they will not be as reactive as the group I metals. As with group I, the reactivity will increase down the group.

E.g. Reaction of group II metals with water

Magnesium will react rapidly with steam.

$$Mg(s) + H_2O(g) \rightarrow MgO(s) + H_2(g)$$

magnesium + water (steam) → magnesium oxide + hydrogen

Calcium reacts with cold water.

$$Ca(s) + 2H_2O(l) \rightarrow Ca(OH)_2(aq) + H_2(g)$$

calcium + water → calcium hydroxide + hydrogen

Barium is stored in oil because it reacts rapidly with cold water.

$$Ba(s) + 2H_2O(l) \rightarrow Ba(OH)_2(aq) + H_2(g)$$

barium + water → barium hydroxide + hydrogen

Table 9.6 shows the arrangement of electrons in the halogen family (group VII of the Periodic Table).

Table 9.6 Electron arrangement in group VII

Element	*Atomic number*	*Arrangement of electrons*
F	9	2,7
Cl	17	2,8,7
Br	35	2,8,18,7
I	53	2,8,18,18,7

Note that each member of the group has seven electrons in the outer energy level. This is just one electron short of the full energy level.

When halogen elements react, they gain an electron to complete that outer energy level. This will form a negative ion.

E.g. $Cl + e^- \rightarrow Cl^-$

As an electron is being gained in the reaction, the most reactive member of the family will be the one where the extra electron is closest to the nucleus, i.e. fluorine. The reactivity decreases down the group.

9.9 Trends within a Period

We have already seen that metallic elements form positive ions and that nonmetals form negative ions. The metallic elements will then be those with only a few electrons in their outer energy level. The most metallic elements will be on the extreme left-hand side of the table, i.e. group I.

The nonmetallic elements will be on the right-hand side of the table. The heavy line in Fig. 9.2 divides metals from nonmetals. In any period of the Periodic Table, there is a gradual change from metallic to nonmetallic from left to right.

E.g. third period:

Na Mg Al Si P S Cl Ar

Also, from left to right in any period the atoms gradually decrease in size. This surprises many people because, going from left to right, each element has one more electron than the previous element. However, this electron goes into the same energy level and the extra positive charge on the nucleus, caused by the extra proton, increases the attraction on the electrons and makes the atom slightly smaller.

9.10 Heavy or Transition Metals

The elements between groups II and III in the Periodic Table are called the heavy or transition metals. These metals are similar in many ways. They generally:

1 Have high melting points, high boiling points and high densities.
2 Form more than one positive ion, e.g. iron forms Fe^{2+} and Fe^{3+} ions.
3 Form many coloured compounds.
4 Are often used as catalysts.

9.11 Summary

The chemical elements can be divided into groups in various ways. In a simple way they could be divided into:

metals and nonmetals
or solids, liquids and gases.

A better way of dividing the elements is into chemical families with elements having similar but not identical properties.

Lithium, sodium and potassium are members of the alkali metal family. Chlorine, bromine and iodine are members of the halogen family. Helium, neon, argon, krypton and xenon are the unreactive gases called noble gases.

The Periodic Table is an arrangement of elements in order of increasing atomic number with elements with similar properties, i.e. in the same family, in the same vertical column. The vertical columns are called groups and the horizontal rows are called periods.

The position of an element in the Periodic Table can tell you a great deal about its properties. Elements on the left-hand side of the table are metals and those on the right-hand side are nonmetals. There is a relationship between the position in the Periodic Table and the arrangement of electrons in an atom of the element.

Much of the chemistry you learn can be understood from the Periodic Table.

10 OXIDATION AND REDUCTION

This unit is written in two parts:

Part A	10.1–10.3	Simple oxidation and reduction
Part B	10.4–10.7	Further oxidation and reduction

Check on pages xvi–xvii to find which parts are required for your syllabus.

10.1 Oxidation

Oxidation can be defined in various ways. Two simple definitions are when:

1 Oxygen is added to a substance.
2 Hydrogen is lost by a substance.

E.g. 1. If magnesium is burnt in oxygen, the magnesium is oxidized.

$$2Mg(s) + O_2(g) \rightarrow 2MgO(s)$$

magnesium + oxygen → magnesium oxide

When any substance is burnt it is oxidized.

E.g. 2. If concentrated hydrochloric acid is oxidized, chlorine gas is produced.

$$MnO_2(s) + 4HCl(aq) \rightarrow MnCl_2(aq) + 2H_2O(l) + Cl_2(g)$$

manganese(IV) oxide + hydrochloric acid → manganese(II) chloride + water + chlorine

The concentrated hydrochloric acid loses hydrogen when being changed to chlorine and is therefore oxidized.

10.2 Reduction

Reduction is the reverse of oxidation. Reduction can be simply defined as reactions where:

1 Oxygen is lost by a substance.
2 Hydrogen is gained by a substance.

E.g. 1. If hydrogen is passed over heated copper(II)oxide, the copper(II) oxide is reduced to copper. The copper(II) oxide loses oxygen.

$$CuO(s) + H_2(g) \rightarrow Cu(s) + H_2O(g)$$
copper(II) oxide + hydrogen → copper + water

E.g. 2. If hydrogen and ethene are passed over a heated catalyst, the ethene is reduced to ethane. Ethene gains hydrogen.

$$C_2H_4(g) + H_2(g) \rightarrow C_2H_6(g)$$
ethene + hydrogen → ethane

10.3 Redox Reactions

Oxidation and reduction processes occur together. If one substance is oxidized, another is reduced. A process where oxidation and reduction are taking place may be called a **redox** (reduction–oxidation) reaction.

E.g. If a mixture of lead(II) oxide and carbon are heated together, the following reaction takes place:

$$PbO(s) + C(s) \rightarrow Pb(s) + CO(g)$$
lead(II) oxide + carbon → lead + carbon monoxide

In this reaction, lead(II) oxide is losing oxygen and forming lead. Lead(II) oxide is, therefore, being reduced. Carbon is gaining oxygen when it forms carbon monoxide: carbon is being oxidized when it forms carbon monoxide: carbon is being oxidized. Oxidation and reduction are both taking place. This is a redox reaction

No reaction would take place if the lead(II) oxide was heated alone. Carbon is the substance which is necessary for the reduction to take place because it removes the oxygen. Carbon is called the **reducing agent**. A reducing agent is a substance which reduces some other substances but it is itself oxidized.

Similarly, lead(II) oxide is the **oxidizing agent**. It supplies oxygen, which is used to oxidize the carbon. An oxidizing agent is a substance which oxidizes some other substance but is itself reduced.

Common reducing agents include hydrogen, carbon, carbon monoxide and metals.

Common oxidizing agents include oxygen, chlorine, concentrated sulphuric acid and concentrated nitric acid.

10.4 Further Oxidation and Reduction

The definitions of oxidation and reduction in 10.1 and 10.2 are not complete. It is often better to define oxidation and reduction in terms of electron loss and gain.

Oxidation is any process where electrons are lost and reduction is any process where electrons are gained.

Remember:

*l*oss of *e*lectrons – *o*xidation (*leo*)

E.g. 1. If chlorine is bubbled into a solution of iron(II) chloride (containing Fe^{2+} ions), the iron(II) chloride is oxidized to iron(III)chloride (containing Fe^{3+} ions). The solution changes from pale green to yellow-brown.

$$2FeCl_2(aq) + Cl_2(g) \rightarrow 2FeCl_3(aq)$$
iron(II) chloride + chlorine → iron(III) chloride

or $$2Fe^{2+}(aq) + Cl_2(g) \rightarrow 2Fe^{3+}(aq) + 2Cl^-(aq)$$

E.g. 2. During the electrolysis of lead(II) bromide (see Unit 12), oxidation and reduction are taking place.

At the cathode, lead ions are reduced to lead:

$$Pb^{2+}(l) + 2e^- \rightarrow Pb(s)$$

At the anode bromide ions are oxidized to bromine:

$$2Br^-(l) \rightarrow Br_2(g) + 2e^-$$

10.5 Common Oxidizing Agents

These are substances that, by their presence, cause other substances to be oxidized.

Oxygen

During the reaction oxygen molecules (O_2) gain electrons and form oxide (O^{2-}) ions.

E.g.

$$2Mg(s) + O_2(g) \rightarrow 2MgO(s)$$

magnesium + oxygen → magnesium oxide

Chlorine

During the reaction, chlorine molecules (Cl_2) gain electrons and form chloride (Cl^- ions).

E.g.

$$Cl_2(g) + 2FeCl_2(aq) \rightarrow 2FeCl_3(aq)$$

chlorine + iron(II) chloride → iron(III) chloride

Potassium manganate (VII) (potassium permanganate) $KMnO_4$ (acidified with dilute sulphuric acid)

During the reaction the solution turns from purple to become colourless as the manganate(VII) ions (MnO_4^-) are reduced to manganese(II) ions.

E.g. with iron(II) sulphate solution

$$MnO_4^-(aq) + 8H^+(aq) + 5e^- \rightarrow Mn^{2+}(aq) + 4H_2O(l)$$

$$Fe^{2+}(aq) \rightarrow Fe^{3+}(aq) + e^-$$

$$MnO_4^-(aq) + 8H^+(aq) + 5Fe^{2+}(aq) \rightarrow Mn^{2+}(aq) + 4H_2O(l) + 5Fe^{3+}(aq)$$

Potassium dichromate(VI) (potassium dichromate) $K_2Cr_2O_7$ (acidified with dilute sulphuric acid)

During the reaction the solution turns from orange to green as the $Cr_2O_7^{2-}$ ion is reduced to Cr^{3+}

E.g. with iron(II) sulphate solution

$$Cr_2O_7^{2-}(aq) + 14H^+(aq) + 6e^- \rightarrow 2Cr^{3+}(aq) + 7H_2O(l)$$

$$Fe^{2+}(aq) \rightarrow Fe^{3+}(aq) + e^-$$

$$Cr_2O_7^{2-}(aq) + 14H^+(aq) + 6Fe^{2+}(aq) \rightarrow 2Cr^{3+}(aq) + 7H_2O(l) + 6Fe^{3+}(aq)$$

Concentrated sulphuric acid H_2SO_4 (usually hot)

When concentrated sulphuric acid acts as an oxidizing agent, sulphur dioxide is always produced.

E.g. with copper

$$Cu(s) + 2H_2SO_4(l) \rightarrow CuSO_4(aq) + 2H_2O(l) + SO_2(g)$$

copper + sulphuric acid → copper(II) sulphate + water + sulphur dioxide

Concentrated nitric acid HNO_3 (usually hot)

When concentrated nitric acid acts as an oxidizing agent, nitrogen dioxide is produced.

E.g. with copper

$$Cu(s) + 4HNO_3(l) \rightarrow Cu(NO_3)_2(aq) + 2H_2O(l) + 2NO_2(g)$$

copper + nitric acid → copper(II) nitrate + water + nitrogen dioxide

10.6 Common Reducing Agents

These are substances that, by their presence, cause other substances to be reduced.

Hydrogen H_2

E.g.

$$C_2H_4(g) + H_2(g) \rightarrow C_2H_6(g)$$

ethene + hydrogen → ethane

Hydrogen sulphide H_2S

When hydrogen sulphide acts as a reducing agent, sulphur is always produced.

E.g. with chlorine

$$H_2S(g) + Cl_2(g) \rightarrow 2HCl(g) + S(s)$$

hydrogen sulphide + chlorine → hydrogen chloride + sulphur

Carbon C

E.g. with lead(II) oxide

$$PbO(s) + C(s) \rightarrow Pb(s) + CO(g)$$
lead(II) oxide + carbon → lead + carbon monoxide

Carbon monoxide CO

When carbon monoxide acts as a reducing agent, carbon dioxide is produced.
E.g. with iron(III) oxide (see 23.6)

$$Fe_2O_3(s) + 3CO(g) \rightarrow 2Fe(l) + 3CO_2(g)$$
iron(III) oxide + carbon monoxide → iron + carbon dioxide

Metals

E.g. iron acts as a reducing agent with copper(II) sulphate solution.

$$Cu^{2+}(aq) + Fe(s) \rightarrow Fe^{2+}(aq) + Cu(s)$$
copper(II) ions + iron → iron(II) ions + copper

10.7 Substances which can act as both Oxidizing and Reducing Agents

Sulphur dioxide (in the presence of water) and hydrogen peroxide can act as both oxidizing and reducing agents.
Sulphur dioxide, in the presence of water, forms sulphurous acid. When this acts as a reducing agent, it is oxidized to sulphuric acid.
E.g. with chlorine

$$SO_2(aq) + 2H_2O(l) + Cl_2(g) \rightarrow H_2SO_4(aq) + 2HCl(aq)$$
sulphur dioxide + water + chlorine → sulphuric acid + hydrochloric acid

In the presence of hydrogen sulphide (a stronger reducing agent), sulphur dioxide acts as an oxidizing agent.

$$2H_2S(g) + SO_2(g) \rightarrow 3S(s) + 2H_2O(l)$$
hydrogen sulphide + sulphur dioxide → sulphur + water

Similarly, hydrogen peroxide can act as an oxidizing agent or as a reducing agent.
E.g. With acidified potassium iodide solution (containing I^- ions), hydrogen peroxide acts as an oxidizing agent and oxidizes the iodide ions to iodine.

$$H_2O_2(aq) + 2H^+(aq) + 2e^- \rightarrow 2H_2O(l)$$
$$2I^-(aq) \rightarrow I_2(s) + 2e^-$$
$$\overline{H_2O_2(aq) + 2H^+(aq) + 2I^-(aq) \rightarrow 2H_2O(l) + I_2(s)}$$

With acidified potassium manganate(VII) (potassium permanganate) (a stronger oxidizing agent than hydrogen peroxide), hydrogen peroxide acts as a reducing agent and reduces the manganate(VII) ions to manganese(II) ions.

$$H_2O_2(aq) \rightarrow O_2(g) + 2H^+(aq) + 2e^-$$
$$MnO_4^-(aq) + 8H^+(aq) + 5e^- \rightarrow Mn^{2+}(aq) + 4H_2O(l)$$
$$\overline{5H_2O_2(aq) + 2MnO_4^-(aq) + 6H^+(aq) \rightarrow 5O_2(g) + 2Mn^{2+} + 8H_2O(l)}$$

10.8 Summary

Oxidation is a reaction where oxygen is gained, hydrogen is lost or electrons are lost.
Reduction is a reaction where oxygen is lost, hydrogen is gained or electrons are gained.
Usually oxidation and reduction occur together in a redox reaction.
An oxidizing agent is a substance which oxidizes another substance and is itself reduced. Common oxidizing agents include oxygen, chlorine, concentrated sulphuric acid and concentrated nitric acid.
A reducing agent is a substance which reduces another substance and is itself oxidized. Common reducing agents include hydrogen, carbon, carbon monoxide and metals.

11 ACIDS, BASES AND SALTS

11.1 Acids

Acids form an important group of chemicals. They are generally thought of as being corrosive and having a sour taste.

Acids can be detected by using an **indicator** such as **litmus**. Indicators are dyes, or mixtures of dyes, which change colour when acids or alkalis are added. A wide range of indicators are available, each having its own characteristic colour in acid and alkali.

E.g. litmus — red in acid, blue in alkali

A very useful indicator is called **Universal Indicator**. This is a mixture of dyes and so gives a greater range of colour changes. The colours of Universal Indicator are shown in Fig. 11.1

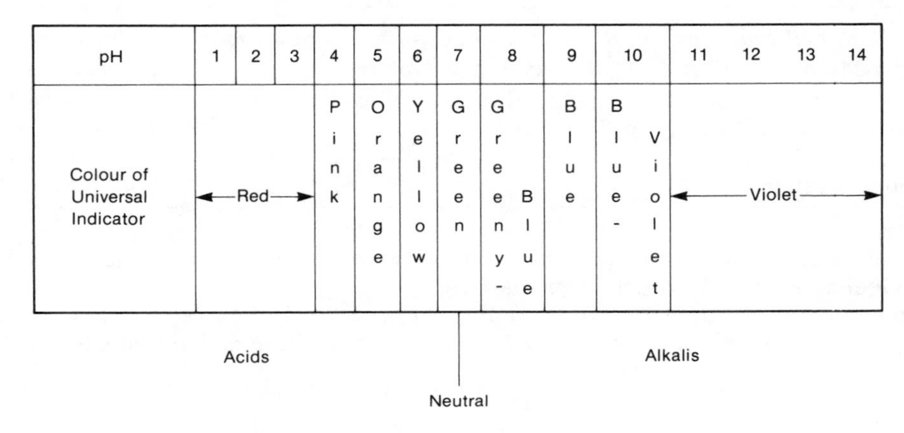

Fig 11.1 Colours of Universal Indicator

Universal Indicator can not only show whether a substance is acid, alkaline or neutral but can also show how strong the acid or alkali is. Table 11.1 gives a list of some common acids.

Table 11.1 Common acids

Acid	*Formula*	
Hydrochloric acid	HCl	Mineral acids
Nitric acid	HNO_3	Mineral acids
Sulphuric acid	H_2SO_4	Mineral acids
Ethanoic acid (acetic acid)	CH_3COOH	Contained in vinegar
Citric acid	$C_6H_8O_7$	Contained in lemon juice

11.2 General Reactions of Acids

Although there are a large number of different acids, there are a number of general chemical reactions common to all acids.

1 Acids turn indicators to their characteristic colour, e.g. litmus turns red.

2 Acids react with a fairly reactive metal to form a salt and evolve hydrogen.

E.g.
$$Mg(s) + H_2SO_4(aq) \rightarrow MgSO_4(aq) + H_2(g)$$
magnesium + sulphuric acid → magnesium sulphate + hydrogen

An exception is nitric acid. This acid tends to release oxides of nitrogen when reacted with metals, although very dilute nitric acid does liberate hydrogen with magnesium.

3 Acids react with metal oxides to form a salt and water. In most cases the acid needs warming.

E.g.
$$CuO(s) + 2HNO_3(aq) \rightarrow Cu(NO_3)_2(aq) + H_2O(l)$$
copper(II) oxide + nitric acid → copper(II) nitrate + water

4 Acids react with metal carbonates to form a salt, carbon dioxide and water.

$$CaCO_3(s) + 2HCl(aq) \rightarrow CaCl_2(aq) + CO_2(g) + H_2O(l)$$

calcium carbonate + hydrochloric acid → calcium chloride + carbon dioxide + water

5 Acids react with alkalis to form a salt and water.

E.g.

$$NaOH(aq) + HCl(aq) \rightarrow NaCl(aq) + H_2O(l)$$

sodium hydroxide + hydrochloric acid → sodium chloride + water

11.3 Importance of Water

Hydrogen chloride gas will dissolve in both water and methylbenzene (toluene) but the solutions have different properties. Table 11.2 compares results of some tests for each solution.

Table 11.2 Testing hydrogen chloride solutions

Test	*Solution of hydrogen chloride in water*	*Solution of dry hydrogen chloride in dry methylbenzene (toluene)*
1 *Dry Universal Indicator paper*	Turns red showing strong acid	Turns green – neutral
2 *Add magnesium ribbon*	Hydrogen evolved	No reaction
3 *Add calcium carbonate*	Carbon dioxide evolved	No reaction
4 *Electrical conductivity*	Good conductor	Nonconductor
5 *Temperature change on forming solution*	Rise in temperature	Little change

These results show that the solution in water behaves as an acid but the solution in methylbenzene shows no acidic properties.

The electrical conductivity of the solution in water indicates the presence of ions.

When hydrogen chloride gas dissolves in water it changes from molecules to ions:

$$HCl(g) \xrightarrow{water} H^+(aq) + Cl^-(aq)$$

This ionization is accompanied by a rise in temperature. **It is the hydrogen ions that cause acidic properties and these are formed in the presence of water.**

You will come across different names and symbols for the hydrogen ion in solution which attempt to show how the hydrogen ion becomes associated with water molecules:

$H^+(aq)$ **hydrated proton** or **hydrated hydrogen ion**

$H_3O^+(aq)$ **oxonium** or **hydronium ion**

Remember that the hydrogen ion is simply a proton. (This is because a hydrogen atom consists of a single proton and a single electron; when H^+ is formed the electron is removed.)

11.4 Definitions of an Acid

An acid can be defined in various ways. These include:

1 An acid is a substance that contains hydrogen which can be wholly or partially replaced by a metal. This hydrogen is called replaceable hydrogen.

2 An acid is a substance which forms hydrogen ions when it is dissolved in water.

3 An acid is a proton donor. It provides protons or H^+ ions.

The general reactions of an acid (see 11.2) can now be seen as reactions of hydrogen ions in solution.

E.g.

$$Mg(s) + 2H^+(aq) \rightarrow Mg^{2+}(aq) + H_2(g)$$
$$O^{2-}(s) + 2H^+(aq) \rightarrow H_2O(l)$$
$$CO_3^{2-}(s) + 2H^+(aq) \rightarrow CO_2(g) + H_2O(l)$$
$$OH^-(aq) + H^+(aq) \rightarrow H_2O(l)$$

11.5 pH Scale

As the reactions of an acid are those of the hydrogen ions in solution, clearly the more hydrogen ions present, the stronger the acid will be. The strength of an acid is measured on the

pH scale. Figure 11.1 relates the colour of Universal Indicator to the pH number. The pH number is a measure of the hydrogen ion concentration.

pH 1	strong acid
pH 2–pH 6	weaker acids
pH 7	neutral
pH 8–pH 13	weaker alkalis
pH 14	strong alkali

The smaller the pH the larger is the concentration of hydrogen ions, and so the stronger the acid is. The pH can be measured with Universal Indicator or by using a pH meter.

11.6 Strong and Weak Acids

Some acids completely ionize when they dissolve in water. These are called **strong acids**, i.e. the solution will contain a high concentration of hydrogen ions.
E.g. sulphuric acid

$$H_2SO_4(l) \xrightarrow{\text{water}} 2H^+(aq) + SO_4^{2-}(aq)$$

Other acids do not completely ionize on dissolving in water, i.e. some of the molecules remain un-ionized in the solution. These are called **weak acids**.
E.g. ethanoic acid (acetic acid)

$$CH_3COOH(l) \underset{}{\overset{\text{water}}{\rightleftharpoons}} H^+(aq) + CH_3COO^-(aq)$$

In a molar solution of ethanoic acid only about four molecules in every thousand change into ions.
Table 11.3 lists some common strong and weak acids.

Table 11.3 Strong and weak acids

Strong acids	*Weak acids*
Hydrochloric acid	Ethanoic acid
Sulphuric acid	Carbonic acid
Nitric acid	Sulphurous acid

11.7 Bases

A **base** is a substance that can accept hydrogen ions, i.e. a proton acceptor. A base will react with an acid to form a salt and water only.

E.g.

$$CuO(s) + H_2SO_4(aq) \rightarrow CuSO_4(aq) + H_2O(l)$$

copper(II) oxide + sulphuric acid → copper(II) sulphate + water

$$O^{2-}(s) + 2H^+(aq) \rightarrow H_2O(l)$$

If a base is soluble in water, the solution is called an **alkali**. Examples of alkalis include sodium hydroxide NaOH, potassium hydroxide KOH and calcium hydroxide $Ca(OH)_2$. An alkali contains a high concentration of hydroxide ions (OH^-) in solution.
When an acid reacts with an alkali, the reaction is between hydrogen ions and hydroxide ions.

$$H^+(aq) + OH^-(aq) \rightarrow H_2O(l)$$

This reaction is called a **neutralization** reaction.

11.8 Strong and Weak Alkalis

If an alkali completely ionizes on dissolving in water, a **strong alkali** is produced.
E.g. sodium hydroxide

$$NaOH(s) \xrightarrow{\text{water}} Na^+(aq) + OH^-(aq)$$

If an alkali does not completely ionize in water, a **weak alkali** is formed.
E.g. ammonium hydroxide (a solution of ammonia gas in water)

$$NH_4OH(aq) \rightleftharpoons NH_4^+(aq) + OH^-(aq)$$

11.9 Salts

A **salt** is a compound formed when the hydrogen of an acid is wholly or partly replaced by a metal.

The number of replaceable hydrogens in an acid is called the **basicity** of the acid. Table 11.4 gives the basicity of some common acids.

Table 11.4 Basicity of common acids

Acid	*Formula*	*Basicity*
Hydrochloric acid	HCl	1
Nitric acid	HNO_3	1
Ethanoic acid	CH_3COOH	1
Sulphuric acid	H_2SO_4	2
Sulphurous acid	H_2SO_3	2
Carbonic acid	H_2CO_3	2
Phosphoric acid	H_3PO_4	3

In the case of an acid containing more than one replaceable hydrogen, salts can be formed when all or some of the hydrogen is replaced. Salts formed by replacing only part of the replaceable hydrogen are called **acid salts**.

Table 11.5 gives examples of salts formed by common acids.

Table 11.5 Salts of common acids

Acid	*Salt*	*Example*
Hydrochloric acid HCl	Chlorides	Sodium chloride $NaCl$
Nitric acid HNO_3	Nitrates	Sodium nitrate $NaNO_3$
Ethanoic acid CH_3COOH	Ethanoates (acetates)	Sodium ethanoate CH_3COONa
Sulphuric acid H_2SO_4	Sulphates	Sodium sulphate Na_2SO_4
	Hydrogensulphates (acid salt)	Sodium hydrogensulphate $NaHSO_4$
Carbonic acid	Carbonates	Sodium carbonate Na_2CO_3
	Hydrogencarbonates (acid salt)	Sodium hydrogencarbonate $NaHCO_3$

11.10 Preparation of Salts

Before we can decide on the method of preparation of a salt, we need to know whether or not it is soluble in water.

The rules are as follows:

1 All nitrates are soluble in water.

2 All sulphates are soluble in water except lead sulphate and barium sulphate. (Calcium sulphate is only slightly soluble in water.)

3 All chlorides are soluble in water except silver chloride, lead chloride and mercury(I) chloride.

4 All carbonates are insoluble in water except sodium carbonate, potassium carbonate and ammonium carbonate.

5 All sulphides are insoluble except sodium sulphide, potassium sulphide and ammonium sulphide.

6 All salts of sodium, potassium and ammonium are soluble in water.

Both soluble and insoluble salts can be prepared by direct combination, e.g. iron(III) chloride, sodium chloride.

11.11 Preparation of Soluble Salts

For soluble salts, there are four general methods of preparation. These are the same reactions as the general reactions of an acid (see 11.2).

1 ACID + METAL

This method is only suitable for fairly reactive metals, e.g. magnesium, zinc and iron. The reaction is too vigorous for the more reactive metals. For metals below hydrogen in the reactivity series (see 8.4) this method is not suitable.

E.g.

$$Mg(s) + H_2SO_4(aq) \rightarrow MgSO_4(aq) + H_2(g)$$

magnesium + sulphuric acid → magnesium sulphate + hydrogen

Magnesium powder is added to warm, dilute sulphuric acid in small amounts until excess magnesium powder remains in the solution. Excess magnesium powder is removed by filtering. Magnesium sulphate crystals are obtained by evaporating the solution until crystals form on the end of a glass rod, which was previously dipped into the hot solution. The solution is then left to cool.

2 ACID + METAL OXIDE

This reaction requires warming to speed up the reaction. The method is the same as in 1.

E.g.

$$H_2SO_4(aq) + CuO(s) \rightarrow CuSO_4(aq) + H_2O(l)$$

sulphuric acid + copper(II) oxide → copper(II) sulphate + water

3 ACID + METAL CARBONATE

This reaction takes place at room temperature. The method is again the same as in 1.

E.g.

$$2HNO_3(aq) + CaCO_3(s) \rightarrow Ca(NO_3)_2(aq) + CO_2(g) + H_2O(l)$$

nitric acid + calcium carbonate → calcium nitrate + carbon dioxide + water

4 ACID + ALKALI

This reaction requires a special technique as both reactants are solutions and so an indicator has to be used to show when reacting quantities of acid and alkali have been used.

Acid is added to a measured volume of alkali until the indicator changes colour. The process is then repeated using the same volumes of acid and alkali but without the indicator. The solution is then evaporated to obtain the salt.

E.g.

$$HCl(aq) + NaOH(aq) \rightarrow NaCl(aq) + H_2O(l)$$

hydrochloric acid + sodium hydroxide → sodium chloride + water

11.12 Preparation of Acid Salts

To prepare an acid salt the exact amount of acid must be added to the alkali. If 25 cm^3 of sulphuric acid is required to produce the normal salt, sodium sulphate, then 50 cm^3 of sulphuric acid is required to produce the acid salt, sodium hydrogensulphate.

$$2NaOH(aq) + H_2SO_4(aq) \rightarrow Na_2SO_4(aq) + 2H_2O(l)$$

sodium hydroxide + sulphuric acid → sodium sulphate + water

$$NaOH(aq) + H_2SO_4(aq) \rightarrow NaHSO_4(aq) + H_2O(l)$$

sodium hydroxide + sulphuric acid → sodium hydrogensulphate + water

11.13 Preparation of Insoluble Salts

There is only one method for preparing insoluble salts. This involves mixing together solutions of two soluble salts each containing half of the required salt. The required insoluble salt is then **precipitated.**

For example the insoluble salt lead carbonate can be prepared by mixing together solutions of a soluble lead salt (lead nitrate) and a soluble carbonate (sodium carbonate).

$$Pb(NO_3)_2(aq) + Na_2CO_3(aq) \rightarrow PbCO_3(s) + 2NaNO_3(aq)$$

lead nitrate + sodium carbonate → lead carbonate + sodium nitrate

Ionic equation:

$$Pb^{2+}(aq) + CO_3^{2-}(aq) \rightarrow PbCO_3(s)$$

This type of reaction is sometimes called **double decomposition** and is represented by the equation:

$$AX + BY \rightarrow AY + BX$$

In order to obtain a pure sample of the insoluble salt it is necessary to filter off the precipitate and then wash it with distilled water and finally dry the precipitate thoroughly.

11.14 Summary

An aqueous solution with a pH less than 7 contains an excess of H^+ ions and contains an acid. The three common mineral acids are sulphuric acid H_2SO_4, nitric acid HNO_3 and hydrochloric acid HCl.

Acids will:

1 Produce hydrogen gas with a reactive metal such as magnesium or zinc.
2 Produce a salt and water with a metal oxide or hydroxide.
3 Produce carbon dioxide with a metal carbonate.
4 Turn indicators to the acid colour.

A strong acid is an acid which is completely ionized in solution. A weak acid is an acid which is only partly ionized in solution. Acids cannot be ionized and show acid properties unless water is present.

A base is a metal oxide, e.g. copper(II) oxide. A base which is soluble in water forms an alkali. Common alkalis include sodium hydroxide NaOH, potassium hydroxide KOH and calcium hydroxide $Ca(OH)_2$. All alkali solutions contain an excess of hydroxide ions OH^-.

The reaction between an acid and an alkali is called a neutralization reaction:

$$H^+(aq) + OH^-(aq) \rightarrow H_2O(l)$$

A salt is formed when all or part of the replaceable hydrogen is replaced by a metal.

Soluble salts are formed by reacting the metal, metal oxide, metal hydroxide or metal carbonate with the correct acid.

Insoluble salts are prepared by mixing suitable aqueous solutions. The insoluble salt is formed by precipitation and can be removed by filtration.

12 THE EFFECT OF ELECTRICITY ON CHEMICALS

This unit is written in two parts:

Part A	12.1–12.5	Qualitative electrolysis
Part B	12.6–12.7	Quantitative electrolysis

Check on pages xvi–xvii to find which parts are required for your syllabus.

12.1 Conductors and Insulators

A substance which allows electricity to pass through it is called a **conductor**. Of the solid elements at room temperature, only metals and graphite (a form of carbon) are good conductors. They conduct electricity because electrons pass freely through the solid.

Substances which do not allow electricity to pass through are called **insulators**. Some substances, e.g. germanium, conduct electricity slightly and are called **semiconductors**. They are important for making transistors.

12.2 Electrolytes

Certain substances do not conduct electricity when solid but do when molten or dissolved in water. They are called **electrolytes**. However, the passage of electricity through the melt or solution is accompanied by a chemical decomposition. The splitting up of an electrolyte, when molten or in aqueous solution, is called **electrolysis**.

Electrolytes include:

acids, metal oxides, metal hydroxides and salts

Electrolytes are composed of **ions** (i.e. ionic bonding, see 7.4) but in the solid state the ions are rigidly held in regular positions and are unable to move to an electrode. Sodium chloride is composed of a regular lattice of sodium Na^+ and chloride Cl^- ions.

Melting the electrolyte breaks down the forces between the ions. The ions are, therefore, free to move in a molten electrolyte. In molten sodium chloride the sodium and chloride ions are able to move freely.

Dissolving an electrolyte in water (or other polar solvent) also causes the breakdown of the lattice and again the ions are free to move.

12.3 Electrolysis of Molten Lead(II) Bromide

Lead(II) bromide is an electrolyte.
The apparatus in Fig. 12.1 could be used for the electrolysis of molten lead(II) bromide.
The bulb does not light up while the lead(II) bromide is solid showing that no electric current is passing through the solid lead(II) bromide. As soon as the lead(II) bromide melts, the bulb lights up. After a while, bromine is seen escaping and, at the end of the experiment, lead can be found inside the crucible.
The reaction is, therefore

$$PbBr_2(l) \rightarrow Pb(l) + Br_2(g)$$

lead(II) bromide → lead + bromine

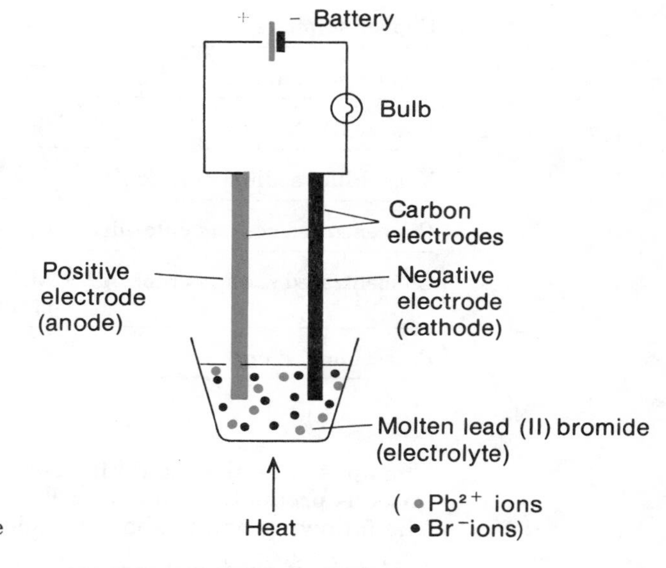

Fig 12.1 Electrolysis of molten lead bromide

The carbon rods are called **electrodes**. The electrode attached to the positive terminal of the battery is the positive electrode (sometimes called the **anode**). The electrode attached to the negative terminal of the battery is the negative electrode (sometimes called the **cathode**). The positive electrode has a shortage of electrons and the negative electrode a surplus of electrons. Electrons are constantly flowing through the wire.

Lead(II) bromide is composed of lead ions Pb^{2+} and bromide ions Br^-. When the lead bromide is molten, the ions move (or migrate) towards the electrode of opposite charge, i.e. Pb^{2+} ions move to the negative electrode and Br^- ions to the positive electrode.

When the ions reach the oppositely charged electrode they are discharged. At the positive electrode (anode), bromide ions lose electrons to the electrode and form bromine molecules.

$$2Br^- \rightarrow Br_2 + 2e^- \quad \text{(oxidation)}$$

At the negative electrode (cathode), lead ions gain electrons from the electrode and form lead atoms.

$$Pb^{2+} + 2e^- \rightarrow Pb \quad \text{(reduction)}$$

The electrolysis of a molten electrolyte is comparatively easy to understand because only one positive and one negative type of ion are present. In a solution it is possible to have two types of positive ions and then these ions can compete to be discharged at the cathode.

12.4 Electrolysis of Aqueous Solutions

In pure water about 1 in every 600 000 000 water molecules ionize to form hydrogen and hydroxide ions:

$$H_2O(l) \rightleftharpoons H^+(aq) + OH^-(aq)$$

This very slight ionization of water molecules explains the very slight electrical conductivity of pure water.

A solution of sodium chloride in water contains the following ions:

$H^+(aq)$	$OH^-(aq)$	from the water
$Na^+(aq)$	$Cl^-(aq)$	from the sodium chloride

Both positive ions migrate to the negative electrode and both negative ions move to the positive electrode. At each electrode one or both of the ions may be discharged.

Table 12.1 below summarizes the results of some electrolysis experiments.

Table 12.1 Examples of electrolysis of solutions

Solution	*Electrodes*	*Ion discharged at positive electrode*	*Ion discharged at negative electrode*	*Product at positive electrode*	*Product at negative electrode*
Dilute sulphuric acid	Carbon	$OH^-(aq)$	$H^+(aq)$	Oxygen	Hydrogen
Dilute sodium hydroxide	Carbon	$OH^-(aq)$	$H^+(aq)$	Oxygen	Hydrogen
Copper sulphate	Carbon	$OH^-(aq)$	$Cu^{2+}(aq)$	Oxygen	Copper
Copper sulphate	Copper	None	$Cu^{2+}(aq)$	None	Copper
Copper(II) chloride	Carbon	$Cl^-(aq)$	$Cu^{2+}(aq)$	Chlorine	Copper
Very dilute sodium chloride	Carbon	$OH^-(aq)$	$H^+(aq)$	Oxygen	Hydrogen
Concentrated sodium chloride	Carbon	$Cl^-(aq)$	$H^+(aq)$	Chlorine	Hydrogen
Concentrated sodium chloride	Mercury cathode	$Cl^-(aq)$	$Na^+(aq)$	Chlorine	Sodium (amalgam)
Potassium iodide	Carbon	$I^-(aq)$	$H^+(aq)$	Iodine	Hydrogen

The apparatus that could be used for the electrolysis of a solution and the collection of the gaseous products is shown in Fig. 12.2.

The following points about the electrolysis of solutions are worth remembering:

1 Metals, if produced, are discharged at the negative electrode.

2 Hydrogen is produced at the negative electrode only.

3 Nonmetals, apart from hydrogen, are produced at the positive electrode.

4 Reactive metals are not formed at the cathode during electrolysis of aqueous solutions. An exception is during the electrolysis of sodium chloride using a mercury cathode.

5 The products obtained can depend upon the concentration of the electrolyte in the solution. For example, electrolysis of concentrated sodium chloride produces chlorine at the anode but electrolysis of dilute sodium chloride can produce oxygen at the anode.

6 Providing the concentrations of the negative ions in solution are approximately the same, the order of discharge is

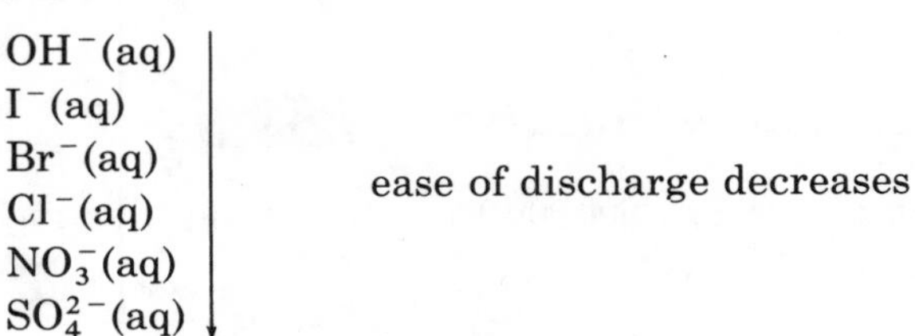

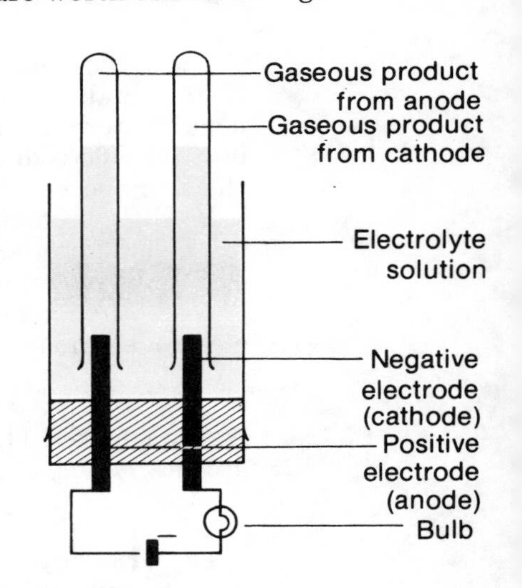

Fig 12.2 Electrolysis of solutions

12.5 Uses of Electrolysis

1 ELECTROPLATING

Electrolysis can be used to form a very thin coating of a metal on the surface of another metal. This can be used for decorative purposes or to prevent corrosion.

If a piece of copper is to be nickel plated the surface must be clean and grease free. The copper is made to be the cathode and a piece of nickel is the anode in an electrolysis experiment with nickel(II) sulphate solution as the electrolyte. A nickel coating is deposited on the negative electrode (the copper) and the nickel anode goes into solution as nickel(II) ions.

2 EXTRACTION OF METALS (Units 23.4 and 23.5)

3 PURIFICATION OF COPPER

4 MANUFACTURE OF SODIUM HYDROXIDE (Unit 28.3)

12.6 Quantitative Electrolysis

If the bulb in Fig. 12.1 is replaced by an ammeter, it is possible to measure the current flowing in the circuit (in amperes, A) and to calculate the quantity of electricity passed in a certain time.

If a current of **1 A** flows for **1 second, s**, the quantity of electricity passed is **1 coulomb c.** During an experiment it is found that, because a large number of coulombs are passed, it is better to work in units of Faradays.

1 Faraday (F) = 96 500 coulombs (approximately)

$$\text{quantity of electricity passed} = \frac{\text{current (in A)} \times \text{time (in s)}}{96\,500}\text{ F}$$

It is found, by experiment, that the quantity of electricity passed determines the mass of products formed.

Faraday's first law of electrolysis states that **the mass of a given element liberated during electrolysis is directly proportional to the quantity of electricity consumed during the electrolysis.**

12.7 Quantity of Different Elements Deposited by the Same Quantity of Electricity

Faraday's second law states that **the masses of different elements liberated by the same quantity of electricity form simple whole number ratios when divided by their relative atomic masses.**

For example, during an electrolysis experiment 1.08 g of silver are deposited and 0.32 g of copper. The relative atomic masses (A_r) of silver and copper are 108 and 64 respectively. Therefore dividing the mass deposited by the appropriate relative atomic mass gives:

Silver	*Copper*
$\frac{1.08}{108} = 0.01$	$\frac{0.32}{64} = 0.005$

The answers obtained are in a simple ratio of 2:1. This is in accordance with Faraday's second law of electrolysis.

It is found that the discharge of one mole of atoms of an element from an ion with a single positive or negative charge requires 1 F of electricity.

∴ 1 F produces 108 g of silver from Ag^+ ions
or 35.5 g of chlorine from Cl^- ions

In a similar way, the discharge of one mole of atoms of an element from an ion with a double positive or negative charge requires 2 F of electricity.

∴ 2 F produce 64 g of copper from Cu^{2+} ions
or 16 g of oxygen from O^{2-} ions

Sample calculation

Calculate the mass of calcium atoms produced when a current of 5 A is passed for 32 min 10 s through molten calcium bromide ($Ca^{2+}.2Br^-$) ($A_r(Ca) = 40$).

$$\text{quantity of electricity passed} = 5 \times 1930 \text{ C}$$
$$= \frac{5 \times 1930}{96\,500} \text{ F}$$
$$= 0.1 \text{ F}$$

Since calcium ions are Ca^{2+}, when 2 F are passed, the number of moles of calcium atoms formed = 1.

$$\therefore 0.1 \text{ F produces } \frac{0.1}{2} \text{ mol of calcium atoms}$$
$$= \frac{0.1}{2} \times 40 \text{ g of calcium}$$
$$= 2 \text{ g of calcium}$$

N.B. One Faraday of electricity contains 1 mole of electrons.

12.8 Summary

Metals and carbon conduct electricity well because electrons can pass freely through them. They are called conductors. Substances which will not let electrons pass through are insulators.

Some substances, called electrolytes, do not conduct electricity when solid but do when they are molten or dissolved in water. This passing of electricity splits up the electrolyte to produce new products. Common electrolytes are metal oxides, acids, alkalis and salts. They are all composed of ions. The ions are not free to move in the solid but become free to move on melting or dissolving in water.

Electrolysis of lead(II) bromide produces lead metal at the negative electrode (cathode) and bromine gas at the positive electrode (anode).

Electrolysis of aqueous solutions can produce a greater variety of products. Hydrogen gas is commonly produced at the cathode and oxygen is commonly produced at the anode. Reactive metals such as sodium and potassium are never produced during electrolysis of aqueous solutions.

Electrolysis is an important industrial process but it is expensive because of the large quantities of electricity needed.

The quantity of products formed during electrolysis depends upon the quantity of electricity used. The quantity of electricity used is measured in coulombs or Faradays.

13 RATES OF REACTION

13.1 Introduction

The **rate** of a chemical reaction is a measure of how fast the reaction takes place. It is important to remember that a rapid reaction is completed in a short time. Some reactions are very fast, e.g. the formation of silver chloride precipitate when silver nitrate and hydrochloric acid solutions are mixed. Other reactions are very slow, e.g. the rusting of iron. For practical reasons, reactions used in the laboratory for studying rates of reaction must not be too fast or too slow.

13.2 Finding Suitable Measurable Changes

Having selected a suitable reaction it is necessary to find a change that can be observed during the reaction. An estimate of the rate of the reaction can be obtained from the time taken for the measurable change to take place. Suitable changes include:

1 Colour.
2 Formation of precipitate.
3 Change in mass (e.g. a gas evolved causing a loss of mass).
4 Volume of gas evolved.
5 Time taken for a given mass of reagent to disappear.
6 pH.
7 Temperature.

13.3 Studying Rates of Reaction

Some of the easiest reactions to study in the laboratory are those in which a gas is evolved. The reaction can be followed by measuring the volume of gas evolved over a period of time using the apparatus in Fig. 13.1.

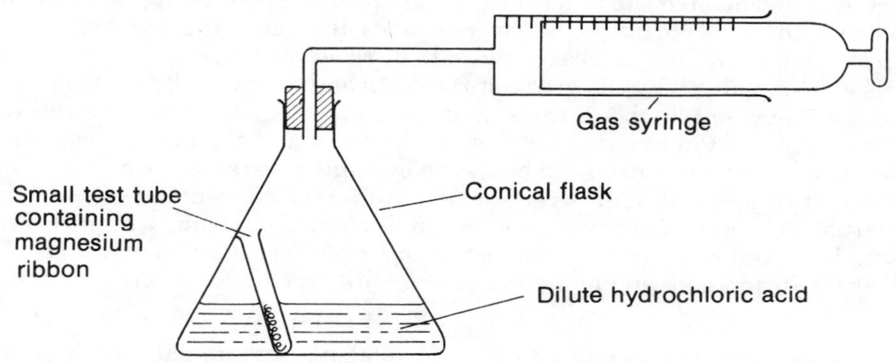

Fig 13.1 Studying the reaction between magnesium and dilute hydrochloric acid

A suitable reaction is:

$$Mg(s) + 2HCl(aq) \rightarrow MgCl_2(aq) + H_2(g)$$

magnesium + hydrochloric acid → magnesium chloride + hydrogen

It is important to keep the reactants separate whilst setting up the apparatus so that the starting time of the reaction can be measured accurately.

Figure 13.2 shows a typical graph obtained for the reaction between dilute hydrochloric acid and magnesium. The dotted line shows the graph for a similar experiment using the same quantities of magnesium and hydrochloric acid but with conditions changed so that the reaction is slightly faster. The rate of the reaction is greatest when the graph is steepest, i.e. at the start of the reaction. The reaction is complete when the graph becomes horizontal, i.e. there is no further increase in the volume of hydrogen. N.B. If you are carrying out an experiment all of the points may not lie on the curve. This is because of experimental error. You should draw the best graph through as many points as possible. It is often possible to follow the course of similar reactions by measuring the loss of mass during the reaction due to escape of gas. However, in this case the loss of mass is very small.

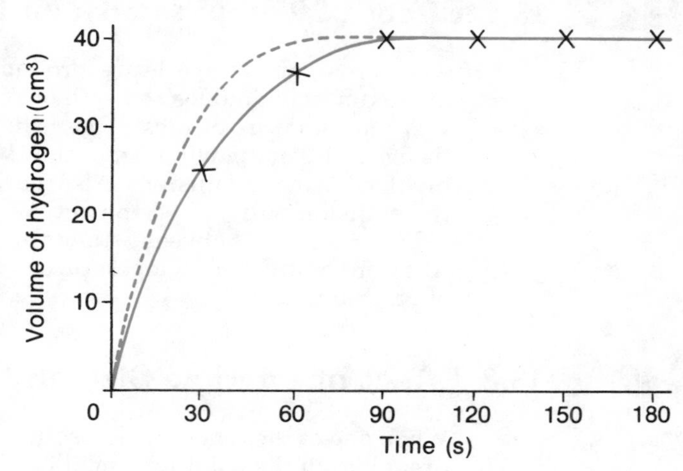

Fig 13.2 A graph of volume of hydrogen collected at intervals

13.4 Collision Theory for Rates of Reaction

Before looking at the factors that can alter the rate of reaction, we must consider what happens when a reaction takes place.

First of all, the particles of the reacting substances must **collide** with each other and, secondly, a fixed amount of energy called the **activation energy (E_a)** must be reached if the reaction is to take place (Fig. 13.3).

If a collision between particles can produce sufficient energy (i.e. if they collide fast enough and in the right direction) a reaction will take place. Not all collisions will result in a reaction.

A reaction is speeded up if the number of suitable collisions is increased.

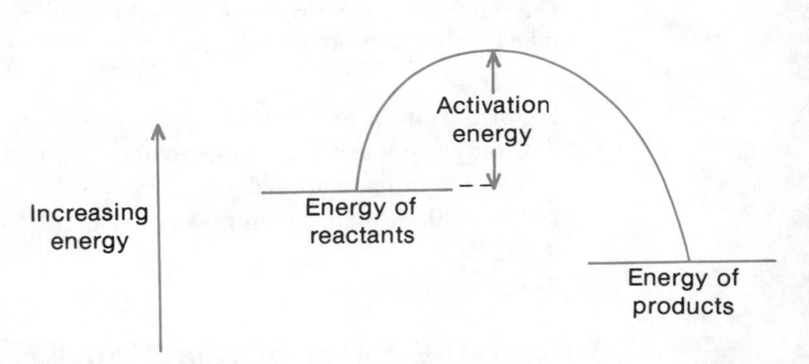

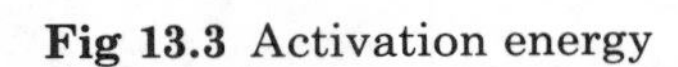
Fig 13.3 Activation energy

13.5 Effect of Concentration on the Rate of Reaction

The more concentrated the reactants, the greater will be the rate of reaction. This is because increasing the concentration of the reactants increases the number of collisions between particles and, therefore, increases the rate of reaction.

This also explains why the greatest rate of reaction is usually as soon as the reactants are mixed i.e. they are both at their highest concentrations. As the reaction proceeds the concentrations of the reacting substances decrease and the rate of reaction decreases.

The effect of concentration can be shown by doing several experiments using equal masses of magnesium ribbon and hydrochloric acid of different concentrations.

Alternatively, a series of experiments can be carried out using a standard solution of sodium thiosulphate and hydrochloric acid solutions of different concentrations. The time is taken until the solution goes so cloudy that a cross disappears when viewed through the solution.

$$Na_2S_2O_3(aq) + 2HCl(aq) \rightarrow 2NaCl(aq) + S(s) + H_2O(l) + SO_2(g)$$

sodium thiosulphate + hydrochloric acid → sodium chloride + sulphur + water + sulphur dioxide

The cloudiness is due to the precipitation of sulphur.

13.6 Effect of Pressure on the Rate of Reaction

When one or more of the reactants are gases an increase in pressure can lead to an increased rate of reaction. The increase in pressure forces the particles closer together. This causes more collisions and increases the rate of reaction.

13.7 Effect of Temperature on the Rate of Reaction

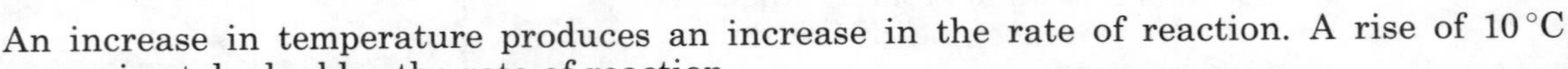

An increase in temperature produces an increase in the rate of reaction. A rise of 10 °C approximately doubles the rate of reaction.

When a mixture of substances is heated, the particles move faster. This has two effects. Since the particles are moving faster they will travel a greater distance in a given time and so will be involved in more collisions. Also, because the particles are moving faster a larger proportion of the collision will exceed the activation energy and so the rate of reaction increases.

The reaction between standard sodium thiosulphate and standard hydrochloric acid solutions at different temperatures can be used to examine the effect of temperature on the rate of reaction.

13.8 Effect of Particle Size on the Rate of Reaction

When one of the reactants is a solid, the reaction must take place on the surface of the solid. By breaking up the solid into smaller pieces, the surface area is increased, giving a greater area for collisions to take place and so causing an increase in the rate of reaction. This explains why mixtures of coal dust and air can cause explosions.

This effect can be examined by reacting equal masses of calcium carbonate with different particle sizes (e.g. chalk and marble chips) with equal volumes of the same hydrochloric acid solution.

$$CaCO_3(s) + 2HCl(aq) \rightarrow CaCl_2(aq) + H_2O(l) + CO_2(g)$$

calcium carbonate + hydrochloric acid → calcium chloride + water + carbon dioxide

13.9 Effect of Light on the Rate of Reaction

The rates of some reactions are increased by exposure to light. Light has a similar effect, therefore, to increasing temperature.

Silver chloride, precipitated by mixing silver nitrate and hydrochloric acid solutions, turns from white to greyish-purple on exposure to sunlight due to the partial decomposition of silver chloride. The effects of light on hydrogen peroxide and concentrated nitric acid explain why they are stored in dark glass bottles.

A mixture of hydrogen and chlorine does not react if kept in the dark but in the presence of light an explosive reaction takes place.

13.10 Effect of Catalysts on the Rate of Reaction

A **catalyst** is a substance which can alter the rate of a reaction but remains chemically unchanged at the end of the reaction. Catalysts usually speed up reactions. A catalyst which slows down a reaction is called a negative catalyst or **inhibitor**.

Catalysts speed up reactions by providing an alternative pathway for the reaction, i.e. one that has a much lower activation energy. More collisions will, therefore, have enough energy for this new pathway (Fig. 13.4).

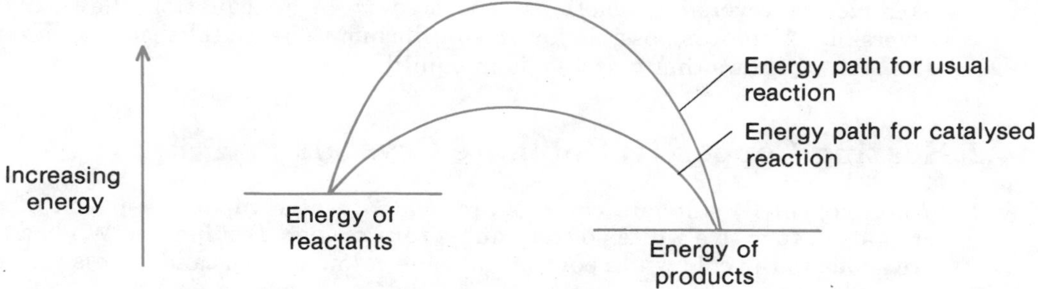

Fig 13.4 Energy diagram for a catalysed reaction

In the laboratory the catalysed decomposition of hydrogen peroxide is usually studied:

$$2H_2O_2(aq) \rightarrow 2H_2O(l) + O_2(g)$$

hydrogen peroxide → water + oxygen

Catalysts for this reaction include manganese(IV) oxide and certain enzymes. **Enzymes** are biological catalysts.

Catalysts are important in many industrial processes. They do not increase the yield of the products but they do increase the rate of production.

Examples include:

1 Iron in the Haber process to produce ammonia (see Unit 25).
2 Vanadium(V) oxide in the Contact process to produce sulphuric acid (see Unit 27).
3 Platinum in the industrial conversion of ammonia to nitric acid (see Unit 25).

Catalysts are usually heavy (transition) metals or compounds of heavy (transition) metals.

13.11 Summary

The factors which affect the rate of a chemical reaction include:

1 Concentration.
2 Particle size.
3 Pressure (for reactions involving gases).
4 Temperature.
5 Light.
6 Presence of a catalyst.

The differences in rate of reaction can be explained in terms of a simple collision theory.

14 REVERSIBLE REACTIONS AND EQUILIBRIUM

This unit is written in two parts:

Part A	14.1–14.4	Reversible reactions
Part B	14.5–14.7	Equilibrium

Check with the Table of Analysis of Examination Syllabuses on p. xvi–xvii to find which part is required for your syllabus.

14.1 Introduction

Most chemical reactions can only go in one direction. For example, when magnesium is reacted with dilute hydrochloric acid, the products are hydrogen and magnesium chloride.

$$Mg(s) + 2HCl(aq) \rightarrow MgCl_2(aq) + H_2(g)$$

magnesium + hydrochloric acid → magnesium chloride + hydrogen

There is no way that the reverse reaction will take place. Hydrogen will not react with magnesium chloride, under any conditions, to produce magnesium and hydrochloric acid.

Some reactions, however, are reversible. A reversible reaction is a reaction that can go in either direction depending on the conditions of the reaction. There are a number of common examples of reversible reactions. The sign $\rightleftarrows$ in an equation shows that the reaction is reversible. Also, it is possible for an equilibrium to be established with a reversible reaction. The sign $\rightleftharpoons$ shows that a system is in equilibrium.

14.2 Heating Copper(II) Sulphate Crystals

When copper(II) sulphate crystals are heated, water vapour is driven off causing the blue crystals to turn to a white powder (anhydrous copper(II) sulphate). When a few drops of water are added to the cold white powder, the blue colour returns and heat is given out, showing the reversible nature of the reaction.

$$CuSO_4.5H_2O(s) \rightleftarrows CuSO_4(s) + 5H_2O(l)$$

copper(II) sulphate crystals ⇄ anhydrous copper(II) sulphate + water

14.3 Heating Ammonium Chloride Crystals

When ammonium chloride crystals are heated, the ammonium chloride dissociates into ammonia gas and hydrogen chloride gas. As the gases cool, they recombine to form solid ammonium chloride.

Similarly if the stopper from a bottle of concentrated ammonia solution is held near a stopper from a bottle of concentrated hydrochloric acid (evolving hydrogen chloride fumes), a dense white smoke of ammonium chloride is formed.

$$NH_4Cl(s) \rightleftarrows NH_3(g) + HCl(g)$$

ammonium chloride ⇄ ammonia + hydrogen chloride

14.4 Formation of Calcium Hydrogencarbonate

Another readily reversible reaction involves the formation of calcium hydrogencarbonate.

If carbon dioxide is bubbled through a solution of calcium hydroxide (limewater), a cloudy white suspension of calcium carbonate is produced. This mixture goes clear again when more carbon dioxide is bubbled through, due to the formation of soluble calcium hydrogencarbonate.

Heating calcium hydrogencarbonate solution causes the reverse reaction and the cloudiness returns as insoluble calcium carbonate is reformed.

$$CaCO_3(s) + H_2O(l) + CO_2(g) \rightleftarrows Ca(HCO_3)_2(aq)$$

calcium carbonate + water + carbon dioxide ⇄ calcium hydrogencarbonate

14.5 Reaction of Iron with Steam

When steam is passed over heated iron, a slow reaction takes place producing hydrogen and an iron oxide. The apparatus suitable for this experiment is shown in Fig. 14.1.

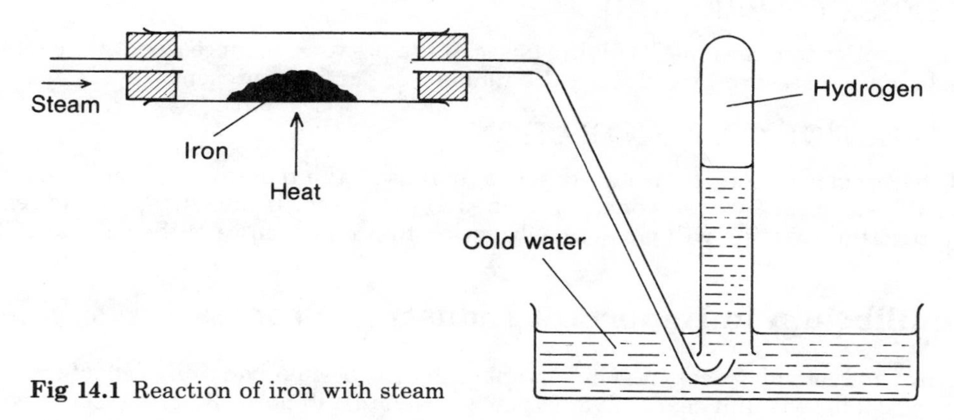

Fig 14.1 Reaction of iron with steam

$$3Fe(s) + 4H_2O(g) \rightarrow Fe_3O_4(s) + 4H_2(g)$$

iron + water (steam) → iron(II) di-iron(III) oxide + hydrogen

In a second experiment dry hydrogen gas is passed over heated iron(II) di-iron(III) oxide (Fig. 14.2). This time the reverse reaction takes place and the iron oxide is reduced to iron.

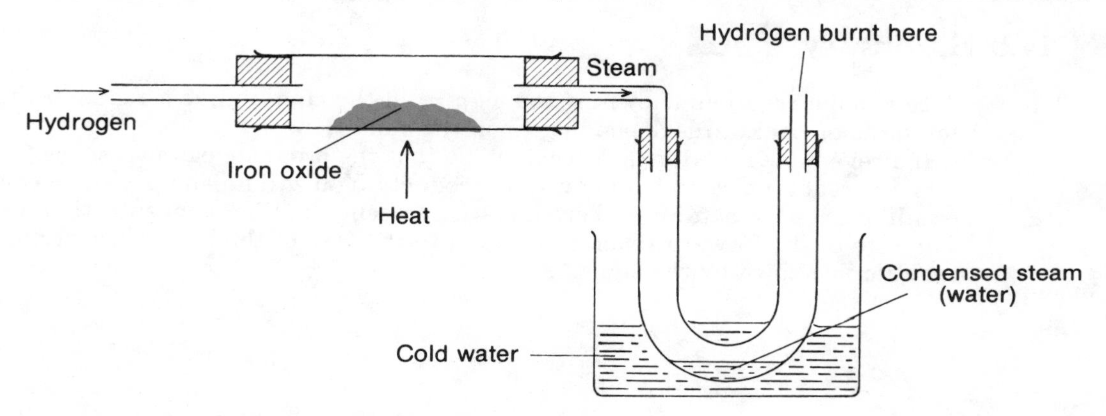

Fig 14.2 Reaction of hydrogen and iron oxide

$$Fe_3O_4(s) + 4H_2(g) \rightarrow 3Fe(s) + 4H_2O(g)$$

An interesting consideration now is what would happen if the four chemicals were together present in a sealed container.

The iron would react with the steam to form hydrogen and iron oxide, but at the same time hydrogen would be reacting with the iron oxide to produce iron and steam.

After a time a state of balance would be set up where the rate of the forward reaction would be equal to the rate of the reverse reaction. This is called a state of **chemical equilibrium.**

$$Fe_3O_4(s) + 4H_2(g) \rightleftharpoons 3Fe(s) + 4H_2O(g)$$

Note that although there will appear to be no overall change in the concentrations of any of the chemicals, the equilibrium is a **dynamic** process i.e. both forward and reverse reactions are taking place but at the same rate.

An equilibrium can be established only in a sealed system, where no chemicals can enter or leave the system. A system in equilibrium is very delicately balanced. Any disturbance of the system, e.g. a change in temperature, may disturb the equilibrium by favouring the forward or reverse reaction.

14.6 Factors Affecting an Equilibrium

In any equilibrium, the position of the equilibrium can be altered by changing the conditions.

Le Chatelier, a French chemist, stated a principle that governs the behaviour of equilibria. The principle states that **for a system in equilibrium, if any change is made to the conditions, the equilibrium will alter so as to oppose the change.**

Changes in conditions include temperature, pressure (if gases are involved) and concentration.

EFFECT OF TEMPERATURE

If the temperature of the system in equilibrium is lowered, the reaction will move in a direction to produce more heat, i.e. the exothermic reaction is favoured.

EFFECT OF PRESSURE

This applies to reactions involving gases. If the pressure is increased, the reaction will move to reduce the pressure by reducing the number of particles present.

EFFECT OF CONCENTRATION

If the concentration of one substance is increased, the reaction will move in a direction to use up the substance whose concentration was increased. If one substance is removed from the system, the reaction will move in a direction to produce more of the substance being removed.

14.7 Equilibrium in Important Industrial Processes

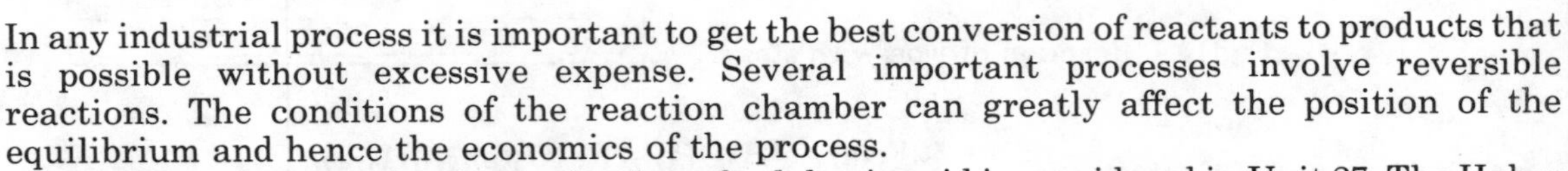

In any industrial process it is important to get the best conversion of reactants to products that is possible without excessive expense. Several important processes involve reversible reactions. The conditions of the reaction chamber can greatly affect the position of the equilibrium and hence the economics of the process.

The Contact process for the production of sulphuric acid is considered in Unit 27. The Haber process for the production of ammonia is considered in Unit 25. In both cases there is a consideration of the best way of achieving a good yield of products economically.

14.8 Summary

A reversible reaction, shown by the sign $\rightleftarrows$ in the equation, is a reaction that can go either forwards or backwards depending upon the conditions.

If a reversible reaction is carried out so that the products cannot escape, it is impossible to turn the reactants completely into the products. You will finish up with an equilibrium. In the equilibrium, all reactants and products are present and their concentrations are not changing. The rate of the forward reaction is equal to the rate of the reverse reaction. An equilibrium reaction is shown by the sign $\rightleftharpoons$.

15 ENERGY CHANGES IN CHEMISTRY

This unit is written in three parts:

Part A	15.1–15.5	Exothermic and endothermic reactions
Part B	15.6–15.7	Calculations involving energy changes
Part C	15.8–15.9	Cells

Check on pages xvi–xvii to find which parts are required for your syllabus.

15.1 Introduction

A consideration of energy, and more particularly energy change, is a fundamental aspect of a chemistry course. In this unit the distinction between physics and chemistry becomes blurred.

A change in energy content of chemicals will be recognized by a change in temperature of the surroundings. There are different forms of energy. The most important forms of energy to a chemist are heat energy and electrical energy.

Energy can neither be created nor destroyed in any process. It is just changed from one form to another. This is called the Law of Conservation of Energy, which is related to the Law of Conservation of Mass. The work of Einstein has shown that mass and energy can be interconverted.

15.2 Energy Possessed by Chemicals

All substances are composed of small particles. At any temperature above absolute zero (−273 °C) the particles are in motion. These particles can possess two types of energy:

1 Kinetic energy

Particles that are moving possess kinetic energy (energy of movement). As the temperature rises, the particles move faster and possess more kinetic energy.

2 Chemical energy (or bonding energy)

When bonds are formed energy is released. If the bonds holding particles together are to be broken, energy has to be supplied.

It is not possible to measure the total energy possessed by any chemical in the laboratory. It is only possible to measure energy changes.

The energy change that accompanies a chemical reaction is due to changes in chemical or bonding energy between reactant and product.

15.3 Exothermic Processes

There are many examples where energy is liberated during the reaction and the temperature of the system rises.

For example, the burning of carbon produces energy. Such a reaction is called an **exothermic reaction**. The quantity of energy produced depends on the quantity of carbon burnt.

$$C(s) + O_2(g) \rightarrow CO_2(g)$$

$$\text{carbon} + \text{oxygen} \rightarrow \text{carbon dioxide} \quad (\Delta H = -393.5\ \text{kJ})$$

This information tells a chemist that if 1 mole of carbon atoms (12 g) is burnt in an adequate supply of oxygen, 393.5 kJ of energy are liberated. ΔH is called the **enthalpy** (or heat) of reaction and is **negative** if the reaction is exothermic.

This process can be summarized in Fig. 15.1.

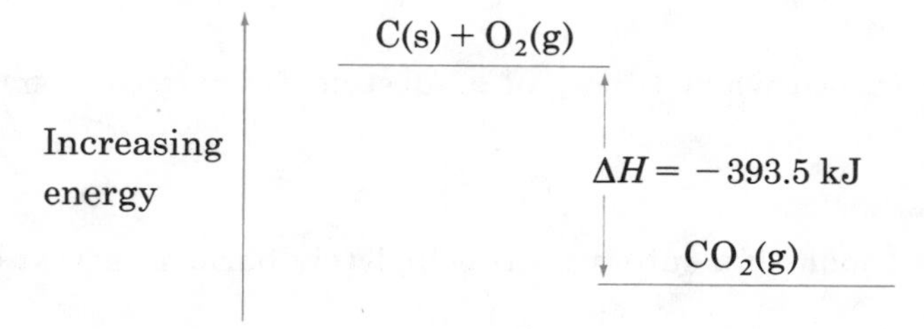

Fig 15.1 Energy diagram for the complete combustion of carbon

During an exothermic reaction, more energy is produced from the formation of new bonds than the energy required to break existing bonds, and therefore there is a surplus of energy, which is lost to the surroundings.

15.4 Endothermic Processes

There are a few reactions where energy is absorbed from the surroundings during the reaction and the temperature falls.

For example, the formation of hydrogen iodide from hydrogen and iodine is an **endothermic reaction**.

$$H_2(g) + I_2(g) \rightleftharpoons 2HI(g)$$
$$\text{hydrogen} + \text{iodine} \rightleftharpoons \text{hydrogen iodide} \quad (\Delta H = +52\ \text{kJ})$$

ΔH is **positive**, in this case, because the reaction is endothermic. This process is summarized in Fig. 15.2.

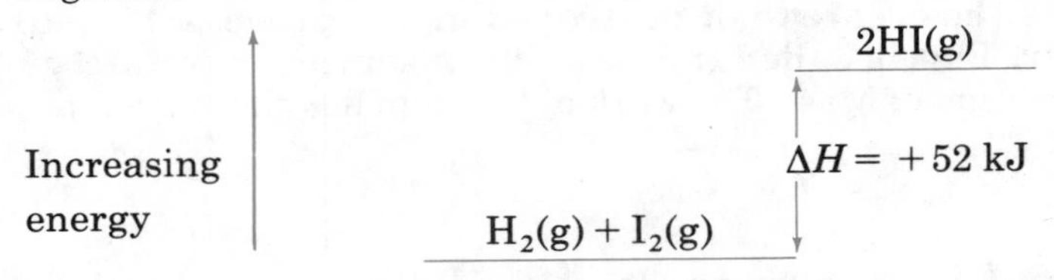

Fig 15.2 Energy diagram for the synthesis of hydrogen iodide

During an endothermic reaction, more energy is required to break bonds than is liberated when new bonds are formed. This deficiency in energy has to be made up from the surroundings.

15.5 Activation Energy

If the products contain less energy than the reactants, it might be expected that the exothermic changes would always occur spontaneously. This is not the case. Before the reaction can take place energy has to be supplied, called the activation energy, to start the reaction off (Fig. 15.3).

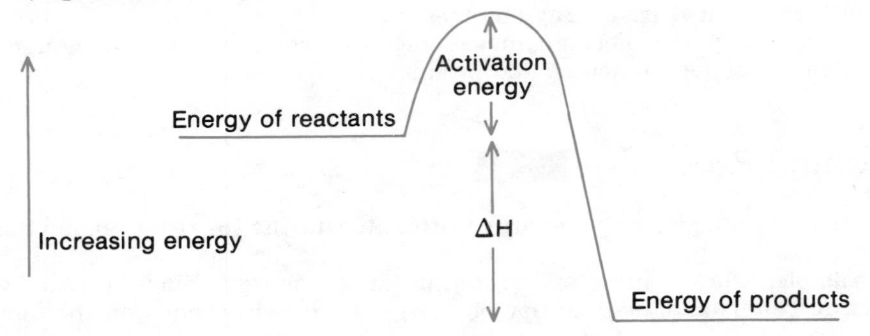

Fig 15.3 Energy changes during a reaction

The function of a catalyst is usually to speed up a reaction by lowering the activation energy barrier (see 13.10).

15.6 Fundamental Definitions

Heat of formation

The heat liberated or absorbed when 1 mole of a substance is formed from its constituent elements.

Heat of combustion.

The heat liberated when 1 mole of a substance is completely burnt in excess oxygen.

Heat of neutralization

The heat liberated when 1 mole of hydrogen ions $H^+(aq)$ reacts with 1 mole of hydroxide ions $OH^-(aq)$.

$$H^+(aq) + OH^-(aq) \rightarrow H_2O(l)$$

The heat of neutralization of many acids and alkalis is -58 kJ/mol.

15.7 To Find the Heat of Combustion of an Alcohol

A small spirit lamp containing the alcohol is weighed accurately. During the weighing the spirit lamp should be covered to prevent evaporation of the alcohol.

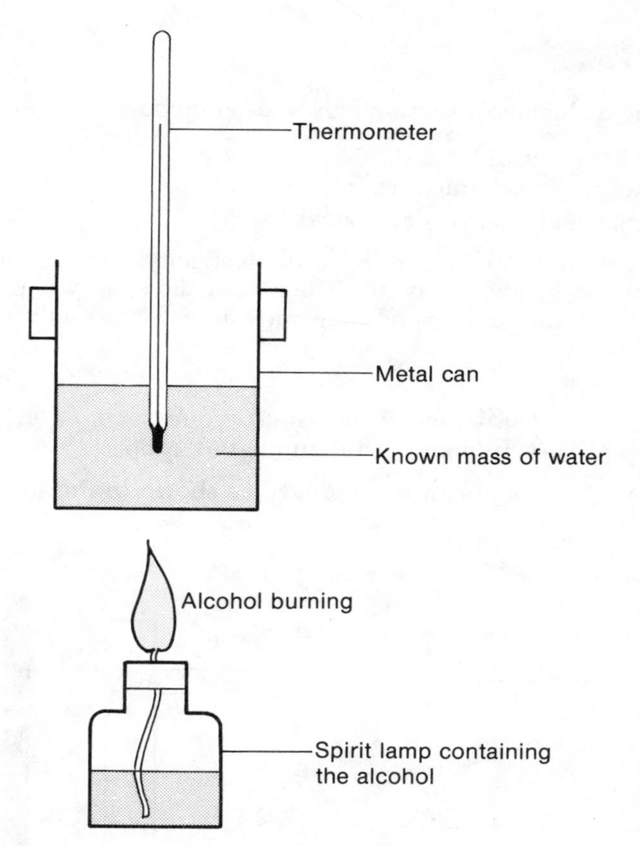

Fig 15.4 Experiment to find the heat of combustion of an alcohol

A known mass of water is taken in a metal can. The temperature of the water is recorded. The spirit lamp is lit and placed underneath the can. The can is heated until the temperature has risen by about 30 °C. The final temperature is recorded. The spirit lamp is then reweighed to find the mass of alcohol burnt.

Sample results

$$\text{mass of spirit lamp + ethanol before burning} = 21.94 \text{ g}$$
$$\text{mass of spirit lamp + ethanol after burning} = 21.10 \text{ g}$$
$$\text{mass of ethanol burnt} = 0.84 \text{ g}$$
$$\text{mass of water in the can} = 100 \text{ g}$$

(You can ignore the heat required to raise the temperature of the can.)

$$\text{Specific heat capacity of the water} = 4.2 \text{ kJ/kg/°C}$$
$$\text{temperature rise} = 30\ °\text{C}$$

$$\text{heat gained by water} = \text{mass (in kg)} \times \text{temp. rise} \times \text{specific heat capacity}$$
$$= \frac{1}{10} \times 30 \times 4.2 \text{ kJ}$$
$$= 12.6 \text{ kJ}$$

It is assumed that all the heat produced when the ethanol burns is used to heat up the water.

12.6 kJ is produced when 0.84 g of ethanol burns

$\frac{12.6}{0.84}$ kJ is produced when 1 g of ethanol burns

mass of 1 mol of ethanol molecules C_2H_2OH = 46 g

heat produced when 46 g of ethanol burns $= \frac{12.6}{0.84}$ kJ

Since the combustion of ethanol is exothermic, the heat of combustion

$$\Delta H = -690 \text{ kJ/mol}$$

However, the heat of combustion of ethanol in the data book is -1370 kJ/mol.

The big difference between the theoretical and practical heats of combustion is due to the heat lost to the surroundings.

15.8 Cells and Batteries

The energy produced during a chemical reaction can be used in various ways. These include:

1 Heat (chemical energy → heat energy).
2 To do work (chemical energy → mechanical energy).
3 To produce electricity (chemical energy → electrical energy).

Not all the energy produced, ΔH, can be converted to electricity. The maximum amount of energy that can be converted to electricity is ΔG. The rest of the energy is wasted.

A simple cell is produced when rods of zinc and copper are dipped into a solution of copper(II) sulphate.

The reaction taking place is:

$$Zn(s) + CuSO_4(aq) \rightarrow ZnSO_4(aq) + Cu(s)$$

zinc + copper(II) sulphate → zinc sulphate + copper

A simple cell does not operate well and can be improved as shown in Fig. 15.5.

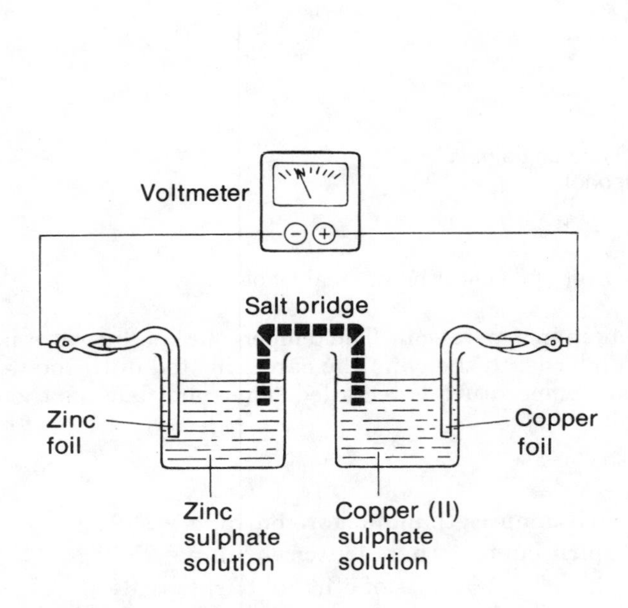

Fig 15.5 A simple cell

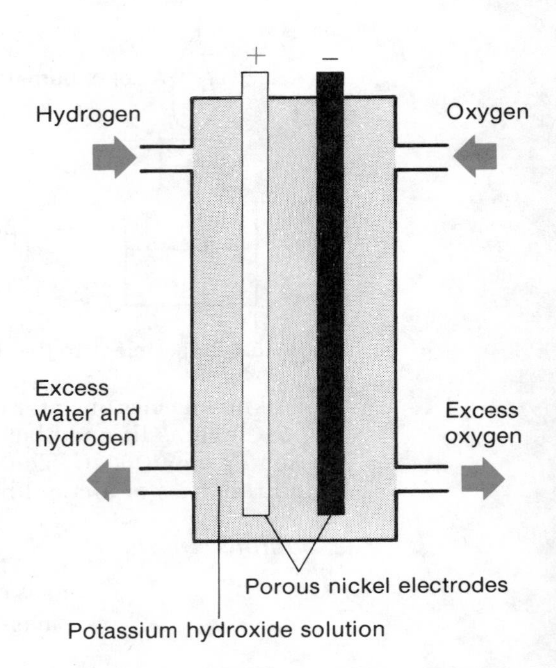

Fig 15.6 A simple hydrogen–oxygen fuel cell

15.9 Fuel Cells

A fuel cell is an efficient means of converting chemical energy into electrical energy. A continuous supply of electrical energy can be obtained if there is a continuous supply of fuel.

In a hydrogen–oxygen fuel cell (Fig. 15.6), hydrogen and oxygen combine to form water. The energy is released as electricity.

Fuel cells have been used in vehicles and space rockets. They produce electricity without pollution.

15.10 Summary

Energy changes are often noticed during chemical reactions.

A reaction where energy is given out to the surroundings is called an **exothermic** reaction. In an exothermic reaction, the substances produced (called products) contain less energy than the reacting substances (called reactants).

In an **endothermic** reaction, energy is taken in from the surroundings. The reactants contain less energy than the products.

There are many examples of exothermic reactions but few endothermic reactions.

A cell is a means of producing the energy from a chemical reaction in an efficient way. The chemical energy is converted into electrical energy.

16 SOCIAL, ECONOMIC AND ENVIRONMENTAL CONSIDERATIONS

16.1 Introduction

All chemistry courses now contain aspects of social, economic and environmental chemistry. Rather than set separate questions to test your understanding of these important factors, they will be incorporated into questions throughout the examination.

Up to date information about these aspects of chemistry can be obtained from newspapers, magazines such as *New Scientist*, and radio and television programmes.

Social, economic and environmental aspects of chemistry are often incorporated into questions on industrial processes. The most important are:

extraction of iron (Unit 23.6)
Contact process (Unit 27.9)
Haber process (Unit 25.3)
extraction of aluminium (Unit 23.5)
nitric acid manufacture (Unit 25.7)
fractional distillation of liquid air (Unit 17.2)
sodium hydroxide manufacture (Unit 28.1)
extraction of sodium (Unit 23.4)
Solvay process (Unit 28.5)
hydrogen manufacture (Unit 5.4)
oil refining (Unit 20.8)
making polymers (Unit 20.10)
making soaps and detergents (Unit 19.5)

As you work through the rest of the book try to see how the five principles below in 16.2–16.6 can be applied.

16.2 Choice of Site for the Process

There are a number of factors which will determine the best place to site a new factory. If bulky raw materials are required it is often advisable to site the factory near the natural deposits or on the coast if the raw materials are to be imported. For this reason, oil refineries are often sited on the coast. If more than one raw material is required, then it is necessary to decide which raw material will be more difficult to transport.

If a process requires a large amount of energy, e.g. extraction of aluminium, it may be advisable to site the factory close to hydroelectric power for cheap electricity.

Another consideration when choosing a site is the location of the customers for the product. The building of a sodium hydroxide factory in north Cheshire (close to salt deposits) can be justified by processes nearby (e.g. soap making) which use large quantities of sodium hydroxide. Another example is the siting of a factory producing oxygen by fractional distillation of liquid air adjacent to a factory producing electric light bulbs. The inert gases produced during the fractional distillation are used to fill electric light bulbs.

A suitable supply of labour and good communications (road, rail etc.) are also important.

If a process needs water for cooling purposes, the factory must be sited near a lake or river.

If the process uses large quantities of toxic or inflammable materials or operates at a high pressure, it would be advisable to site the factory away from housing areas.

16.3 Using the Minimum Amount of Energy

In order to reduce energy costs, it is necessary to consider ways of re-using energy or reducing working temperatures. In the extraction of sodium from sodium chloride, the melting point of sodium chloride is reduced from about 800°C to about 550°C by adding an impurity of calcium chloride. This does not affect the products but reduces the fuel required to keep the sodium chloride molten.

In the Contact process, reducing the temperatures of the catalyst chamber saves energy but also causes a greater conversion of sulphur dioxide to sulphur trioxide. It also, however, slows down the process. The slowing down of the process when temperature is reduced has always to be remembered.

Often, following an exothermic reaction, the products are at a higher temperature than is required. Under these circumstances, a heat exchanger can be used to remove heat from the products to be re-used elsewhere. For example in the first stage of the conversion of ammonia and air to nitric acid, the reaction is exothermic and the temperature of the gases has to be reduced considerably. A heat exchanger or a series of heat exchangers can be used to transfer the energy to some other part of the process. The energy possessed by hot gases that are being produced and expelled through a chimney is removed by heat exchangers. Often waste gases contain carbon monoxide and other gases that can be burnt to produce energy rather than being allowed to escape into the atmosphere.

Energy needs can also be reduced by re-using waste products. For example, glass bottles can be collected, crushed and re-used to make new glass. Apart from the amount of energy required to make 'recycled glass' being less, there is a saving in raw materials. Recycling becomes important as raw materials become expensive and energy costs increase.

16.4 Continuity of a Process

A process that works continuously rather than in batches is usually more economic. A blast furnace producing iron continuously can produce 1800 tonnes of iron per day. It can work continuously because the impurities in the iron are being removed in the slag rather than accumulating in the furnace.

Often a process that appears to be continuous is not in practice. For example, the manufacture of nitric acid would appear to be continuous but a run of production of more than a few days is unusual because the platinum catalyst gauzes have to be replaced. In order to do this, the plant has to be shut down.

16.5 Re-use of Byproducts or Production of Valuable Byproducts

The economics of a process depend heavily upon the usefulness of the byproducts of the process. In the manufacture of sodium hydroxide by electrolysis, the byproducts (chlorine and hydrogen) are very valuable and can be easily sold.

When ammonia is being produced industrially from nitrogen and hydrogen only about 10 per cent of the gases are converted to ammonia. The unused nitrogen and hydrogen are then recycled.

16.6 Pollution Considerations (see Units 17.12, 19.11)

The control of waste from a factory is monitored by inspectors. It is necessary to treat wastes correctly before they are allowed to escape into the air or into water.

One of the major impurities in waste gases, which needs to be removed, is sulphur dioxide. When this gets into the air it becomes oxidized to sulphuric acid. In addition to polluting the atmosphere, the escape of sulphur dioxide is wasting valuable sulphur that could be re-used.

It is necessary to treat waste water to remove toxic materials like cyanides and heavy metals. It is usual to precipitate these substances to facilitate their removal. Apart from these impurities, it is necessary to monitor the temperature, pH and nitrogen content of the water. If ammonia is present in the water, bacteria will oxidize the ammonia to nitrates and remove oxygen from the water. This is called **eutrophication** and can be recognized by green algae on the water. These green algae prevent light entering water, and further oxygen is removed from the water when underwater vegetation dies and decays.

16.7 Summary

It is important to consider more than the chemicals used and produced in our chemical factories. Factors which should be considered include:

1 Choice of site for the process.
2 Using the minimum amount of energy.
3 Continuity of the process.
4 Re-use of byproducts or production of valuable byproducts.
5 Pollution.

We must understand that a chemical factory must make a profit. We must consider ways of increasing the profit. We must also consider the effects, good and bad, of having chemical factories. It is important in all cases to try and get a balanced view on every issue. You will always find differences of opinion, even among chemists!

17 OXYGEN AND THE AIR

17.1 The Composition of Air

The composition of air varies from place to place because air is a mixture of gases. The composition, by volume, of a typical sample of air is as follows:

nitrogen	78 per cent
oxygen	21 per cent
carbon dioxide	0.03 per cent
argon	0.9 per cent
helium	0.0005 per cent
neon	0.002 per cent
krypton	0.00001 per cent
xenon	0.000001 per cent

Air also contains varying amounts of water vapour. Air does not usually contain hydrogen. Argon, helium, neon, krypton and xenon belong to the family of noble gases (see Unit 9.4).

17.2 Separating Air into its Constituent Gases

Separating air into its constituent gases is a very difficult process. It cannot be done satisfactorily in the laboratory.

However, it is an important process in industry because all of the gases are valuable as pure gases. The separation involves the fractional distillation (Unit 2.3) of liquid air.

The air is first cooled in a refrigeration plant to separate carbon dioxide and water vapour. Both of these gases solidify in the cooler and the solid can be removed. If these gases were not removed at this stage, they would later solidify in the pipes and block them.

The compressed air is then allowed to expand through a small hole. This expansion causes the gas to cool rapidly. The compressing and expanding cycles are repeated until the temperature is about −200°C. At this temperature most of the air has liquified.

When liquid air warms up the different compounds boil off at different temperatures. Nitrogen boils at −196°C and oxygen later, at −183°C.

17.3 Percentage of Oxygen in Air by Volume

An approximate value of the percentage of oxygen in air can be obtained by burning a piece of phosphorus in a fixed volume of air trapped over water.

Figure 17.1 shows the apparatus that can be used to find the percentage of oxygen accurately.

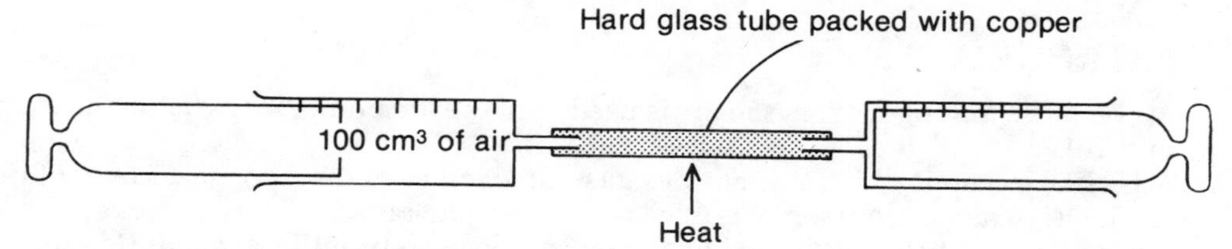

Fig 17.1 Percentage of oxygen in air

One hundred cubic centimetres of air is trapped in one of the syringes. This air is passed backwards and forwards over heated copper turnings in a hard glass tube. The copper reacts with oxygen in the sample of air producing solid copper(II) oxide. The heating is continued until there is no further reduction in the volume of air. The apparatus is left to cool to room temperature and the volume of gas remaining (i.e. air minus the oxygen) is measured.

E.g.

$$\text{volume of gas before} = 100\ \text{cm}^3$$
$$\text{volume of gas after} = 80\ \text{cm}^3$$
$$\therefore\ 100\ \text{cm}^3 \text{ of air contains } 20\ \text{cm}^3 \text{ of oxygen}$$
$$\text{percentage of oxygen} = 20 \text{ per cent}$$

17.4 Laboratory Preparation of Oxygen

Oxygen can be prepared in the laboratory by the decomposition of hydrogen peroxide solution using manganese(IV) oxide (manganese dioxide) as a catalyst.

$$2H_2O_2(aq) \rightarrow 2H_2O(l) + O_2(g)$$

hydrogen peroxide → water + oxygen

The oxygen produced can be collected over water (Fig. 17.2).

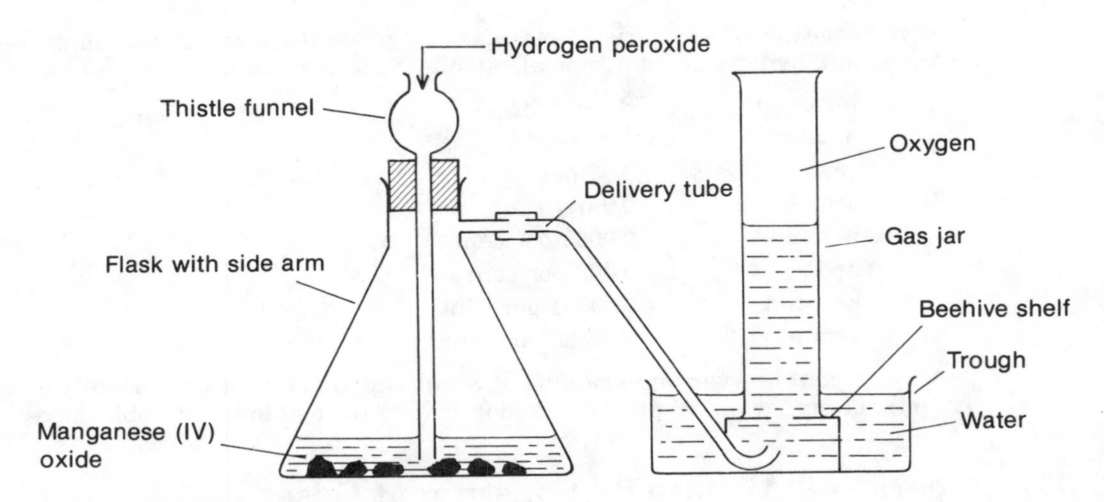

Fig 17.2 Preparation of oxygen

17.5 Uses of Oxygen and Nitrogen

OXYGEN

Pure oxygen is very important for helping breathing. 'Medical grade' oxygen, the purest grade, can often revive a patient who is having problems breathing – perhaps after an accident or during an operation. Climbers who are going to climb high mountains, pilots of high-flying fighter planes and divers require cylinders of oxygen for their breathing.

Oxygen which is less pure can be used in industry in large amounts. Oxygen is used in steelmaking to oxidize unwanted impurities. Ethyne (still called acetylene) is a hydrocarbon which burns in air. This burning, however, produces a higher temperature if oxygen is used in place of air. Oxyacetylene flames produce high temperatures which are sufficient to cut metals or weld them together.

Oxygen in liquid form is carried in rockets to enable the rocket fuel to burn.

Oxygen is also used in sewage treatment. Blowing air through raw sewage assists its breaking down.

Oxygen is supplied in cylinders that are black in colour.

NITROGEN

Much of the nitrogen from the air is used to make ammonia, nitric acid and fertilizers (Units 25 and 26).

Unlike oxygen, which is an active gas, nitrogen is an inactive gas. Many of its uses rely upon its inactive nature. Nitrogen is used to fill food packages, e.g. crisp packets. Oxygen spoils food and so air in contact with food has to be avoided. Light bulbs also contain gases which include nitrogen.

Liquid nitrogen (boiling point −196°C) is an inexpensive liquid that is used for a wide range of 'cooling' uses. Small quantities of food can be 'deep frozen' by dipping them into liquid nitrogen. It is also useful for transporting biological tissue for transplant surgery and samples of semen to a farm for artificial insemination.

17.6 Processes Involving Gases in the Air

There are various processes that involve the gases in the air. These are summarized in Table 17.1.

Table 17.1 Processes that involve the gases in the air

Gas	*Combustion*	*Rusting*	*Respiration*	*Photosynthesis*
Nitrogen	Usually not involved	Not involved	Not involved	Not involved
Oxygen	Usually necessary	Necessary	Necessary	Produced
Carbon dioxide	Formed when carbon and carbon compounds burn	Speeds up rusting but it is not essential	Produced	Necessary
Noble (inert) gases	Not involved	Not involved	Not involved	Not involved
Water vapour	Formed when hydrogen and hydrogen compounds burn	Necessary	Produced	Necessary

17.7 Combustion or Burning

The combustion or burning of a substance is the combination of the substance with oxygen. During the combustion, heat and light are usually given out. Substances burn better in pure oxygen than in air.

E.g. sulphur burns in air or oxygen with a blue flame:

$$S(s) + O_2(g) \rightarrow SO_2(g)$$

sulphur + oxygen → sulphur dioxide

E.g. methane burns in excess air or oxygen:

$$CH_4(g) + 2O_2(g) \rightarrow CO_2(g) + 2H_2O(g)$$

methane + oxygen → carbon dioxide + water

and in a limited supply of air or oxygen:

$$2CH_4(g) + 3O_2 \rightarrow 2CO(g) + 4H_2O(g)$$

methane + oxygen → carbon monoxide + water

Providing all the products are weighed and are not allowed to escape, a substance increases in mass during combustion.

17.8 Rusting of Iron

The rusting of iron is discussed in Unit 24.4.

17.9 Respiration

Oxygen is transported from the lungs to the muscles by the haemoglobin in the blood. The oxygen is used to oxidize carbohydrates to form carbon dioxide.

E.g.
$$C_6H_{12}O_6(aq) + 6O_2(g) \rightarrow 6CO_2(g) + 6H_2O(l) + \text{energy}$$

glucose + oxygen → carbon dioxide + water + energy

The carbon dioxide is returned to the lungs and expelled from the body. Carbon monoxide is poisonous because it destroys the oxygen-carrying capacity of the haemoglobin.

17.10 Photosynthesis

Green plants absorb carbon dioxide from the air through the underside of their leaves. The carbon dioxide combines with water in the presence of sunlight and certain enzymes to produce carbohydrates. The oxygen produced escapes into the air.

$$6CO_2(g) + 6H_2O(l) + \text{energy} \rightarrow C_6H_{12}O_6(aq) + 6O_2(g)$$

carbon dioxide + water + energy → glucose + oxygen

(N.B. This is the reverse of the equation in 17.9.)

17.11 Carbon Cycle

The carbon cycle is shown in Fig. 17.3. Photosynthesis by plants ensures that the percentages of various gases in the air remain fairly constant (carbon dioxide is used up while oxygen is produced – this is the opposite of respiration and the burning of fuels).

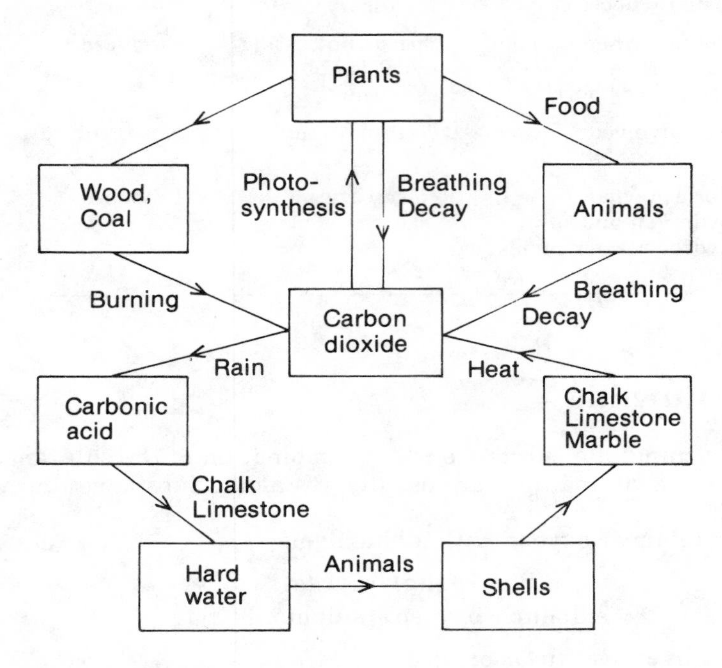

Fig 17.3 The carbon cycle

17.12 Air Pollution

Apart from the gases normally found in air, other gases such as sulphur dioxide, oxides of nitrogen and carbon monoxide can be present. These gases can cause air pollution and are called pollutants.

Pollutants can cause a variety of problems to the environment.

17.13 Sulphur Dioxide as a Pollutant

Sulphur dioxide is a major cause of air pollution. Coal and fuel oil contain about 2 per cent sulphur. When they are burnt this sulphur is turned into sulphur dioxide. Sulphur dioxide is a colourless gas with a strong smell. It dissolves in water to form sulphurous acid.

$$\text{sulphur dioxide} + \text{water} \rightarrow \text{sulphurous acid}$$
$$SO_2(g) + H_2O(l) \rightarrow H_2SO_3(aq)$$

In the early 1950s sulphur dioxide and the smoke which is always with it caused great problems. Apart from blackening buildings and producing long term fogs they caused serious health problems. In 1952, for example, over 4000 people died because of air pollution in London alone. Most of these deaths were due to lung illnesses including bronchitis.

The Clean Air Act 1956 set up 'smokeless zones' in large towns and cities. Within these areas coal could not be burned and smokeless fuels had to be used. The result of this Act, and other measures taken, has been to improve greatly the conditions in towns and cities.

The burning of coal in household fires was regarded as the major problem. Even today considerable quantities of sulphur dioxide are lost from factories and power stations. The chimneys of factories and power stations are much higher than household chimneys and this gives more chance for the sulphur dioxide to disperse. Even today there are links between the emission of sulphur dioxide and **acid rain**.

17.14 Oxides of Nitrogen as Pollutants

About 30–40 per cent of the oxides of nitrogen in the air come from car exhausts. In the car engine nitrogen and oxygen combine together to form oxides of nitrogen. Other sources of nitrogen oxide pollution are factories and fires.

Even small concentrations of oxides of nitrogen can cause serious environmental problems. Like sulphur dioxide, oxides of nitrogen dissolve in water to form an acid.

17.15 Carbon Monoxide as a Pollutant

Carbon monoxide is a poisonous gas produced by the partial combustion of fuels. Much of the carbon monoxide comes from the incomplete combustion of petrol in the car engine.

Around heavy traffic the concentration of carbon monoxide in the air can reach 10–20 parts per million (p.p.m.). Levels as high as 200 p.p.m. have been recorded. Levels above 50 p.p.m. can be harmful to adults. Carbon monoxide is poisonous because it forms a stable compound with the haemoglobin in the blood called carboxyhaemoglobin, which prevents the binding of haemoglobin to oxygen and so stops the transport of oxygen around the body. Low concentrations are still harmful as they affect the proper working of the brain.

17.16 Lead Compounds as Pollutants

A lead compound called tetraethyl lead is added in small quantities to petrol. This increases the octane number of petrol (the number of stars on the petrol pump), lubricates the valves in the engine and enables more litres of petrol to be made from a given amount of petroleum.

A car covering 9000 miles per year consumes about 200 g of lead; most of this escapes into the air through the exhaust system. Recent studies have emphasized how serious the emission of lead into the air can be. Lead is an intoxicant in much the same way as alcohol is, and some studies have linked it with antisocial behaviour. A study in a London school has shown that children with high lead levels in their blood were less intelligent and found it more difficult to concentrate. There is also considerable evidence that vegetables such as lettuces and cabbages grown near busy roads contain high levels of lead.

Producing 'lead-free' petrol is possible but this adds about 2 pence to the price per litre. Also most of today's cars cannot use it without expensive changes. However, there are plans to reduce the levels of lead in petrol and 'lead-free' petrol is now produced.

17.17 Problems Caused by Air Pollution

Exhaust gases from cars are causing problems throughout the world. Strong sunlight acting on a mixture of oxides of nitrogen, carbon monoxide and hydrocarbon vapour (unburnt petrol) produces a new and more unpleasant mixture which includes ozone. Its acrid fumes collect in valleys and cause many problems. These fumes are sometimes called 'photochemical smog'. Apart from health problems these fumes:

1 Cause the breaking down of rubber and plastics.

2 Fade dyes.

Much of the problem of air pollution could be minimized by cleaning up car exhausts. A converter containing a platinum catalyst fitted to the exhaust system would convert all of the harmful gases into carbon dioxide, nitrogen, oxygen and water vapour. These gases are all normally present in air and therefore do not cause pollution.

Pure rain-water has a pH of approximately 5. Oxides of nitrogen and sulphur dioxide dissolve in water to form acids. The presence of these pollutants reduces the pH making the rain-water more acidic and causing certain problems. The effects of **'acid rain'** include:

1 Stonework on buildings is damaged. St Paul's Cathedral and Westminster Abbey are just two of the buildings that show the damage caused by acid rain. First a black skin appears on the surface. This then blisters and cracks, causing the stonework to be seriously disfigured.

2 Rivers and lakes over a wide area become more acid. This is said to be affecting wildlife, e.g. otters. There are many lakes in Sweden and Norway that now have no life.

3 Forests are seriously damaged. Forests in Scandinavia and Germany especially are being damaged by acid rain. Trees are stunted, needles and leaves drop off and the trees die. It has been estimated that acid rain is costing the German forestry industry about £150 million each year.

4 Human life can be affected. Acid conditions can alter levels of copper, lead and aluminium in the body. These changes have been linked with diarrhoea in small babies, breathing disorders and mental deterioration in old age (senile dementia).

5 Metalwork is damaged. Acid rain can speed up corrosion of metals. Wrought iron railings in city areas can show considerable damage.

17.18 Summary

Air is a mixture of gases. Approximately four fifths of air is composed of the inactive gas nitrogen. Approximately one fifth of the air is composed of the active gas oxygen. Other gases present in small amounts include carbon dioxide, helium, neon, argon, krypton, xenon and water vapour.

The constituent gases in air can be separated by fractional distillation of liquid air. This process relies upon the different boiling points of the gases present.

Combustion, rusting and respiration all use up oxygen from the air. Fortunately photosynthesis removes carbon dioxide from the air and replaces it with oxygen.

Air pollution is caused by other gases present in the air. These include sulphur dioxide, oxides of nitrogen, carbon monoxide and lead compounds. Pollutants such as these can cause considerable environmental and social problems, including acid rain.

18 CHALK, LIMESTONE AND MARBLE

18.1 Introduction

Chalk, limestone and marble are three forms of the same chemical compound – calcium carbonate. It has been estimated that there are about 60 000 000 000 000 000 tonnes of these minerals in the rocks of the earth. Large amounts of these minerals are used by the chemical industry.

18.2 How were these Minerals Formed?

Millions of years ago the seas were filled with sea creatures with shells. When the animals died the shells fell to the sea bed and built up a deposit up to 600 metres thick. These deposits were compressed and became **chalk** or **limestone**. As the earth moved the rocks were pushed upwards and formed hills. The Downs of the south-east of England are examples of chalk hills.

When limestone is heated to high temperatures and compressed further, it re-crystallizes forming **marble**. This is harder than chalk or limestone. It is often found particularly in mountainous regions, e.g. Italy.

It is possible to find the remains of sea creatures in chalk or limestone as **fossils**.

18.3 Reactions of Calcium Carbonate

All forms of calcium carbonate – chalk, limestone and marble – react in similar ways.

Calcium carbonate does not dissolve in pure distilled water. However, it does dissolve slightly in water in the presence of carbon dioxide. Calcium hydrogencarbonate is formed in solution and this causes temporary hardness in water (Unit 19.3).

$$\text{calcium carbonate} + \text{water} + \text{carbon dioxide} \rightleftharpoons \text{calcium hydrogencarbonate}$$
$$CaCO_3(s) + H_2O(l) + CO_2(g) \rightleftharpoons Ca(HCO_3)_2(aq)$$

Dissolving of calcium carbonate leads to the formation of underground caverns. Stalagmites and stalactites are formed when calcium hydrogencarbonate decomposes into insoluble calcium carbonate, carbon dioxide and water.

Calcium carbonate does not easily decompose on heating. A temperature of about 900°C is needed to decompose it. When calcium carbonate is strongly heated it decomposes forming calcium oxide and carbon dioxide gas.

$$\text{calcium carbonate} \rightarrow \text{calcium oxide} + \text{carbon dioxide}$$
$$CaCO_3(s) \rightarrow CaO(s) + CO_2(g)$$

This reaction is accompanied by a dim white light. Calcium oxide is sometimes called **quicklime**.

In industry calcium carbonate is converted into calcium oxide in a lime kiln.

When cold water is added to cold calcium oxide, a violent reaction takes place. A great deal of steam is produced and the mixture becomes very hot. The white solid remaining is calcium hydroxide, sometimes called **slaked lime**.

$$\text{calcium oxide} + \text{water} \rightarrow \text{calcium hydroxide}$$
$$CaO(s) + H_2O(l) \rightarrow Ca(OH)_2(s)$$

This reaction is exothermic and the heat produced causes some of the water to boil.

When solid calcium hydroxide is added to water it forms a creamy-coloured suspension called 'milk-of-lime'. If this suspension is filtered a clear solution of calcium hydroxide is produced. This is called **limewater**.

When carbon dioxide is bubbled through a solution of limewater the solution first goes milky white and then goes clear again. The milkiness is caused by the formation of insoluble calcium carbonate. Calcium hydrogencarbonate, which is soluble in water, is formed in the final clear solution.

$$\text{calcium hydroxide} + \text{carbon dioxide} \rightarrow \text{calcium carbonate} + \text{water}$$
$$Ca(OH)_2(aq) + CO_2(g) \rightarrow CaCO_3(s) + H_2O(l)$$

$$\text{calcium carbonate} + \text{water} + \text{carbon dioxide} \rightarrow \text{calcium hydrogencarbonate}$$
$$CaCO_3(s) + H_2O(l) + CO_2(g) \rightarrow Ca(HCO_3)_2(aq)$$

Some of the relationships between calcium compounds are shown in Fig. 18.1.

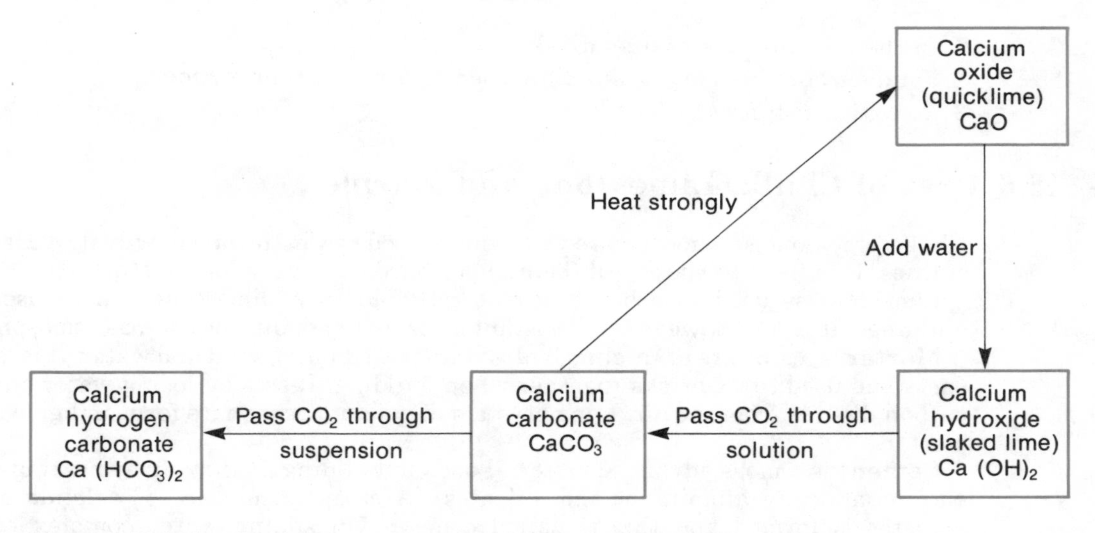

Fig 18.1 Relationship between common calcium compounds

18.4 Laboratory Preparation of Carbon Dioxide

Chalk, limestone and marble are attacked by acids. Carbon dioxide is a product of all reactions between calcium carbonate and an acid.

E.g. calcium carbonate + hydrochloric acid → calcium chloride + water + carbon dioxide

$$CaCO_3(s) + 2HCl(aq) \rightarrow CaCl_2(aq) + H_2O(l) + CO_2(g)$$

Carbon dioxide can be prepared by the reaction between marble chips (calcium carbonate) and dilute hydrochloric acid using the apparatus in Fig. 18.2.

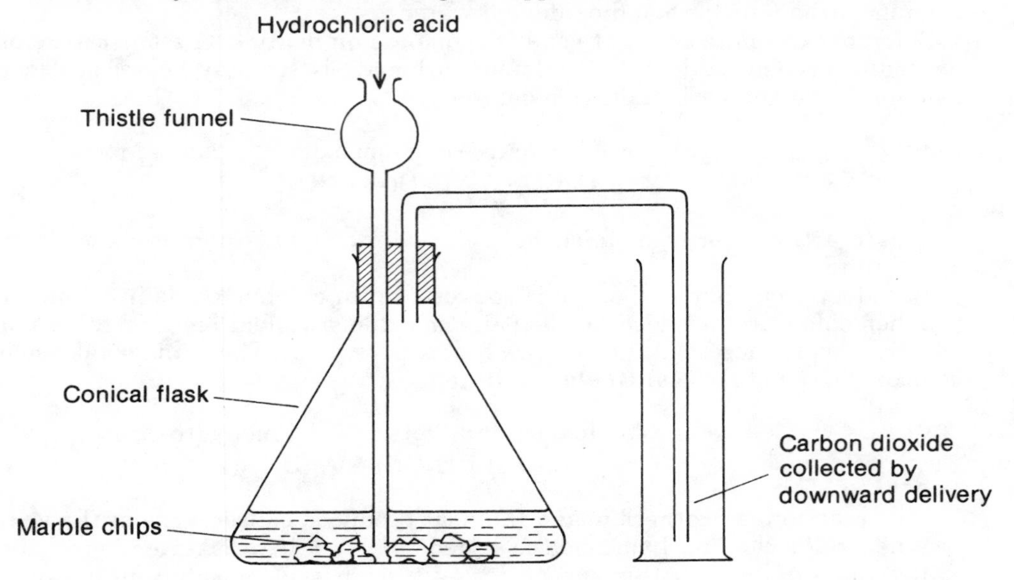

Fig 18.2 Preparation of carbon dioxide

The carbon dioxide can be collected by downward delivery (because carbon dioxide is much denser than air) or over water (although it is quite soluble in water). A solution of calcium chloride remains in the flask.

If a pure, dry sample of carbon dioxide is required it may be passed through a solution of potassium hydrogencarbonate (to remove hydrochloric acid spray which may become suspended in the gas), dried by passing it through concentrated sulphuric acid and collected by downward delivery.

The reaction between calcium carbonate and sulphuric acid is much slower because the product, calcium sulphate, is not very soluble.

18.5 Uses of Carbon Dioxide

1 In the manufacture of fizzy drinks.
2 Solid carbon dioxide ('Drikold') is used as a refrigerant for food.
3 To extinguish fires.

18.6 Uses of Chalk, Limestone and Marble

Chalk, limestone and marble are very widely used raw materials in industry. Marble is used for statues. It is hard to shape but, being very hard, it is very long lasting.

Limestone is used as a building material. Blocks of limestone can be used to construct buildings. It is not, however, as resistant as brick to conditions such as atmospheric pollution.

Mortar is a mixture of calcium hydroxide (slaked lime), sand and water. It is mixed to a thick paste and used to fix bricks together when building. It sets by losing water and by absorbing carbon dioxide from the air. Long crystals of calcium carbonate form and give strength to the mortar.

Cement is a more advanced material used in building. It is made by heating limestone with clay (containing aluminium and silicates). A complex mixture of calcium and aluminium silicates is formed and this is called cement. On adding water complex reactions occur producing calcium hydroxide. The setting of cement is similar to the setting of mortar.

Cement is used with sand, small stones and water to make **concrete**. The properties of concrete depend upon the proportions of the different ingredients. Concrete is not very strong but it can be strengthened by rods of steel or steel meshing (reinforced concrete).

Ordinary **glass** is made by mixing calcium carbonate, silicon dioxide (sand) and sodium carbonate together and melting them. The resulting mixture of sodium and calcium silicates produces glass on cooling. This type of glass is used for windows. Hardened glass, such as 'Pyrex', contains boron and is called 'borosilicate' glass. This can be cooled quickly without cracking. Lead added to glass makes it very hard and suitable for making 'cut glass'.

Glass is often coloured and this is due to the presence of impurities such as metal oxides. When glass is collected in 'bottle banks' for recycling it is necessary to collect glass of different colours in different containers.

Calcium carbonate is used as a raw material in the manufacture of a number of important industrial chemicals. Sodium hydrogencarbonate and sodium carbonate are produced in the Solvay process (Unit 28.5). Calcium carbide CaC_2 is also produced from calcium carbonate. Large quantities of calcium carbonate are used in iron extraction (Unit 23.6) to remove unwanted materials from the furnace.

Calcium carbonate and calcium hydroxide are used widely in agriculture. They neutralize excess acidity in the soil.

18.7 Summary

Chalk, limestone and marble are all forms of calcium carbonate. They are the remains of sea shells deposited in the seas millions of years ago.

Calcium carbonate is decomposed on strong heating to form calcium oxide and carbon dioxide.

When water is added to calcium oxide, calcium hydroxide is formed.

When carbon dioxide is bubbled through a solution of calcium hydroxide (limewater) the solution first turns cloudy (due to the formation of calcium carbonate) and then clear again (due to the formation of calcium hydrogencarbonate). This reaction is used as a test for carbon dioxide.

When dilute hydrochloric acid is added to calcium carbonate, effervescence is seen and colourless carbon dioxide gas is produced. This reaction is used for the laboratory preparation of carbon dioxide.

Large amounts of chalk, limestone and marble are used in industry. Mortar, cement, concrete, iron and glass all require limestone in their making.

19 WATER

19.1 Introduction

Water is a compound of hydrogen and oxygen (hydrogen oxide H_2O). It is a colourless, odourless liquid at room temperature and pressure.

The **water cycle** explains the regular supply of rain which provides the water essential for life. Evaporation of water from rivers, lakes and the sea provides water vapour which is held in the atmosphere in clouds. When the clouds cool the water vapour condenses and falls as rain.

The water supply to our homes comes from unpolluted rivers, lakes or suitable underground sources. It is not pure but contains a range of dissolved substances depending on the rocks through which the water has passed. The water is filtered and treated with a small quantity of chlorine to kill bacteria.

19.2 Water as a Solvent

Water dissolves a wide range of different substances. Water is said to be a good **solvent** and the substances dissolved are called **solutes**. Water is a **polar solvent**, i.e. it contains small positive and negative charges caused by the slight movement of electrons in the covalent bonds. Polar solvents dissolve compounds containing ionic bonds, e.g. sodium chloride. **Nonpolar solvents** (e.g. tetrachloromethane CCl_4) are poor at dissolving ionic compounds but are good solvents for molecular compounds.

A solution which contains as much solute as can be dissolved at a particular temperature is called a **saturated solution**. If any more solute is added to a saturated solution, the extra solute remains undissolved. The **solubility** of a solute is the mass of solute (in grams) which dissolves in 100 g of solvent at a particular temperature to form a saturated solution.

Generally the solubility of ionic compounds in water (e.g. potassium nitrate) increases as the temperature rises but the solubility of gases in water decreases as temperature rises. A graph of solubility versus temperature for a solute is called a **solubility curve**. Figure 19.1 shows the solubility curves for some common solutes.

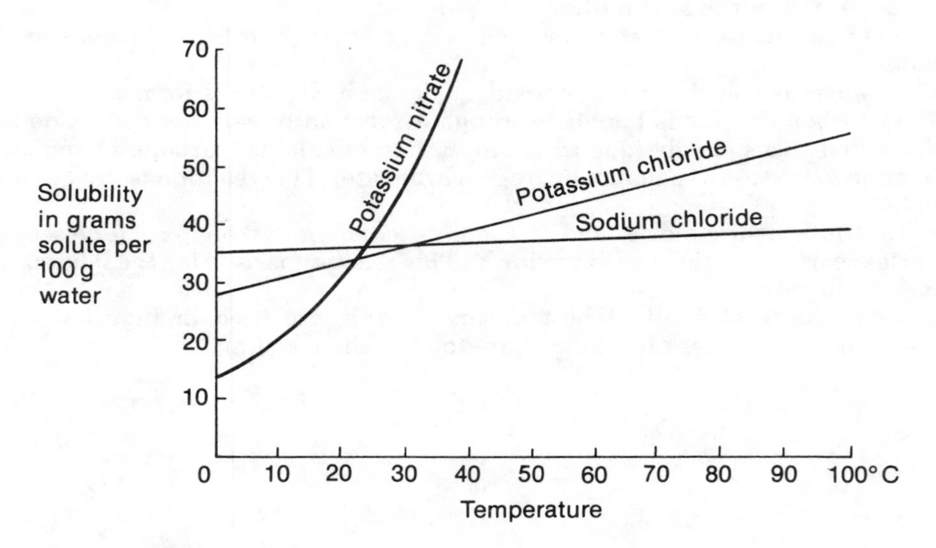

Fig 19.1 Solubility curves

19.3 Hard and Soft Water

Distilled water (pure water) contains no dissolved solid impurities. It lathers well with soap. Rain water quite closely resembles distilled water.

Some water samples do not lather well with soap but form scum. These water samples are said to be hard. **Hard water** is caused by certain dissolved substances in water. These substances become dissolved as the water trickles through the ground. Water (containing dissolved carbon dioxide from the air) trickling through chalk (calcium carbonate) dissolves some of the chalk forming calcium hydrogencarbonate.

$$CaCO_3(s) + H_2O(l) + CO_2(g) \rightleftharpoons Ca(HCO_3)_2(aq)$$

calcium carbonate + water + carbon dioxide ⇌ calcium hydrogencarbonate

Hard water is caused by dissolved calcium and magnesium compounds in water.

There are two types of hardness in water – permanent hardness and temporary hardness. **Permanent hardness** is caused by dissolved **calcium sulphate** and **magnesium sulphate**. This type of hardness is not removed by boiling. It has to be softened (i.e. the hardness removed) by chemical reaction. **Temporary hardness** is caused by dissolved **calcium hydrogencarbonate**.

Apart from using more soap than would otherwise be required, using hard water has other effects. It causes deposits in kettles, boilers and hot water pipes. Hard water is, however, better for brewing beer and supplies calcium to the human body.

19.4 Removal of Hardness

Temporary hardness is removed by boiling due to the decomposition of calcium hydrogencarbonate forming calcium carbonate, which is insoluble.

$$Ca(HCO_3)_2(aq \rightleftharpoons CaCO_3(s) + H_2O(l) + CO_2(g)$$

calcium hydrogencarbonate ⇌ calcium carbonate + water + carbon dioxide

It is this deposit of calcium carbonate which forms the scale or 'fur' in a kettle.

Permanent and temporary hardness can be removed by adding washing soda crystals (sodium carbonate crystals).

$$Ca(HCO_3)_2(aq) + Na_2CO_3(aq) \rightarrow CaCO_3(s) + 2NaHCO_3(aq)$$

calcium hydrogencarbonate + sodium carbonate → calcium carbonate + sodium hydrogencarbonate

$$CaSO_4(aq) + Na_2CO_3(aq) \rightarrow CaCO_3(s) + Na_2SO_4(aq)$$

calcium sulphate + sodium carbonate → calcium carbonate + sodium sulphate

$$MgSO_4(aq) + Na_2CO_3(aq) \rightarrow MgCO_3(s) + Na_2SO_4(aq)$$

magnesium sulphate + sodium carbonate → magnesium carbonate + sodium sulphate

In each case the calcium or magnesium ions in solution are precipitated and no longer cause problems. Other substances, e.g. calcium hydroxide, 'Calgon' (sodium metaphosphate) and sodium sesquicarbonate, work in a similar way by precipitating the substances which cause hardness. Sodium sesquicarbonate is better than sodium carbonate for household use because it is less alkaline.

Hardness can be removed by using an **ion exchange column**. A column is filled with a suitable resin in small granules. The resin contains an excess of sodium ions. When the hard water passes through the column, the calcium and magnesium ions in the water (causing hardness) are exchanged for sodium ions. When the sodium ions on the column have all been removed, the column is recharged.

19.5 Soaps and Soapless Detergents

Soap is produced by treating vegetable or animal fats with concentrated sodium hydroxide solution, and precipitating the soap with salt solution. This reaction is an example of **saponification.**

Soapless detergents are produced from residues from crude oil distillation (see Unit 20.8). These hydrocarbons are treated with concentrated sulphuric acid.

Soap and soapless detergent molecules are very similar in structure despite the different methods of production. Both have a long hydrocarbon chain (e.g. $C_{17}H_{35}$—) attached to an ionic group (—CO_2^- in soap or —SO_3^- in soapless detergents). The hydrocarbon 'tail' will dissolve in fats and grease while the ionic 'head' will dissolve readily in water. The cleansing actions of soap or soapless detergents are summarized in Fig. 19.2. The soap or soapless detergent molecules are represented as 'tadpoles'. When the detergent (soap or soapless) is added to water the molecules are in clusters in the solution. The tails of the detergent molecules stick into the greasy dirt and attraction between the water molecules and the detergent molecules lifts the dirt from the fibre. Agitation of the solution helps to lift the dirt. The grease is then suspended in the solution, with repulsive forces between detergent molecules preventing grease from returning to the material.

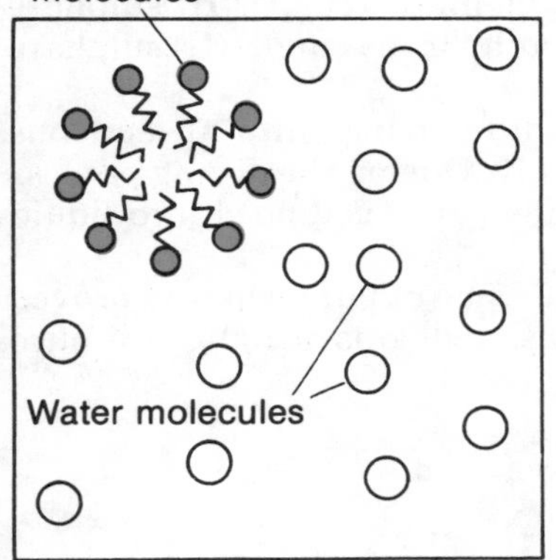

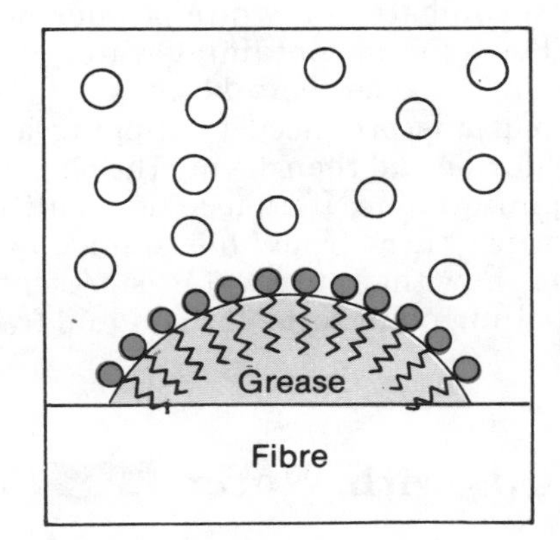

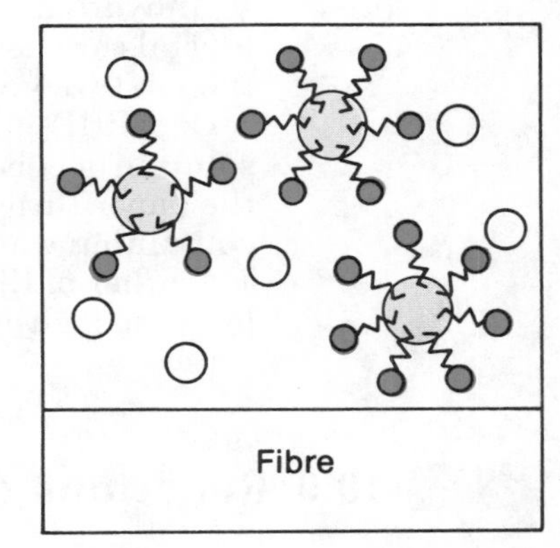

Fig 19.2 Cleansing action of soap

Whereas soap forms scum with calcium and magnesium compounds in hard water, soapless detergents do not. This is an advantage of soapless detergents.

$$2C_{17}H_{35}CO_2^-Na^+(aq) + Ca^{2+}(aq) \rightarrow (C_{17}H_{35}CO_2)_2Ca(s) + 2Na^+(aq)$$

sodium stearate (soap) + calcium ions → calcium stearate (scum) + sodium ions

N.B. Sodium stearate is sometimes called sodium octodeconate.

19.6 Water of Crystallization

Certain substances crystallize (form crystals) with a fixed number of molecules of water contained within the crystal. This water is called **water of crystallization**. For example, copper(II) sulphate crystallizes into copper(II) sulphate crystals $CuSO_4.5H_2O$. They are blue in colour and are said to be **hydrated**. On heating, the water of crystallization is evolved and white anhydrous copper(II) sulphate is formed. Other examples of substances containing water of crystallization are:

hydrated sodium sulphate	$Na_2SO_4.10H_2O$
hydrated sodium carbonate	$Na_2CO_3.10H_2O$
hydrated magnesium sulphate	$MgSO_4.7H_2O$

There is no way of predicting the number of molecules of water of crystallization in a particular substance.

19.7 Changes Taking Place when Chemicals are Exposed to the Air

Some substances absorb water vapour from the air and are not appreciably changed. They are said to be **hygroscopic**. For example, copper(II) oxide absorbs small amounts of water vapour and should always be thoroughly dried before use in experiments involving weighing.

Other substances absorb water vapour to a much greater extent and the substance dissolves in the water absorbed. These substances are said to be **deliquescent**. An example of deliquescence is anhydrous calcium chloride. When left exposed to the atmosphere, the anhydrous calcium chloride absorbs water vapour from the air and after a couple of days a pool of calcium chloride solution results.

Hygroscopic substances increase in mass by a small amount but deliquescent substances increase in mass considerably when left exposed to the atmosphere.

Some compounds containing water of crystallization can lose some of this water on standing exposed to the air. This is called **efflorescence** (not to be confused with effervescence). Efflorescent compounds include hydrated sodium carbonate $Na_2CO_3.10H_2O$ and hydrated sodium sulphate $Na_2SO_4.10H_2O$. Hydrated sodium carbonate crystals are colourless, transparent crystals which on standing in the air turn to a white, opaque powder $Na_2CO_3.H_2O$. The mass of efflorescent compounds decreases on standing in air.

19.8 Testing for Water

Water is frequently produced during chemical reactions. However, you cannot assume that any colourless, odourless liquid that you come across is water. Either anhydrous copper(II) sulphate or cobalt(II) chloride paper can be used to show that water is present in a liquid.

Anhydrous copper(II) sulphate is a white powder produced when blue copper(II) sulphate crystals are heated. When a liquid containing water is added to anhydrous copper(II) sulphate the mixture goes blue and becomes very hot.

Cobalt(II) chloride paper is produced by dipping a piece of filter paper into an aqueous solution of cobalt(II) chloride and then drying the paper thoroughly. During the drying process the paper turns from pink to blue. If a piece of cobalt(II) chloride paper is dipped into liquid containing water the paper turns from blue to pink.

Neither of these tests show that pure water is present. The presence of pure water is proved by doing melting and boiling point tests. If the liquid freezes at 0°C and boils at 100°C it is pure water.

19.9 Reactions of Metals with Water

The difference in reactivity of metals can be seen in the reactions of metals with water.

1 POTASSIUM

A small piece of potassium reacts violently with cold water. It floats on the water in a small molten ball and reacts with the water producing hydrogen gas, which catches alight and burns with a pinkish flame (see Unit 32.3). An alkali (potassium hydroxide) remains in the solution.

$$2K(s) + 2H_2O(l) \rightarrow 2KOH(aq) + H_2(g)$$

potassium + water → potassium hydroxide + hydrogen

2 SODIUM

Sodium reacts with cold water in a similar way to potassium but more slowly. The hydrogen does not usually ignite.

$$2Na(s) + 2H_2O(l) \rightarrow 2NaOH(aq) + H_2(g)$$

sodium + water → sodium hydroxide + hydrogen

3 CALCIUM

Calcium reacts slowly with cold water and usually sinks.

$$Ca(s) + 2H_2O(l) \rightarrow Ca(OH)_2(aq) + H_2(g)$$

calcium + water → calcium hydroxide + hydrogen

During the experiment the solution produced turns cloudy due to the formation of calcium hydroxide, which is not very soluble and precipitates out.

4 MAGNESIUM

Magnesium reacts very slowly with cold water. It takes several weeks to produce a test tube full of hydrogen from a short length of magnesium ribbon. It reacts well, however, when strongly heated in steam (Fig. 19.3).

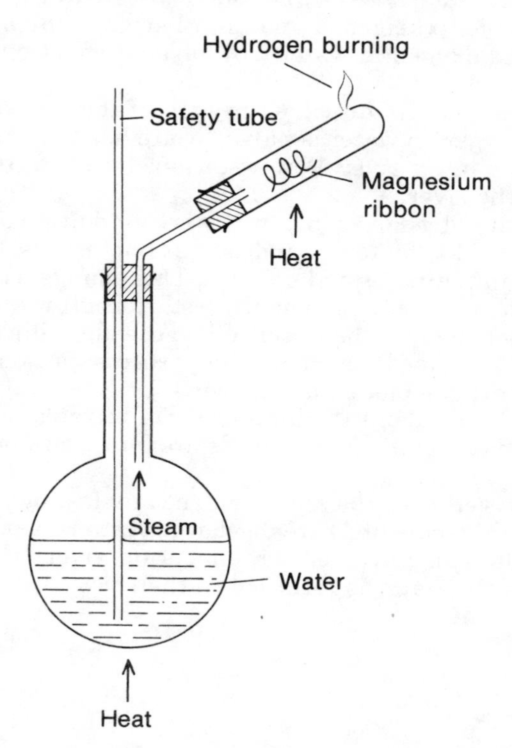

Fig 19.3 Reaction of magnesium with steam

$$Mg(s) + H_2O(g) \rightarrow MgO(s) + H_2(g)$$

magnesium + water(steam) → magnesium oxide + hydrogen

(N.B. Magnesium oxide is formed because, at the temperature of the experiment, magnesium hydroxide decomposes.)

5 ZINC

Zinc reacts with steam in a similar way.

$$Zn(s) + H_2O(g) \rightarrow ZnO(s) + H_2(g)$$

zinc + water(steam) → zinc oxide + hydrogen

6 IRON (see Unit 14.5)

Iron, being less reactive, reacts only reversibly with steam.

$$3Fe(s) + 4H_2O(g) \rightleftharpoons Fe_3O_4(s) + 4H_2(g)$$

iron + water(steam) ⇌ iron(II)di-iron(III) oxide + hydrogen

Other metals lower in the reactivity series (e.g. lead, copper) do not react with water.

19.10 Importance of Water to Industry

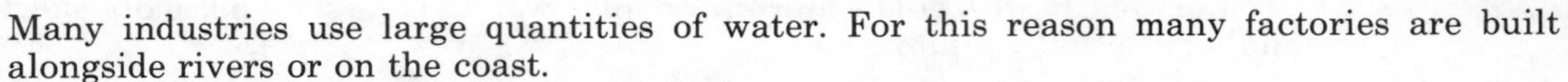

Many industries use large quantities of water. For this reason many factories are built alongside rivers or on the coast.

The water used by industry can be used for various reasons. These include:

1 As an essential ingredient in the product, e.g. beer making, whisky production.

2 For water to cool parts of the process, e.g. making electricity in a oil- or coal-fired power station.

3 As a source of energy, e.g. making electricity in a hydroelectric power station.

4 As a raw material which is removed during the process, e.g. paper making.

19.11 Water Pollution

Over 50 million gallons of waste water enter rivers in Great Britain every day from sewage works, factories and farms. Great care has to be taken to ensure that unwanted impurities, called pollutants, are not in the water.

At room temperature only about 30 cm^3 of oxygen will dissolve in 1 litre of water. This small amount of oxygen is all that is available to keep fish and all other river life alive. Anything which reduces the amount of dissolved oxygen in water will have an effect on the life of the river.

A temperature rise will reduce the amount of dissolved oxygen. If the temperature rises to about 40°C, the volume of dissolved oxygen in water is halved. Care has to be taken therefore not to pump large amounts of hot water into a river. Power stations return water to rivers but it is allowed to cool before it enters the river.

A major reason for the reduction of dissolved oxygen in water is the addition of ammonia and ammonium compounds from sewage works, factories and farms. Bacteria in the water turn ammonia and ammonium compounds into nitrites and nitrates. This change uses up dissolved oxygen. Because nitrates encourage plant growth, green algae start to grow on the surface of the water. These algae prevent light getting into the water. When the algae die they decay and this again reduces the amount of dissolved oxygen. This process is sometimes called **eutrophication** and the result can be a completely dead river.

There is considerable concern at present about the high levels of nitrates in rivers and the effects that these high levels can have on health. These levels are four times larger than they were 20 years ago and are still increasing.

A fast-flowing river can dissolve oxygen from the air to replace any lost by pollution. More serious problems are caused when metal compounds are discharged into rivers. Compounds of lead, cadmium and iron, for example, can have serious effects on river life. These metal compounds should be removed before the water is returned to the river.

19.12 Summary

Water is a very important compound of hydrogen and oxygen. It is a very good solvent dissolving a wide range of substances. For this reason it is very difficult to get pure water.

The solubility of most solutes increases with rise in temperature.

Water that does not lather well with soap but forms scum is called hard water. Hard water is caused by dissolved calcium and magnesium compounds. There are two types of hardness. Temporary hardness is caused by calcium hydrogencarbonate. Permanent hardness is caused by dissolved calcium and magnesium sulphates.

When hardness is removed by a softening process, soft water is produced. Softening can be done by adding chemicals such as sodium carbonate crystals or by using an ion exchange column.

Soapless detergents produce a lather with hard water without forming scum. They are easier to use, therefore, in hard water areas.

Some chemicals absorb water vapour from the atmosphere and dissolve in this water. These substances are said to be deliquescent. Other hydrated chemicals lose water vapour to the atmosphere. They are said to be efflorescent.

The presence of water in a substance can be shown with anhydrous copper(II) sulphate (turns white to blue) or cobalt(II) chloride paper (turns blue to pink). Pure water boils at 100 °C and freezes at 0 °C.

Water pollution can cause considerable problems. These problems often involve reduction of the small volume of dissolved oxygen in the water.

20 CHEMICALS FROM PETROLEUM

20.1 Petroleum

Petroleum (also called **crude oil**) is a most important mineral. It is found in various parts of the world including the Middle East, USA (including Alaska and Texas), Venezuela and the North Sea. Apart from producing valuable fuels, it is also the source of a wide range of chemicals and useful everyday materials.

Petroleum was formed by the decomposition of animal and plant material over millions of years. It is a fossil fuel (see Unit 22.5). It is found trapped in permeable rock layers between layers of impermeable rock (Fig. 20.1). It is extracted by drilling deep holes into the earth. Often the oil will escape from the ground under its own pressure at first but pumping may be required later.

Petroleum is a complex mixture of hydrocarbons. **Hydrocarbons** are compounds of carbon and hydrogen only. Most of the compounds present in petroleum are alkanes.

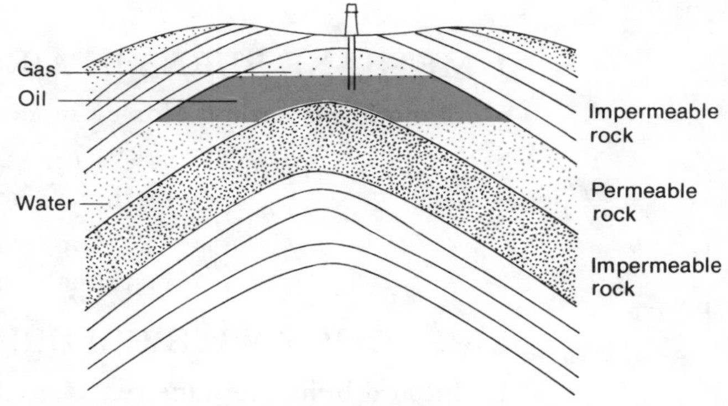

Fig 20.1 Formation of petroleum

20.2 Alkanes

The **alkanes** are a series of hydrocarbons (compounds of carbon and hydrogen) with the general formula C_nH_{2n+2} where $n = 1,2,3,...$, for successive members of the series.

The first member of the series ($n = 1$) is methane CH_4 and the second member ($n = 2$) is ethane C_2H_6.

Table 20.1 summarizes some information about the simplest alkanes.

Table 20.1 The simplest alkanes

Alkane	*Molecular formula*	*Structural formula*	*Melting point °C*	*Boiling point °C*	*State at room temperature and pressure*
Methane	CH_4	CH_4 (H—C—H with H above and below)	−183	−162	Gas
Ethane	C_2H_6	H—C—C—H (each C with H above and below)	−183	−89	Gas
Propane	C_3H_8	H—C—C—C—H (each C with H above and below)	−188	−42	Gas
Butane	C_4H_{10}	H—C—C—C—C—H (each C with H above and below)	−135	−0.5	Gas
Pentane	C_5H_{12}	H—C—C—C—C—C—H (each C with H above and below)	−130	36	Liquid

Alkanes contain only single bonds and are said to be **saturated**. Organic compounds containing double or triple bonds are said to be **unsaturated**.

A series of compounds which are related to each other (e.g. the alkanes) is called a **homologous series**. Each member is called a **homologue**. In each homologous series, each member has the same general formula but differs from the next in the series by a unit of CH_2.

The physical properties of the members show a gradual change with increasing relative molecular mass. Thus, for example, in the alkanes the melting points and boiling points rise with increasing relative molecular mass. In a similar way, the densities increase and the solubility in water decreases with rising relative molecular mass.

Natural gas which is piped to our homes and used for all gas appliances is almost pure methane. Calor gas which is bought in cylinders for camping is propane. Butane is used in gas cigarette lighters. Higher alkanes are used in petrol and other fuels.

20.3 Chemical Reactions of Methane

1 METHANE BURNS IN AIR OR OXYGEN

Methane forms carbon dioxide in excess air and carbon monoxide in a limited supply of air.

$$CH_4(g) + 2O_2(g) \rightarrow CO_2(g) + 2H_2O(g)$$

methane + oxygen → carbon dioxide + water

$$2CH_4(g) + 3O_2(g) \rightarrow 2CO(g) + 4H_2O(g)$$

methane + oxygen → carbon monoxide + water

2 REACTION WITH CHLORINE

In diffused light, methane reacts giving a series of products by the successive replacement of hydrogen atoms. Each of the reactions is a **substitution reaction**.

$$CH_4(g) + Cl_2(g) \rightarrow HCl(g) + CH_3Cl(g)$$

methane + chlorine → hydrogen chloride + chloromethane

$$CH_3Cl(g) + Cl_2(g) \rightarrow HCl(g) + CH_2Cl_2(g)$$

chloromethane + chlorine → hydrogen chloride + dichloromethane

$$CH_2Cl_2(g) + Cl_2(g) \rightarrow HCl(g) + CHCl_3(g)$$

dichloromethane + chlorine → hydrogen chloride + trichloromethane

$$CHCl_3(g) + Cl_2(g) \rightarrow HCl(g) + CCl_4(g)$$

trichloromethane + chlorine → hydrogen chloride + tetrachloromethane

20.4 Isomerism

There is only one possible structure for each of the first three alkanes but the four carbon atoms and the ten hydrogen atoms in a molecule of butane, C_4H_{10}, can be linked in two different ways as shown in Fig. 20.2.

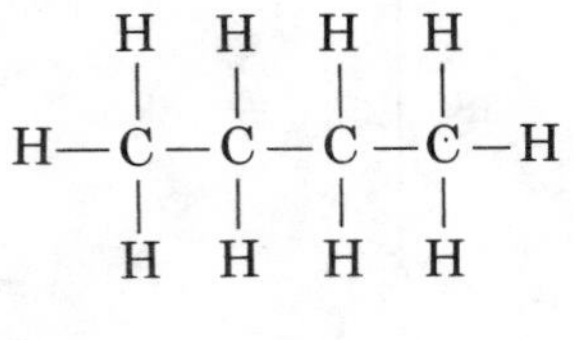

```
    H   H   H   H               H   H   H
    |   |   |   |               |   |   |
H — C — C — C — C — H       H — C — C — C — H
    |   |   |   |               |   |   |
    H   H   H   H               H   |   H
                                    |
                                H — C — H
                                    |
                                    H
```

Butane
(or n-Butane)

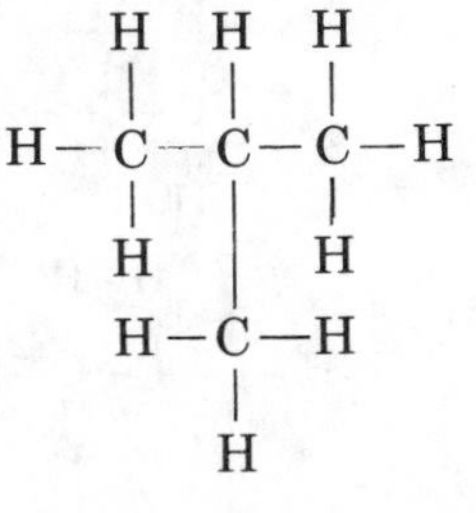

2-methylpropane
(or iso-butane)

Fig 20.2 Isomers of butane

This is an example of **isomerism**, i.e. the existence of two or more compounds (called **isomers**) having the same molecular formula but different structural formulae.

Isomerism becomes more common in the higher alkanes, e.g. there are 75 isomers of decane $C_{10}H_{22}$. Isomerism also occurs within other homologous series and also with compounds in different homologous series, e.g. C_2H_5OH (an alcohol) and CH_3OCH_3 (an ether).

20.5 Alkenes

The **alkenes** are a series of hydrocarbons with the general formula C_nH_{2n}, where $n = 2, 3, \ldots$. Table 20.2 summarizes some information about the simplest alkenes.

Table 20.2 The simplest alkenes

Alkene	*Molecular formula*	*Structural formula*	*Melting point °C*	*Boiling point °C*	*State at room temperature and pressure*
Ethene	C_2H_4	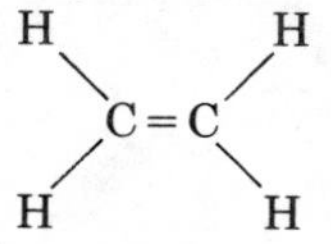 $H_2C{=}CH_2$	−169	−102	Gas
Propene	C_3H_6	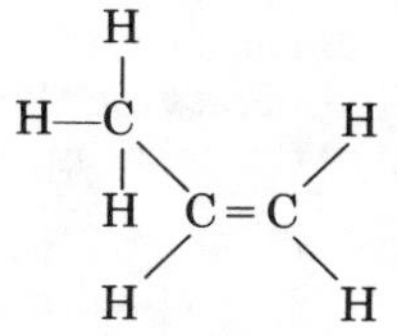 $H_3C{-}CH{=}CH_2$	−185	−48	Gas
Butene	C_4H_8	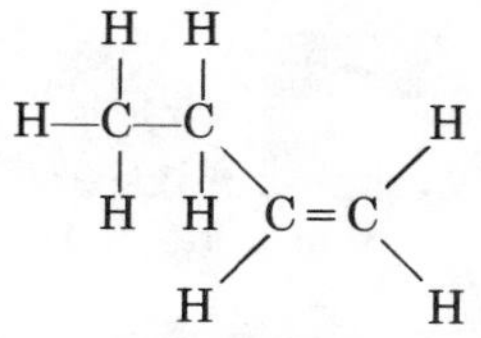 $H_3C{-}CH_2{-}CH{=}CH_2$	−185	−6	Gas
Pentene	C_5H_{10}	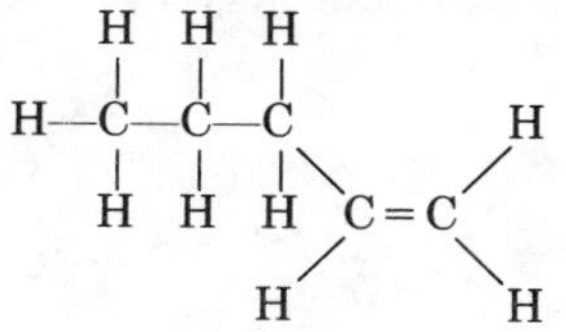 $H_3C{-}CH_2{-}CH_2{-}CH{=}CH_2$	−138	30	Liquid

It should be noted that all alkenes contain a double bond between two carbon atoms. They are, therefore, **unsaturated**. Alkenes with four or more carbon atoms can exist as different isomers.

20.6 Reactions of Ethene

Other alkenes react in a similar way.

1 COMBUSTION

Ethene burns in air or oxygen if ignited. In excess air carbon dioxide and water are produced and in a limited supply of air carbon monoxide and water are produced.

2 ADDITION REACTIONS

Addition reactions are common with all unsaturated compounds. In such a reaction two substances combine to produce a single new substance. The reaction between ethene and bromine is most frequently mentioned. Ethene reacts rapidly with bromine vapour to form colourless oily drops of 1,2-dibromoethane (Fig. 20.3).

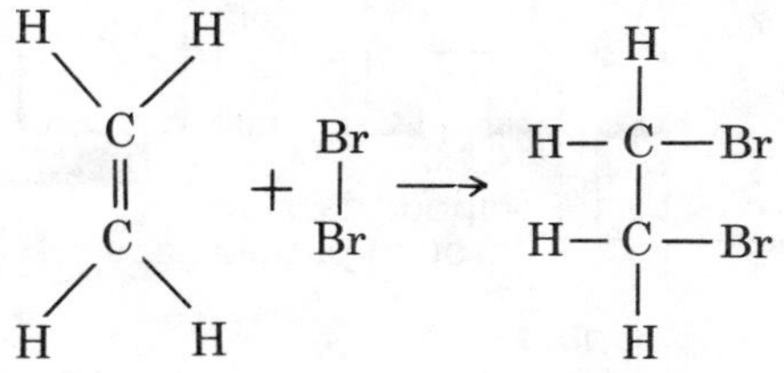

$$H_2C{=}CH_2 + Br{-}Br \longrightarrow H{-}CHBr{-}CHBr{-}H$$

Fig 20.3 Addition of bromine to ethene

This reaction is used to detect compounds containing double or triple bonds (unsaturated). The compound is shaken with bromine dissolved in a suitable solvent (e.g. tetrachloromethane). Unsaturated compounds remove the reddish colour of the bromine. This test does not, of course, distinguish between compounds containing double and triple bonds.

Ethene also reacts with hydrogen at 200 °C, in the presence of a finely divided nickel catalyst, to form ethane.

$$C_2H_4(g) + H_2(g) \rightarrow C_2H_6(g)$$
$$\text{ethene} + \text{hydrogen} \rightarrow \text{ethane}$$

This reaction is similar to the reaction in which natural fats and oils are converted to margarine by the addition of hydrogen at about 5 atmospheres pressure and 180 °C.

20.7 Alkynes

The **alkynes** are a series of hydrocarbons with a general formula C_nH_{2n-2}. The simplest member is ethyne (sometimes called acetylene) C_2H_2, which has the structural formula H—C≡C—H. All alkynes contain a triple bond between two carbon atoms and, like the alkenes, are unsaturated.

20.8 Refining of Petroleum

Petroleum consists mainly of a complex mixture of hydrocarbons, the greater proportion of which are alkanes. The petroleum can be split up into various fractions by **fractional distillation**.

Fractional distillation of petroleum may be carried out on a small scale in the laboratory using the apparatus in Fig. 20.4.

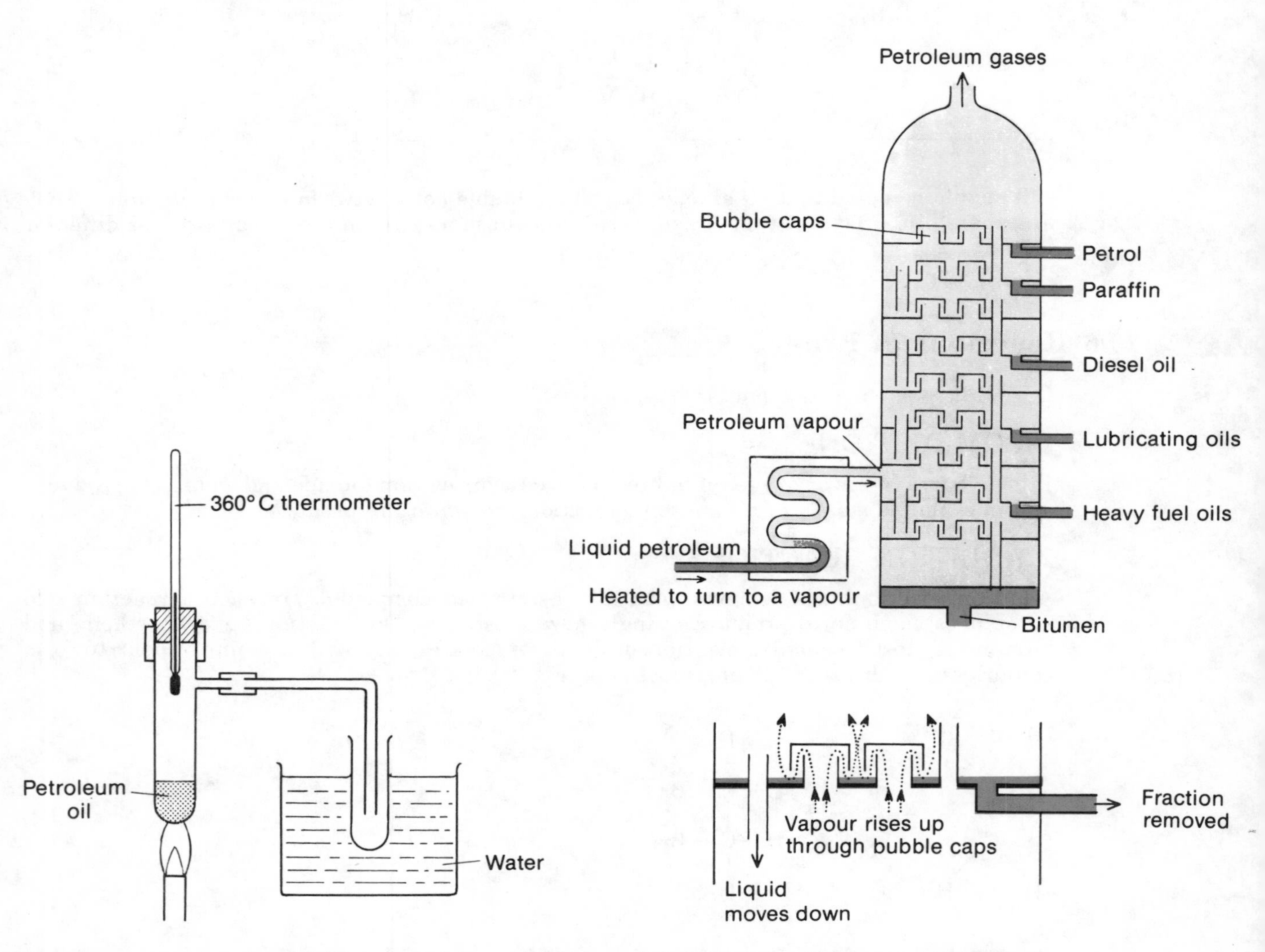

Fig 20.4 Laboratory distillation of petroleum

Fig 20.5 Industrial distillation of petroleum

Several fractions may be obtained corresponding to different boiling point ranges (e.g. first fraction up to 70°C, second fraction 70–120°C, third fraction 120–70°C, fourth fraction 170–220 °C etc.). On examining the fractions obtained, certain definite changes in properties of the fractions with increasing boiling point can be seen. In particular, the yellow colour and viscosity (i.e. the ease with which the liquid pours, or stickiness) increase with increasing boiling point whilst the flammability decreases.

Industrially, the fractional distillation of petroleum is carried out on a large scale in an oil refinery. The fractionation is carried out in a fractional distillation column (Fig. 20.5).

The main fractions include:

1 Petrol (6–12 carbon atoms).
2 Paraffin (11–16 carbon atoms).
3 Lubricating oil (more than 20 carbon atoms).

20.9 Cracking

Petroleum contains a complex mixture of hydrocarbons. After refining, many of the higher boiling point fractions containing long chain molecules are difficult to sell profitably. Oil companies have developed cracking and reforming processes to convert these fractions into shorter chain molecules which are very easy to sell to the chemical industry. They are used for making chemicals and making polymers.

Cracking involves the breaking of long chain alkanes (Fig. 20.6) and can be carried out in two ways:

1 Thermal cracking – carried out simply by heating.
2 Catalytic cracking – carried out using a catalyst.

```
    H  H  H  H  H  H          H  H       H  H  H  H
    |  |  |  |  |  |          |  |       |  |  |  |
H — C— C— C— C— C— C—H  ——→   C==C  + H—C— C— C— C—H
    |  |  |  |  |  |          |  |       |  |  |  |
    H  H  H  H  H  H          H  H       H  H  H  H

        Hexane               Ethene         Butane
```

Fig 20.6 Cracking of hexane

Cracking always produces some unsaturated products.

Reforming involves processes where the shapes of the molecules are changed but the sizes of the molecules are not. It is used, for example, to produce petrol with a high octane rating.

A simple cracking experiment can be carried out in the laboratory using the apparatus in Fig. 20.7. Liquid paraffin vapour is passed over strongly heated broken china where the cracking takes place. The product, ethene gas, can be collected over water.

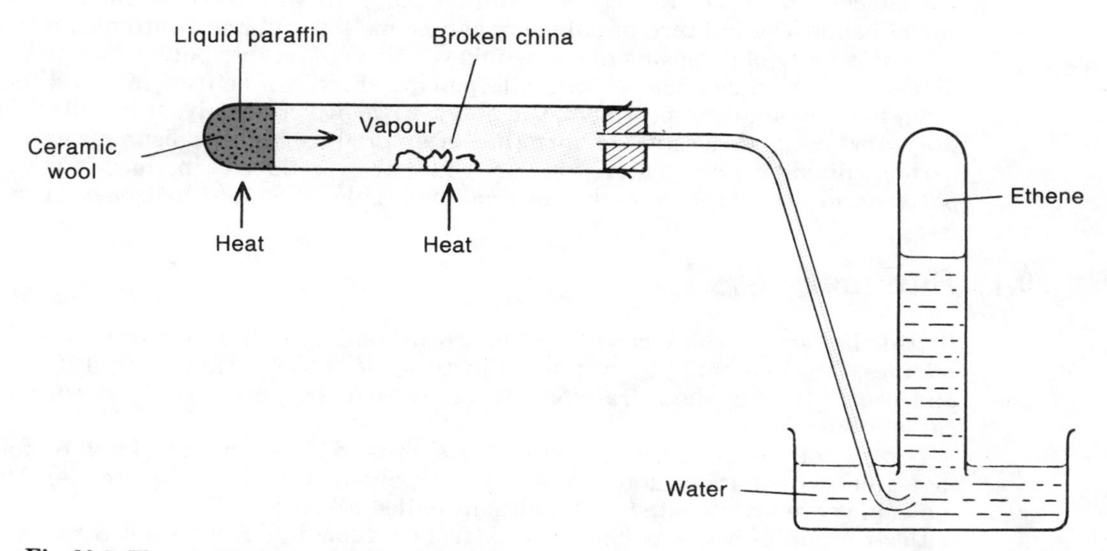

Fig 20.7 Thermal cracking of liquid paraffin

20.10 Addition Polymers

Polymerization is the linking together of relatively small and simple molecules to form large units called **polymers**. The individual small molecules are called **monomers**.

Addition polymerization is one type of polymerization. It may be represented as follows:

$$nM \rightarrow (M)n$$

where M is a monomer unit.

This type of polymerization can take place only if the monomer is an unsaturated molecule. Many monomers, in fact, contain a double bond between two carbon atoms.

E.g. **Ethene** produces the polymer called **polyethene** (polythene) as shown in Fig. 20.8.

$$n\left[\begin{array}{cc} H & H \\ | & | \\ C & = C \\ | & | \\ H & H \end{array}\right] \longrightarrow \; ---\begin{array}{c} H \; H \; H \; H \; H \; H \\ | \;\; | \;\; | \;\; | \;\; | \;\; | \\ C-C-C-C-C-C \\ | \;\; | \;\; | \;\; | \;\; | \;\; | \\ H \; H \; H \; H \; H \; H \end{array}---$$

Ethene — Part of polymer molecule

Fig 20.8 Polymerization of ethene

In the resulting polymer unit produced there are usually hundreds of ethene units combined together.

The polymer can be produced in two ways:

1 Ethene gas is bubbled into a hydrocarbon solvent containing a complex catalyst. The temperature is below 100 °C and the pressure of the gas is atmospheric. The polythene produced is called high density (HD) polythene.

2 Ethene is heated at high pressure (1000 atmospheres) and a temperature of 180 °C with a little oxygen present as initiator. The polythene produced is called low density (LD) polythene.

Table 20.3 gives details of some common addition polymers and their uses.

20.11 Advantages and Disadvantages of Addition Polymers

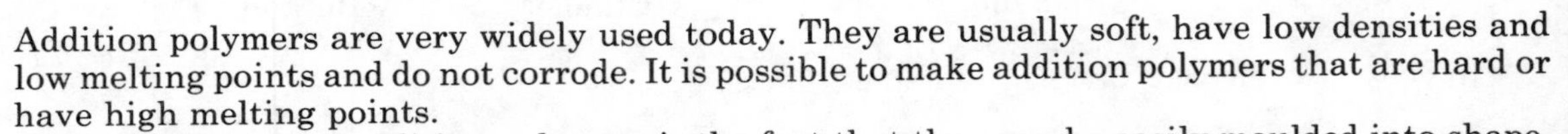

Addition polymers are very widely used today. They are usually soft, have low densities and low melting points and do not corrode. It is possible to make addition polymers that are hard or have high melting points.

One advantage of addition polymers is the fact that they can be easily moulded into shape, making the production of complicated items such as a car dashboard easy.

The biggest disadvantage of addition polymers comes from the fact that they do not decompose. They are extremely difficult to dispose of, for example in household refuse. If tipped onto an open tip they do not rot away.

If household rubbish is screened until only a mixture of waste paper and waste polymers remains it is possible to separate the polymers. Thorough wetting with water will make the paper sink to the bottom of a water tank. Polymers are not wetted and continue to float. They can be scooped off. The result is a mixture of polymers which cannot easily be separated into its components. The mixture of polymers can be melted and made into cheap insulation blocks.

Another way of disposing of household rubbish containing polymers is to burn it. In several places household rubbish is made into pellets which can be used for solid fuel boilers.

On heating addition polymers, the polymers melt very easily. It is called a **thermoplastic polymer**. When these start to burn they often produce highly poisonous gases, along with the carbon dioxide and water vapour that are produced in large quantities. Burning poly(acrylonitrile) for example can produce highly poisonous hydrogen cyanide.

20.12 Summary

Petroleum is a complex mixture of hydrocarbons, especially saturated hydrocarbons called alkanes. It is separated by fractional distillation into fractions with different boiling points and uses. Many of these fractions are used as fuels. These fractions contain hydrocarbons called alkanes.

Long chain hydrocarbon fractions are broken down by cracking to form short chain molecules which are used to make addition polymers such as poly(ethene). These short chain molecules are unsaturated hydrocarbons called alkenes.

Destruction of waste polymers is extremely difficult. They do not rot away easily and can produce poisonous gas when they burn.

Table 20.3 Examples of addition polymers

Monomer Formula/name	Polymer Name	Trade name	Formula	Uses
$H_2C{=}CH_2$ ETHENE	Poly(ethene)	Polythene	$(-CH_2-CH_2-)_n$	Plastic sheets, pipes, plastic bags
$H_2C{=}CH(CH_3)$ PROPENE	Poly(propene)	Propathene	$(-CH_2-CH(CH_3)-)_n$	Plastic sheets, electric insulators, washing-up bowls
$H_2C{=}CHCl$ VINYL CHLORIDE (CHLOROETHENE)	Poly(vinylchloride) or poly(chloroethene)	PVC	$(-CH_2-CHCl-)_n$	Records, clothes, electrical wire insulators
$H_2C{=}CH(C_6H_5)$ STYRENE (PHENYLETHENE)	Polystyrene or poly(phenylethene)	—	$(-CH_2-CH(C_6H_5)-)_n$	Packaging materials, ceiling tiles, plastic model kits
$H_2C{=}CH(COOCH_3)$ METHYL METHACRYLATE	Poly(methyl-methacrylate)	Perspex	$(-CH_2-CH(COOCH_3)-)_n$	Substitute for glass
$H_2C{=}CH(CN)$ ACRYLONITRILE	Poly(acry-lonitrile)	Orlon, Courtelle, Acrilan	$(-CH_2-CH(CN)-)_n$	Synthetic fibre
$F_2C{=}CF_2$ TETRAFLUOROETHENE	Poly(tetra-fluoroethene)	Teflon PTFE	$(-CF_2-CF_2-)_n$	Coating for nonstick saucepans; bridge bearings

21 ETHANOL AND ETHANOIC ACID

This unit is written in four parts

Part A:	21.1–21.5	Ethanol
Part B:	121.6–21.7	Ethanoic acid
Part C:	21.8	Condensation polymerization
Part D:	21.9–21.11	Carbohydrates

Check the Table of Analysis of Examination Syllabuses on pp. xvi–xvii to find which parts are required for your syllabus.

21.1 Introduction

Ethanol C_2H_5OH is an organic chemical of great importance. It belongs to the homologous series of alcohols. **Alcohols** have a general formula $C_nH_{2n+1}OH$. They may be regarded as being derived from an alkane by replacing a hydrogen atom by a hydroxyl (—OH) group.

Ethanoic acid (which used to be called acetic acid) is a weak acid that is closely related to ethanol. Both compounds contain two carbon atoms.

Both ethanol and ethanoic acid are widely used in industry.

21.2 Laboratory Preparation of Ethanol

Ethanol can be prepared by the fermentation of glucose (or any other sugar solution) using enzymes in yeast. The apparatus (Fig. 21.1) is kept at about 30 °C for several days. The fermentation lock allows the escape of carbon dioxide gas without the entry of oxygen, which could oxidize the ethanol produced.

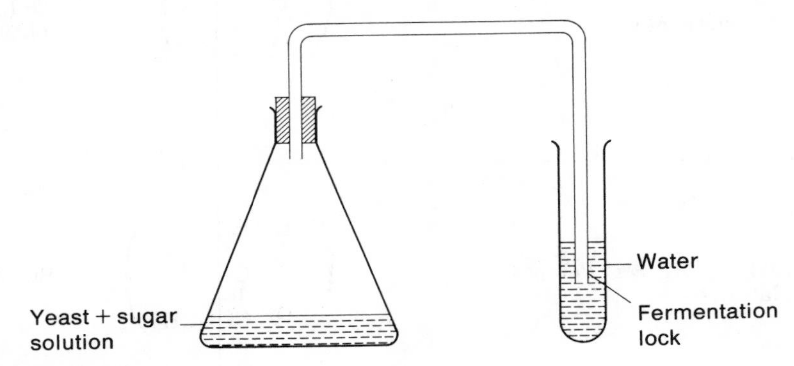

Fig 21.1 Fermentation

During the fermentation the glucose is converted to ethanol by the enzyme zymase in the yeast.

$$C_6H_{12}O_6(aq) \rightarrow 2C_2H_5OH(aq) + 2CO_2(g)$$
$$\text{glucose} \rightarrow \text{ethanol} + \text{carbon dioxide}$$

The final solution will be a dilute solution of ethanol. This process is used in wine and beer making.

A more concentrated ethanol solution can be produced by fractional distillation of the resulting solution (see Unit 2.3). This is the basis of industries producing whisky, gin and other spirits.

The production of large quantities of ethanol for industry by fermentation is possible with surplus wine production in the EEC. In Brazil factories exist for the production of ethanol on a large scale by fermentation of sugar cane for use as a petrol substitute.

21.3 Industrial Production of Ethanol

Ethanol is produced in large quantities from ethene; it is produced by cracking (see Unit 20).

Ethene is mixed with steam and passed over a phosphoric acid catalyst at 300 °C.

$$C_2H_4(g) + H_2O(g) \rightarrow C_2H_5OH(g)$$
ethene + water(steam) → ethanol

21.4 Uses of Ethanol

Ethanol is a widely used solvent. It is, however, very flammable. It is used in the manufacture of varnishes, inks, glues and paints. It evaporates quickly and its evaporation explains the cooling effect on the skin when deodorants, perfumes, etc. are used.

Ethanol is the alcohol present in alcoholic drinks. There is more and more evidence for the harmful effects of ethanol, especially on the liver. The adverse effects of ethanol on the nervous system – impaired coordination and slower responses – are responsible for many road accidents. Alcoholic drinks are very heavily taxed.

Pure ethanol cannot be purchased in shops or people would drink it and avoid taxes. We usually buy 'methylated spirits'. This is ethanol but with added methanol. Methanol is highly poisonous. Other substances are added to make it undrinkable and a purple dye is added as a warning.

21.5 Reactions of Ethanol

1 COMBUSTION

When ethanol is ignited in a plentiful supply of air or oxygen the products are carbon dioxide and water.

The ethanol burns with an almost invisible blue flame.

2 DEHYDRATION OF ETHANOL

This is the reverse of the reaction used to produce ethanol from ethene. It can be carried out using concentrated sulphuric acid or by passing ethanol vapour over heated aluminium oxide.

$$C_2H_5OH(g) \rightarrow C_2H_4(g) + H_2O(g)$$
ethanol → ethene + water

3 REACTION WITH SODIUM

Ethanol reacts with sodium metal to produce hydrogen gas. This reaction is similar to the reaction of sodium with water (see Unit 19.9) but slower.

$$2C_2H_5OH(l) + 2Na(s) \rightarrow 2C_2H_5O^-.Na^+ \text{ (ethanol)} + H_2(g)$$

4 OXIDATION

If ethanol is allowed to come into contact with air it can be oxidized by the oxygen in the air. The result of this atmospheric oxidation is ethanoic acid. This souring of ethanol was the original method of making vinegar.

If oxidation is carried out in the laboratory using an acidified solution of potassium dichromate(VI), it is possible to produce ethanal (an aldehyde) and then ethanoic acid.

$$C_2H_5OH + [O] \rightarrow CH_3CHO + H_2O$$
ethanol + [oxygen] → ethanal + water

$$CH_3CHO + [O] \rightarrow CH_3COOH$$
ethanol + [oxygen] → ethanoic acid

5 ESTER FORMATION

(See Unit 21.7.)

21.6 Ethanoic Acid

Ethanoic acid is a weak acid produced by the oxidation of ethanol. In aqueous solution it ionizes slightly.

$$CH_3COOH(aq) \rightleftharpoons CH_3COO^-(aq) + H^+(aq)$$

21.7 Ester Formation

Ethanol reacts reversibly with ethanoic acid, in the presence of concentrated sulphuric acid, to form ethyl ethanoate (ethyl acetate). Ethyl ethanoate is an **ester**.

$$CH_3COOH(l) + C_2H_5OH(l) \rightleftharpoons CH_3COOC_2H_5(l) + H_2O(l)$$

ethanoic acid + ethanol ⇌ ethyl ethanoate + water

ACID + ALCOHOL ⇌ ESTER + WATER

Esters are sweet-smelling liquids found naturally in flowers and fruit to which they give scent and flavour.

Many naturally occurring fats and oils consist largely of esters. When an ester is boiled with an alkali solution the ester is split up into the constituent acid and alcohol. This is called **saponification**.

21.8 Condensation Polymerization

In 20.10 the topic of addition polymerization was discussed. There is another type of polymerization possible; this type is called **condensation polymerization**.

When two molecules react together to form a larger molecule and lose a small molecule (e.g. water) this is called a **condensation reaction**. An example is the esterification reaction in 21.7.

In condensation polymerization reactions the monomer molecules join together with the elimination of small molecules. Each monomer unit must contain **two reactive groups** otherwise no polymer is possible.

Nylon is an important man-made condensation polymer. Although there are various types of nylon, the commonest is nylon-6,6 (so-called because both starting materials contain six carbon atoms). For nylon-6,6, the starting materials are hexane-1,6-diamine and hexanedioic acid. They may be represented as follows:

$H_2N—□—NH_2$
Hexane-1,6-diamine
(reactive group—NH_2)

HOOC—●—COOH
Hexanedioic acid
(reactive group—COOH)

One of the reactive —NH_2 groups on the hexane-1,6-diamine molecule reacts with one of the reactive —COOH groups on the hexanedioic acid molecule with the elimination of a molecule of water. The product still contains two reactive groups and a series of similar reactions take place resulting in the formation of a polymer (Fig. 21.2).

$H_2N—□—NH_2$ + HOOC—●—COOH
↓
$H_2N—□—NHOC—●—COOH + H_2O$
↓ + $H_2N—□—NH_2$
$H_2N—□—NHOC—●—COHN—□—NH_2 + H_2O$
↓ + HOOC—●—COOH
$H_2N—□—NHOC—●—COHN—□—NHOC—●—COOH + H_2O$
↓ etc.

Fig 21.2 Formation of nylon-6,6 (a condensation polymer)

Another condensation polymer is **polyester**. This is widely used for making clothes and home furnishings.

Many natural products are condensation polymers. These include starch, cellulose and proteins. **Starch** and **cellulose** are condensation polymers with glucose as the monomer.

Proteins are long chain condensation polymers which resemble nylon in structure. They are formed by the polymerization of a limited number of amino acids, e.g.:

```
      H
      |
H2N—C—COOH
      |
      R
```

21.9 Carbohydrates

Carbohydrates are compounds containing carbon, hydrogen and oxygen. The last two elements are present in the same proportion as in water. All carbohydrates have the general molecular formula $C_x(H_2O)_y$.

Carbohydrates may be divided into **monosaccharides** (e.g. glucose and fructose, both $C_6H_{12}O_6$), **disaccharides** (e.g. sucrose and maltose, both $C_{12}H_{22}O_{11}$) and **polysaccharides** (e.g. starch). Monosaccharides and disaccharides are often referred to as sugars.

21.10 Tests for Starch and Reducing Sugars

Starch produces a dark blue coloration with iodine solution. This is used as a test for starch.

Certain sugars will reduce hot **Fehling's** (or **Benedict's**) **solution** to a brick red precipitate of copper(I) oxide. These sugars are called **reducing sugars**. Examples of reducing sugars are glucose, fructose, and maltose.

21.11 Hydrolysis of Starch

Starch may be broken down into simpler carbohydrates by **hydrolysis** in aqueous solution. The breakdown involves the reaction of the starch with water but unless a catalyst is present the reaction is extremely slow. The reaction can be catalysed in two ways:

1 Acid-catalysed hydrolysis

This is carried out by heating starch solution with dilute acid. The product is **glucose**.

2 Enzyme-catalysed hydrolysis

The hydrolysis reaction may also be catalysed by the enzyme α-amylase which is present in saliva. This reaction proceeds at room temperature to produce **maltose**.

21.12 Summary

Ethanol C_2H_5OH is a member of the alcohol family. It can be produced by fermentation of glucose or sugar solution with enzymes present in yeast.

Ethanol is prepared industrially by the addition of water to ethene.

Ethanol is oxidized, first to ethanal, and then to ethanoic acid, if it comes into contact with air.

Ethanol and ethanoic acid react together to form an ester, ethyl ethanoate. Esters can be split with alkali in a saponification reaction.

Nylon, polyester, starch and cellulose are condensation polymers. Starch is a carbohydrate condensation polymer and can be split up or hydrolysed with dilute acid or enzymes in saliva.

22 FUELS

22.1 Introduction

A fuel is a substance which can be used to produce energy. Most fuels, and certainly the ones considered in this section, liberate energy when burnt in air or oxygen. These fuels are compounds of carbon and hydrogen, the elements carbon and hydrogen and carbon monoxide.

Fuels can be classified as solid fuels, liquid fuels and gaseous fuels.

22.2 Solid Fuels

The most important solid fuel is **coal**. Coal is a fossil fuel composed largely of carbon. A typical coal sample contains 80 per cent carbon, 6 per cent hydrogen, 6 per cent nitrogen, 5 per cent oxygen and 3 per cent ash.

Coal was formed by the action of heat and pressure on plants over millions of years. Anthracite is a form of coal consisting of a larger percentage of carbon resulting from a more complete decomposition of the plants.

When coal is burnt in excess air, carbon dioxide gas is produced. However, other products formed include water vapour and sulphur dioxide. The sulphur dioxide produced causes pollution problems and for this reason smokeless fuel is preferred to coal.

DESTRUCTIVE DISTILLATION OF COAL

When coal is heated to about 1000 °C in the absence of air it does not burn. It splits up and forms four products – coal gas, ammonia solution, coal tar and coke (or smokeless fuel). Suitable apparatus is shown in Fig. 22.1.

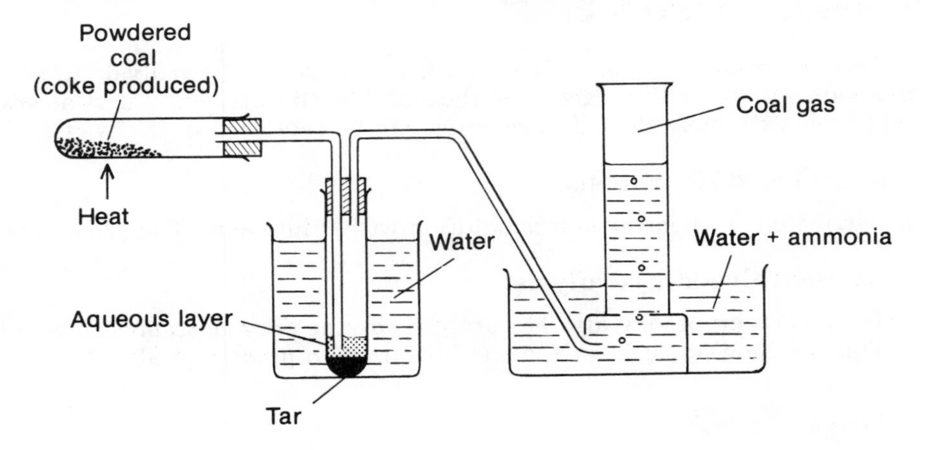

Fig 22.1 Destructive distillation of coal

The coal gas produced used to be used for household gas supply but it has now been replaced by natural gas. **Coal gas** consists chiefly of hydrogen with smaller amounts of methane and carbon monoxide.

The ammonia produced collects in the water. It can be used to make fertilizers. Coal tar is a complicated mixture of hydrocarbons and closely related compounds. It can be separated by fractional distillation into valuable chemicals, e.g. benzene, toluene.

22.3 Liquid Fuels

Liquid fuels include petrol and paraffin. They are often obtained from petroleum. Petrol is a mixture of alkanes including octane. Compounds including tetraethyl lead $Pb(C_2H_5)_4$ are added to the petrol to improve its combustion properties.

Other liquid fuels include the family of alcohols (including methanol and ethanol).

22.4 Gaseous Fuels

The most important gaseous fuel is methane (CH_4). It is found as **natural gas**, either alone or associated with crude oil. Natural gas and oil were formed by the action of heat and pressure on the remains of small sea creatures. Natural gas and oil are trapped below ground by layers of rock that do not allow the natural gas to escape (Fig. 20.1).

Other gaseous fuels include **water gas** and **producer gas**.

Producer gas is formed when air is passed over red-hot coke (carbon).

$$2C(s) + O_2(g) + N_2(g) \rightarrow 2CO(g) + N_2(g)$$
carbon + air → carbon monoxide + nitrogen

This reaction is exothermic and the producer gas yielded is a mixture of carbon monoxide and nitrogen. Producer gas is not a good fuel because it contains a large proportion of nitrogen which does not burn. It is, however, cheap to produce.

Water gas is produced when steam is passed over white-hot coke (carbon).

$$C(s) + H_2O(g) \rightleftharpoons CO(g) + H_2(g)$$
carbon + water ⇌ carbon monoxide + hydrogen

This reaction is endothermic and the coke has to be reheated if the reaction is to continue. Water gas is a mixture of hydrogen and carbon monoxide. It is a good gaseous fuel because both carbon monoxide and hydrogen burn.

22.5 Fossil Fuels and Replaceable Fuels

Fuels such as coal, petroleum and natural gas are present in the earth in limited amounts. As we use them they cannot be replaced. They are called **fossil fuels**.

Hydrogen and ethanol are not fossil fuels. They are replaceable forms of energy.

22.6 Which is the Best Fuel?

There are a number of factors that determine which fuel is best for a particular purpose. These include:

1 Price.
2 Ease of storage.
3 Energy value of fuel.
4 Ease of transport.
5 Ease of ignition.
6 Percentage of impurities.
7 Pollution products.

22.7 Summary

A fuel is a substance which can be used to produce energy. Fuels can be classified as solid fuels, liquid fuels and gaseous fuels.

Coal is the most important solid fuel. It is also a source of chemicals. Destructive distillation of coal will produce very important chemicals.

Fossil fuels must be used carefully as they cannot be replaced.

There are a number of factors to be considered when looking for the 'best' fuel.

Petrol and paraffin are common liquid fuels and natural gas (methane) and butane are gaseous fuels.

23 EXTRACTION OF METALS

23.1 Introduction

Metals are widely used in everyday life. Very rarely are pure metals found free in the ground. Usually they are found in compounds with other materials in the form of an ore. This unit is concerned with methods of extracting metals from their ores. It is important to relate this section to the reactivity series (Unit 8).

23.2 Recycling of Metals

Since the Second World War there has been a threefold increase in the amount of copper used each year and a sixfold increase in the amount of aluminium. Obviously with this increasing use of metals supplies of suitable metal ores will eventually run out. It has been estimated that all aluminium ores will be used up by the year 2100 and all supplies of copper ores will be used up by the year 2000!

One way of making these ores last is to re-use or recycle metals. Copper is being recovered from scrap copper wires and pipes. Aluminium is being recovered from soft drink cans and kitchen foil. Tin is being removed from scrap food cans to be re-used in making new food cans. As metals become more expensive recycling becomes more likely.

23.3 Treatment of Ores

Some of the least reactive metals, e.g. gold, can be found in the form of the unreacted metal in the earth. Other metals can be found as compounds with other unwanted material in the form of an **ore**. Table 23.1 shows the ores of some common metals and the chief chemical constituent of the ore. (The examples in bold type are the ones most frequently asked for by examiners.)

Table 23.1 Common ores

Metal	*Ore*	*Chief chemical constituent*
Sodium	Rock salt	Sodium chloride
Calcium	Chalk, limestone, marble	Calcium carbonate
Magnesium	Magnesite (also in sea water)	Magnesium carbonate magnesium chloride
Aluminium	**Bauxite**	**Aluminium oxide**
Zinc	**Zinc blende**	**Zinc sulphide**
Iron	**Haematite**	**Iron(III) oxide**
Copper	Malchite	Basic copper(II) carbonate
Mercury	Cinnabar	Mercury(II) sulphide

Before the metal is extracted from the ore, the ore is frequently concentrated or purified.

Bauxite (aluminium oxide) is purified by adding the ore to sodium hydroxide solution. The aluminium oxide reacts and forms soluble sodium aluminate. Impurities such as iron(III) oxide can be removed by filtration. The aluminium oxide is then precipitated in a pure form (in fact aluminium hydroxide is precipitated and this is heated to give the pure oxide).

Zinc blende and galena can be concentrated by froth flotation. The ore is added to a detergent bath that is agitated. By careful control of the conditions in the bath it is possible to cause the metallic sulphide to float and the impurities to sink.

The method used to extract the metal from the ore depends on the position of the metal in the reactivity series. If a metal is high in the reactivity series its ores are stable and the metal can be obtained only by electrolysis. Metals that can be obtained by electrolysis include potassium, sodium, calcium, magnesium and aluminium.

Metals in the middle of the reactivity series do not form very stable ores and they can be extracted by reduction, often with carbon. Examples of metals extracted by reduction are zinc, iron and lead.

Metals low in the reactivity series, if present in ores, can be extracted simply by heating because the ores are unstable. For example, mercury can be extracted by heating cinnabar.

Questions on extraction of metals often appear on examination papers and common examples are detailed below.

23.4 Extraction of Sodium

Sodium is extracted by the electrolysis of molten sodium chloride in the Downs cell. Calcium chloride is added to the sodium chloride to lower the melting point to about 600 °C. A Downs cell is shown in Fig. 23.1.

The cathode (negative electrode) is made of iron and the cylindrical anode (positive electrode) is made of graphite (carbon).

During the electrolysis the following reactions take place at the electrodes:

cathode $Na^+(l) + e^- \rightarrow Na(l)$

anode $2Cl^-(l) \rightarrow Cl_2(g) + 2e^-$

The sodium and chlorine produced are kept apart to prevent them reacting and reforming sodium chloride. The chlorine produced is a valuable byproduct.

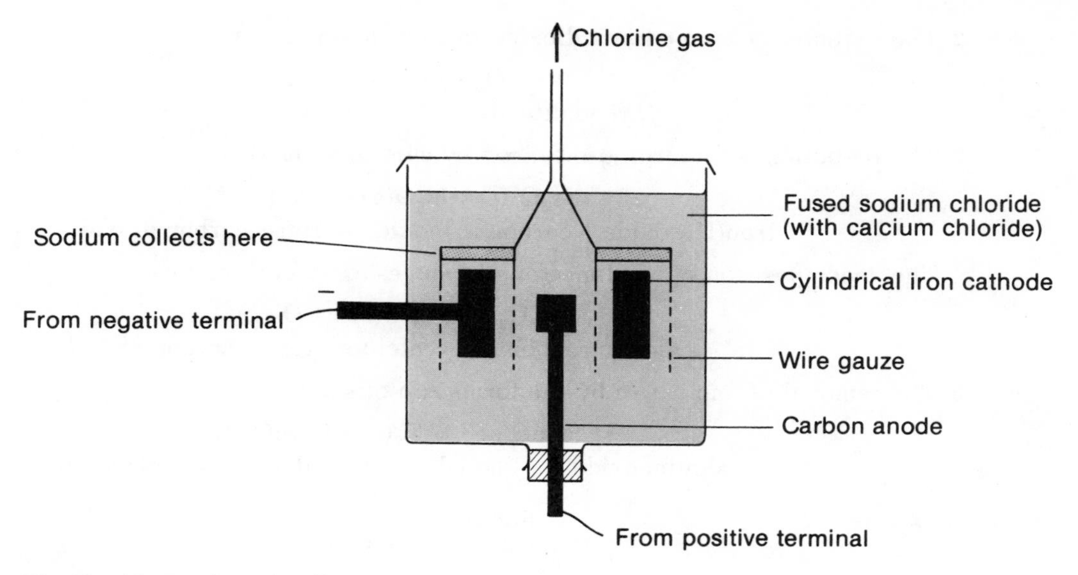

Fig 23.1 Extraction of sodium

23.5 Extraction of Aluminium

Aluminium is extracted from purified aluminium oxide by electrolysis. However, aluminium oxide has a high melting point and is not readily soluble in water but it does dissolve in molten cryolite (Na_3AlF_6) and this produces a suitable electrolyte. An appropriate cell (Hall's cell) is shown in Fig. 23.2.

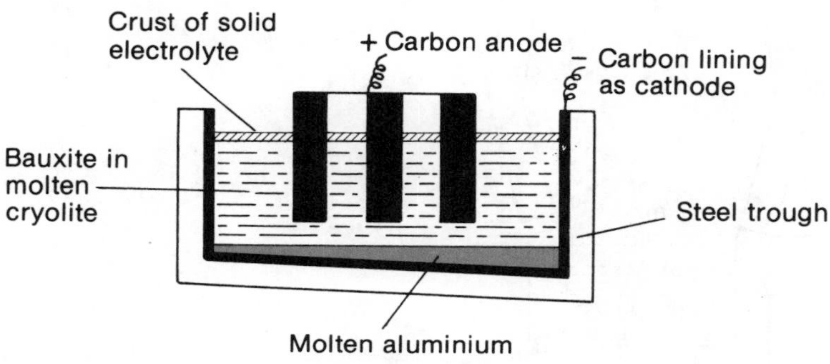

Fig 23.2 Extraction of aluminium

The electrodes are made of carbon. The reactions taking place at the electrodes are as follows:

cathode	$Al^{3+} + 3e^- \rightarrow Al$	overall reaction
anode	$2O^{2-} \rightarrow O_2 + 4e^-$	$4Al^{3+} + 6O^{2-} \rightarrow 4Al + 3O_2$

At the working temperature of the cell, the oxygen reacts with the carbon of the anode to produce carbon dioxide. The anode has, therefore, to be replaced frequently. As this process requires a large amount of electricity an inexpensive source, e.g. hydroelectric power, is an advantage.

23.6 Extraction of Iron

Iron is extracted from iron ore, in large quantities, by reduction in a blast furnace (Fig. 23.3). The furnace is loaded with iron ore, coke and limestone and is heated by blowing hot air into the base from the tuyères. Inside the furnace the following reactions take place raising the temperature to about 1500 °C:

1 The burning of the coke in the air:

$$C(s) + O_2 \rightarrow CO_2(g)$$

carbon + oxygen → carbon dioxide

2 The reduction of the carbon dioxide to carbon monoxide:

$$CO_2(g) + C(s) \rightarrow 2CO(g)$$

carbon dioxide + carbon → carbon monoxide

3 The reduction of the iron ore to iron by carbon monoxide:

$$Fe_2O_3(s) + 3CO(g) \rightarrow 2Fe(l) + 3CO_2(g)$$

iron(III) oxide + carbon monoxide → iron + carbon dioxide

4 The decomposition of the limestone produces extra carbon dioxide:

$$CaCO_3(s) \rightarrow CaO(s) + CO_2(g)$$

calcium carbonate → calcium oxide + carbon dioxide

5 The removal of impurities by the formation of slag:

$$CaO(s) + SiO_2(s) \rightarrow CaSiO_3(l)$$

calcium oxide + silicon dioxide → calcium silicate ('slag')

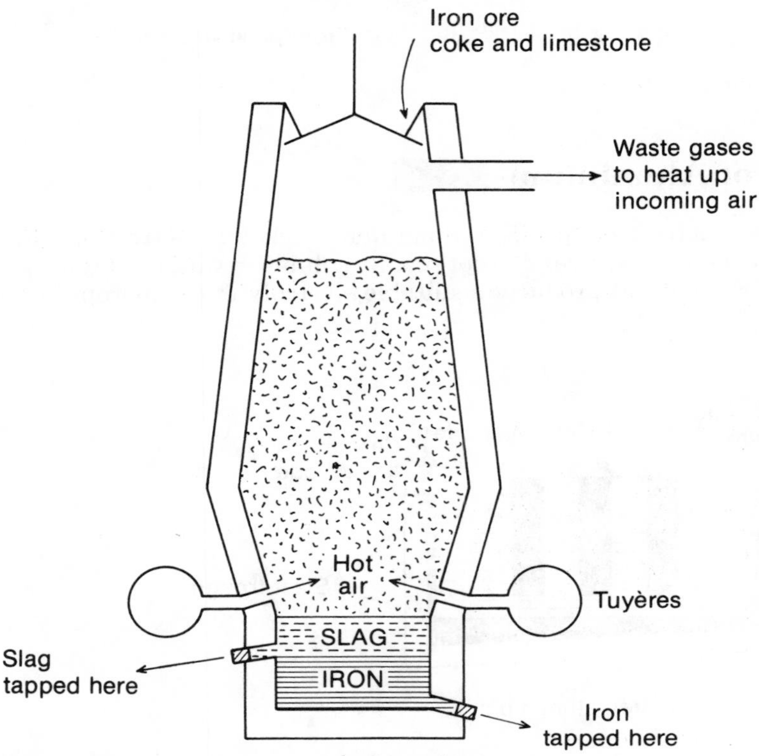

Fig 23.3 Extraction of iron in a blast furnace

The molten iron sinks to the bottom of the furnace and the slag floats on the surface of the molten iron. The iron and slag can be tapped off separately at intervals. The iron produced is called **pig iron** and contains about 4 per cent carbon. The slag, once discarded, is used as a phosphorus fertilizer and for road building.

Most of the pig iron produced is converted into the alloy, steel. The pig iron is melted and scrap steel is added (another example of recycling). Air or oxygen is bubbled through the molten iron. All impurities are oxidized and pure iron remains. The required amounts of carbon are added together with any other metals necessary.

23.7 Extraction of Zinc

Following the concentration of the ore (see Unit 23.3), the ore is heated strongly in air to convert the zinc sulphide to zinc oxide.

$$2ZnS(s) + 3O_2(g) \rightarrow 2ZnO(s) + 2SO_2(g)$$

zinc sulphide + oxygen → zinc oxide + sulphur dioxide

(The sulphur dioxide can be used to manufacture sulphuric acid, see 25.2.)

The zinc oxide and coke are mixed together in a cylindrical furnace and heated strongly. The carbon reduces zinc oxide to zinc.

$$ZnO(s) + C(s) \rightarrow Zn(g) + CO(g)$$

zinc oxide + carbon → zinc + carbon monoxide

At the temperature of the furnace, zinc distils off and is condensed. The carbon monoxide produced is used to heat the furnace.

23.8 Extraction of Titanium

Titanium occurs as titanium(IV) oxide in the ore rutile. Titanium is above carbon in the reactivity series and cannot, therefore, be reduced with carbon.

Titanium is extracted by reduction using the metal sodium as the reducing agent.

Titanium(IV) oxide is mixed with carbon and heated in a stream of chlorine to produce titanium(IV) chloride.

$$TiO_2(s) + 2C(s) + 2Cl_2(g) \rightarrow TiCl_4(l) + 2CO(g)$$

titanium(IV) oxide + carbon + chlorine → titanium(IV) chloride + carbon monoxide

Titanium chloride is then heated with sodium to produce titanium.

$$TiCl_4(l) + 4Na(l) \rightarrow Ti(s) + 4NaCl(s)$$

titanium(IV) chloride + sodium → titanium + sodium chloride

Because sodium is very expensive, the metal titanium will be very expensive.

23.9 Summary

The method used to extract a metal from its ore depends upon the position of the metal in the reactivity series.

Metals high in the reactivity series e.g. sodium and aluminium, form stable compounds (Unit 8) and the metal must be extracted by electrolysis. This will be expensive because of the large quantities of electricity required.

Metals lower in the reactivity series, e.g. iron, can be extracted by reduction using carbon as a reducing agent.

Metals low in the reactivity series may be found uncombined or 'native' in the earth. If they are in compounds, the compounds will be easy to split up.

Shortage of suitable deposits of ores will encourage:

1 Recycling of metals.
2 Exploitation of low quality deposits.
3 The use of alternative materials.

24 CORROSION OF METALS

24.1 Introduction

In Unit 8, metals were arranged in order of reactivity and a reactivity series was established. Many metals react with oxygen and water in the air and are said to **corrode**. Corrosion can be related to the reactivity series. In this unit the topic of corrosion of metals will be considered.

24.2 Corrosion of Reactive Metals

Potassium and sodium are extremely reactive metals. They react rapidly with oxygen and water. To prevent their corrosion they are usually stored in liquid paraffin.

A similar method of storage is used for other alkali metals (see Unit 9.2) and also for barium. Calcium is not usually stored under paraffin oil but it rapidly corrodes.

24.3 Corrosion of Aluminium

Aluminium is not as reactive as its position in the reactivity series might suggest. For example it reacts only after a long delay, with warm, dilute hydrochloric acid.

$$2Al(s) + 6HCl(aq) \rightarrow 2AlCl_3(aq) + 3H_2(g)$$

aluminium + hydrochloric acid → aluminium chloride + hydrogen

The initial delay is caused by a layer of insoluble and unreactive aluminium oxide on the surface, which has to be removed before the reaction can start. The aluminium oxide can be removed by wiping with a cloth soaked in mercury(II) chloride or by dipping the metal in mercury. When the oxide layer is removed, the aluminium is much more reactive. It is possible to make the oxide layer thicker by electrolysis, with the aluminium as the anode. Then the layer can be dyed using a suitable dye to give a very attractive finish for aluminium. This process is called **anodizing**.

24.4 Corrosion of Iron

Iron and steel are very prone to corrosion. The corrosion of iron and steel is called **rusting**. Rusting costs many hundreds of millions of pounds each year in Great Britain.

Rust, formed during the rusting process, is chemically very complicated. It is a hydrated iron(III) oxide.

Figure 24.1 shows an experiment frequently used to show the conditions necessary for rusting.

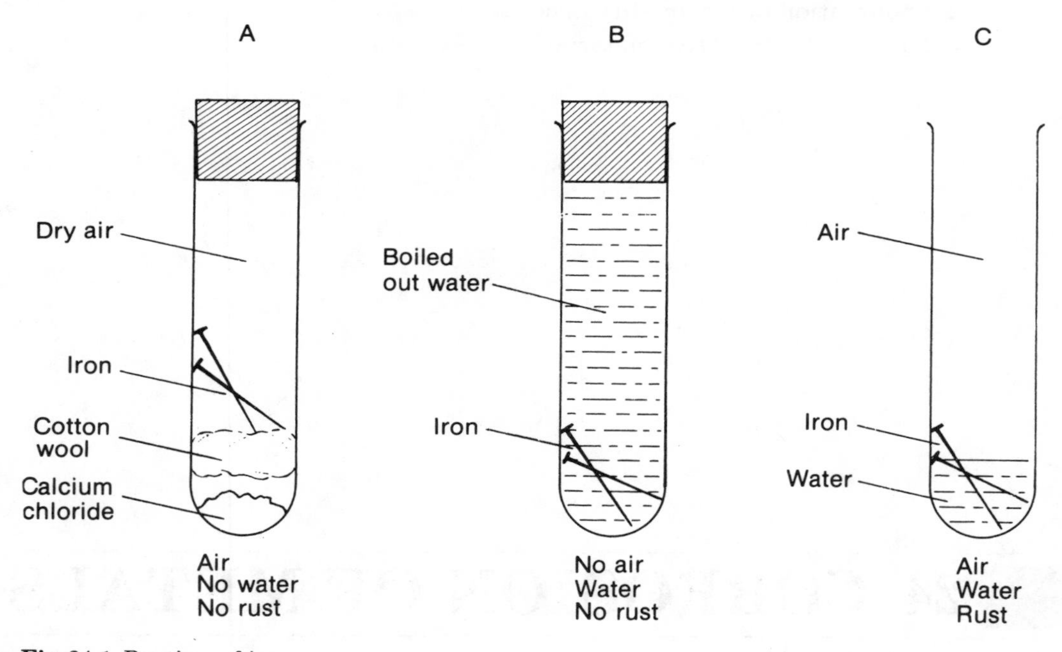

Fig 24.1 Rusting of iron

If three test tubes are set up as shown in Fig. 24.1 and left for a few days, rusting takes place only in test tube C. From this we can conclude that both air and water are necessary for rusting to take place. In fact, it is the oxygen in the air which, along with water, is vital (see Unit 17.8). Rusting is speeded up by the presence of acid (carbon dioxide or sulphur dioxide are examples) or salt.

Rusting can be prevented by excluding air and/or water. It can be prevented by the following methods:

1 Iron can be painted by spraying, dipping or brushing. Providing the paint surface is not broken, rusting will not take place. The paint coating prevents oxygen and water from coming into contact with the iron. When the paint coating is broken rusting will take place. This type of protection is used to protect cars from rusting and for protecting large bridges (e.g. Forth Bridge) and iron railings.

2 Iron can be coated with a layer of plastic. This layer again prevents oxygen and water from coming into contact with the iron. This type of protection is used for draining racks in the kitchen. However, eventually the plastic coating starts to split and rapid rusting occurs.

3 Iron can be coated with oil or grease. Again oxygen and water cannot come into contact with the iron and rusting does not take place. This form of protection is very easy and very effective. A smearing of oil onto a saw blade will prevent its rusting. It is a very good method for treating moving parts.

4 Iron can be coated with a layer of zinc in a process called **galvanizing**. The zinc coating can be put on by dipping or spraying. If the zinc coating is deeply scratched to expose bare iron rusting will not take place.

If a car body repair is being carried out, rust has first to be removed. The bare steel may then be treated with a zinc-based primer paint before spraying with the coloured paint.

Zinc coating cannot be used for food cans because zinc compounds are poisonous.

5 Iron can be coated with a thin coating of tin to make tinplate. This can be used for making food cans. If the tin coating is scratched to show the bare steel, rusting of the steel will take place.

Aluminium is more and more being used to replace tinplate in making food cans.

Modern processes of refuse disposal are now removing tinplate from household refuse, then removing the tin from the cans. The tin obtained is purified by electrolysis and re-used for tin can manufacture.

6 Iron can be protected by **sacrificial protection**. This is used for protecting steel hulls of ships and steel piers. Rusting takes place even faster in the presence of carbon dioxide or salt water. To prevent a steel hull from rusting, blocks of a suitable metal are strapped to the steel hull. The metal used must be more reactive than iron. Zinc or magnesium would be suitable metals to use as they are higher than iron in the reactivity series.

The zinc or magnesium blocks corrode in preference to iron. As long as they remain no rusting will take place. These blocks can be easily replaced when they have corroded away.

7 Steel can be protected by electroplating. Using electrolysis a thin coating of nickel is deposited on the steel to prevent rusting. Finally a very thin coating of chromium is electroplated on top of the nickel. The chromium gives a nice shiny surface. Chromium plating of bicycle handlebars, car bumpers and electric kettles is a very expensive process and is being replaced by aluminium alloys or stainless steel. Stainless steel is an alloy of iron which is less inclined to rusting than iron or other types of steel.

There is a big difference between the rusting of iron and the corrosion of aluminium. When aluminium corrodes it forms a layer of aluminium oxide which seals the surface and prevents further corrosion. When iron rusts the rust is flaky. It flakes off and reveals a new surface which rusts again.

24.5 Corrosion of Copper, Silver and Gold

Copper is extremely slow to corrode. A copper roof on a building can become an extremely pleasant green colour after many years. This green corrosion is a form of hydrated copper carbonate.

For statues, bronze (a copper alloy) is used rather than copper as it is less likely to corrode.

Silver does not really corrode but blackens when the atmosphere contains sulphur compounds.

Gold does not corrode and this is one of the reasons why it is so highly valued.

24.6 Summary

Generally, the higher a metal is in the reactivity series the more likely it is to corrode. However, the corrosion will appear greater if it flakes off to reveal an uncorroded surface.

The corrosion of iron and steel, which we call rusting, is of special economic importance. Rusting is caused by the action of oxygen and water vapour on iron and steel. Methods of preventing rusting include oiling, painting, coating with plastic, galvanizing, plating and sacrificial protection.

Copper, silver and gold (low in the reactivity series) do not readily corrode.

25 AMMONIA AND NITRIC ACID

This unit is written in two parts

Part A	25.1–25.6	Ammonia
Part B	25.7–25.10	Nitric acid

Check in the Table of Analysis of Examination Syllabuses on pp. xvi–xvii to find which part is required for your syllabus.

25.1 Introduction

Ammonia NH_3 is a compound of nitrogen and hydrogen. It is a very important industrial chemical. Much of the ammonia produced in industry is converted into nitric acid HNO_3.

This unit deals with the chemistry of these two nitrogen compounds and other related compounds.

25.2 Laboratory Preparation of Ammonia

Ammonia gas is prepared in the laboratory by heating a mixture of an ammonium salt and an alkali. For example, a mixture of ammonium chloride and sodium hydroxide could be used.

$$NH_4Cl(s) + NaOH(s) \rightarrow NaCl(s) + H_2O(g) + NH_3(g)$$

ammonium chloride + sodium hydroxide → sodium chloride + water + ammonia

or

$$(NH_4)_2SO_4(s) + Ca(OH)_2(s) \rightarrow CaSO_4(s) + 2H_2O(g) + 2NH_3(g)$$

ammonium sulphate + calcium hydroxide → calcium sulphate + water + ammonia

The underlying reaction taking place in each case can be represented by the same ionic equation:

$$NH_4^+ + OH^- \rightarrow NH_3 + H_2O$$

Suitable apparatus for preparing dry ammonia is shown in Fig. 25.1.

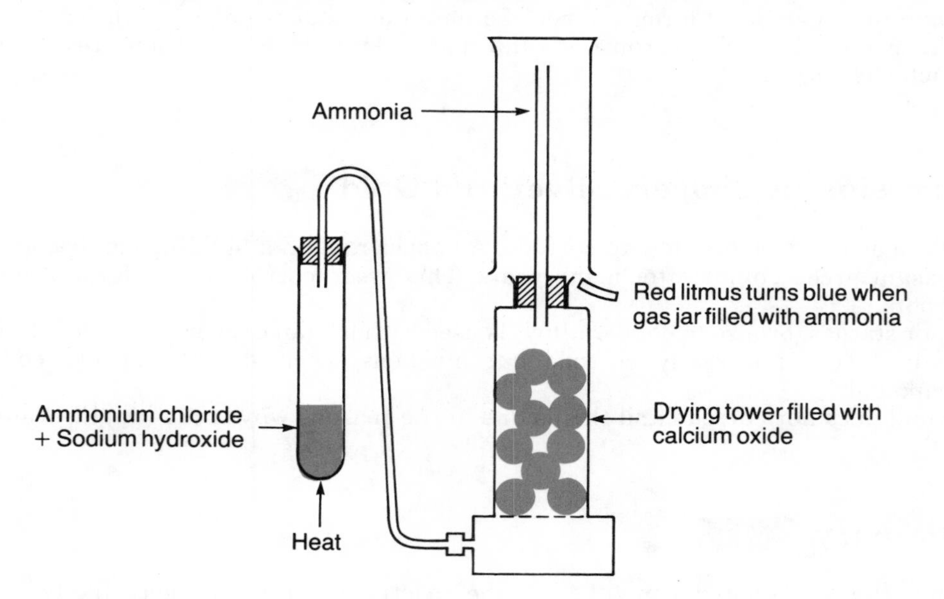

Fig 25.1 Preparation of ammonia

Ammonia gas is dried by passing it through a tower containing calcium oxide (quicklime).

$$CaO(s) + H_2O(l) \rightarrow Ca(OH)_2(s)$$

calcium oxide + water → calcium hydroxide

(Concentrated sulphuric acid cannot be used because it reacts with ammonia gas:

$$2NH_3(g) + H_2SO_4(l) \rightarrow (NH_4)_2SO_4(s)$$

ammonia + sulphuric acid → ammonium sulphate)

Ammonia gas is collected by upward delivery (downward displacement of air) because it is much less dense than air and is readily soluble in water.

25.3 Industrial Production of Ammonia

Ammonia is produced industrially in large amounts from nitrogen and hydrogen using the **Haber process**.

1 SOURCES OF NITROGEN AND HYDROGEN

Nitrogen is obtained by fractional distillation of liquid air (see Unit 17.2). Hydrogen can be obtained from water (see Unit 5.1) and also from oil or natural gas (see Unit 5.4).

2 REACTION BETWEEN NITROGEN AND HYDROGEN

The mixture of nitrogen (1 part) and hydrogen (3 parts) is compressed to 200 atmospheres and passed through the catalyst chamber at 500 °C. The catalyst chamber contains finely divided iron as the catalyst. About 10 per cent of the nitrogen and hydrogen mixture is converted to ammonia.

$$N_2(g) + 3H_2(g) \rightleftharpoons 2NH_3(g)$$

nitrogen + hydrogen ⇌ ammonia

(Forward reaction exothermic, i.e. ΔH is negative.)

Effect of temperature

The forward reaction is exothermic. If the system is cooled, the equilibrium will move to oppose this change, i.e. move to the right producing more ammonia.

Effect of pressure

From the equation it can be seen that 1 mole of nitrogen molecules reacts with 3 moles of hydrogen molecules to give 2 moles of ammonia molecules. This means that the formation of ammonia is accompanied by a reduction in the number of molecules. Increasing the pressure favours the conversion to ammonia.

Effect of concentration

If the concentration of one of the chemicals is altered, the equilibrium will move to oppose the change. This means that if the ammonia is removed from the system as it is formed the equilibrium will move to the right to produce more ammonia.

Industrial conditions for the Haber process

1 Very high pressure (about 200 atmospheres although some modern plants use pressures up to 1000 atmospheres).

2 Temperature of about 500 °C. Note that a lower temperature would cause a greater proportion of ammonia to be formed but the rate of reaction is too slow.

3 The iron catalyst enables the equilibrium to be established more quickly but does not produce *more* ammonia.

3 REMOVAL OF AMMONIA FROM THE MIXTURE OF GASES

When the mixture of gases leaving the catalyst chamber is cooled, only ammonia liquefies and can be removed. The unreacted nitrogen and hydrogen are recycled.

25.4 Testing for Ammonia

1 Ammonia turns red litmus blue and does not burn in air when a lighted splint is applied.

2 Dense white fumes of ammonium chloride are formed when ammonia gas comes into contact with hydrogen chloride gas (e.g. the stopper from a bottle of concentrated hydrochloric acid).

$$NH_3(g) + HCl(g) \rightleftharpoons NH_4Cl(s)$$

25.5 Properties of Ammonia

Ammonia is a colourless gas with a pungent and characteristic odour (smelling salts). It turns red litmus blue and is less dense than air.

1 SOLUBILITY IN WATER

Ammonia dissolves very readily in cold water to form ammonia solution $NH_3(aq)$. This is sometimes called ammonium hydroxide NH_4OH. This solution is alkaline.

$$NH_3(g) + H_2O(l) \rightleftharpoons NH_4^+(aq) + OH^-(aq)$$

The solubility of ammonia in water can be demonstrated by the fountain experiment. A dry flask is filled with dry ammonia and set up as shown in Fig. 25.2. The end of the tube is dipped into water. The water level rises in the tube as the ammonia dissolves and the water fountains into the flask. A similar experiment can be used to demonstrate the solubility of hydrogen chloride in water.

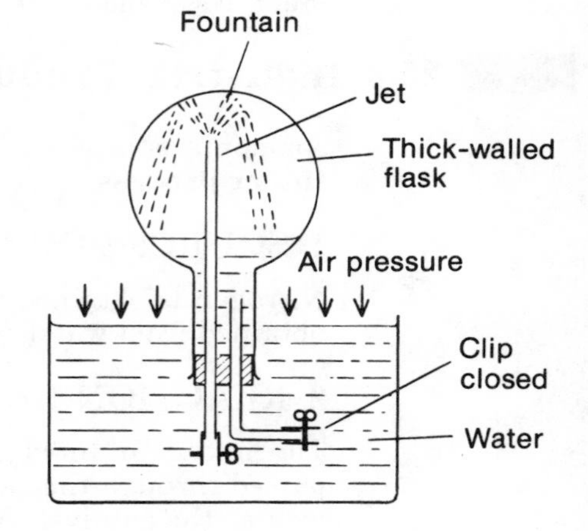

Fig 25.2 Fountain experiment

2 REACTIONS WITH ACIDS

Ammonia gas is absorbed by acids to form ammonium salts.

$$NH_3(g) + HCl(aq) \rightarrow NH_4Cl(aq)$$
ammonia + hydrochloric acid → ammonium chloride

$$NH_3(g) + HNO_3(aq) \rightarrow NH_4NO_3(aq)$$
ammonia + nitric acid → ammonium nitrate

$$2NH_3(g) + H_2SO_4(aq) \rightarrow (NH_4)_2SO_4(aq)$$
ammonia + sulphuric acid → ammonium sulphate

3 COMBUSTION OF AMMONIA

Ammonia does not burn in air but it burns in oxygen (or air enriched with oxygen).

$$4NH_3(g) + 3O_2(g) \rightarrow 2N_2(g) + 6H_2O(g)$$
ammonia + oxygen → nitrogen + water

Substances do not burn in ammonia gas.

It is possible to oxidize ammonia with oxygen in the presence of a heated platinum catalyst to form nitrogen monoxide. A heated platinum wire glows brightly in a mixture of ammonia and oxygen because this reaction is very exothermic.

$$4NH_3(g) + 5O_2(g) \rightarrow 4NO(g) + 6H_2O(g)$$
ammonia + oxygen → nitrogen monoxide + water

This is the basis of the industrial manufacture of nitric acid from ammonia (see Unit 25.7).

4 AMMONIA AS A REDUCING AGENT

When ammonia gas is passed over heated copper(II) oxide, the ammonia is oxidized to nitrogen and the copper(II) oxide is reduced to copper (Fig. 25.3).

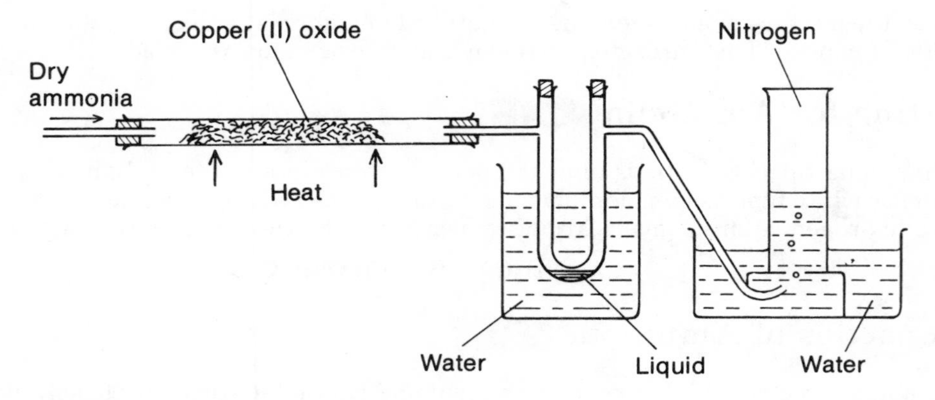

Fig 25.3 Production of nitrogen by the passing of ammonia over copper(II) oxide

$$2NH_3(g) + 3CuO(s) \rightarrow 3Cu(s) + 3H_2O(g) + N_2(g)$$
ammonia + copper(II) oxide → copper + water + nitrogen
(black) (pinkish-brown)

A similar reaction takes place with oxides of lead and iron. This experiment can be used to demonstrate that ammonia contains nitrogen.

25.6 Uses of Ammonia

1 Ammonia is used to manufacture fertilizers, e.g. ammonium sulphate.
2 Ammonia is used to manufacture nitric acid.
3 Ammonia is used to manufacture plastics.

25.7 Industrial Manufacture of Nitric Acid

Nitric acid is manufactured by the catalytic oxidation of ammonia and dissolving the products in water.

1 A mixture of ammonia vapour and excess air is passed over a heated platinum gauze catalyst at 900 °C. An exothermic reaction takes place producing nitrogen monoxide and steam.

$$4NH_3(g) + 5O_2(g) \rightarrow 4NO(g) + 6H_2O(g)$$
ammonia + oxygen → nitrogen monoxide + water

2 The mixture of gases is cooled and nitrogen monoxide reacts with further oxygen in the air to produce nitrogen dioxide.

$$2NO(g) + O_2(g) \rightarrow 2NO_2(g)$$
nitrogen monoxide + oxygen → nitrogen dioxide

3 The mixture of gases is then passed through a tower containing a flow of cold water. Nitric acid is produced by the reaction of nitrogen dioxide with water in the presence of oxygen.

$$4NO_2(g) + O_2(g) + 2H_2O(l) \rightarrow 4HNO_3(l)$$
nitrogen dioxide + oxygen + water → nitric acid

25.8 Properties of Concentrated Nitric Acid

1 ACTION OF HEAT ON CONCENTRATED NITRIC ACID

Thermal decomposition of concentrated nitric acid produces a mixture of nitrogen dioxide, steam and oxygen. If the gas produced is collected over water, only oxygen is collected.

$$4HNO_3(l) \rightarrow 2H_2O(g) + 4NO_2(g) + O_2(g)$$
nitric acid → water + nitrogen dioxide + oxygen

2 OXIDIZING PROPERTIES OF CONCENTRATED NITRIC ACID

Concentrated nitric acid is a strong oxidizing agent. When it acts as an oxidizing agent it is reduced to nitrogen dioxide and water. The following are examples of this property.

Copper

Concentrated nitric acid reacts with copper to form a blue solution of copper(II) nitrate. The nitric acid is reduced to nitrogen dioxide and water. The reaction is exothermic.

$$Cu(s) + 4HNO_3(l) \rightarrow Cu(NO_3)_2(aq) + 2H_2O(l) + 2NO_2(g)$$
copper + nitric acid → copper nitrate + water + nitrogen dioxide

The reaction takes place without heating. Concentrated nitric acid reacts with all common metals except gold and platinum.

Carbon

Carbon is oxidized by warm, concentrated nitric acid producing carbon dioxide. The nitric acid is reduced to nitrogen dioxide and water.

$$C(s) + 4HNO_3(l) \rightarrow CO_2(g) + 4NO_2(g) + 2H_2O(g)$$
carbon + nitric acid → carbon dioxide + nitrogen dioxide + water

25.9 Properties of Dilute Nitric Acid

1 WITH INDICATORS

Dilute nitric acid turns blue litmus red.

2 WITH METALS

Most dilute acids will react with some metals to produce hydrogen. Hydrogen is only produced using very dilute nitric acid. If magnesium is reacted with cold, very dilute nitric acid hydrogen is produced.

$$Mg(s) + 2HNO_3(aq) \rightarrow Mg(NO_3)_2(aq) + H_2(g)$$
magnesium + nitric acid → magnesium nitrate + hydrogen

In other cases, dilute nitric acid is still a sufficiently strong oxidizing agent to oxidize the hydrogen to water.

$$3Cu(s) + 8HNO_3(aq) \rightarrow 3Cu(NO_3)_2(aq) + 4H_2O(l) + 2NO(g)$$
copper + nitric acid → copper(II) nitrate + water + nitrogen monoxide

The nitrogen monoxide forms nitrogen dioxide in contact with air.

$$2NO(g) + O_2(g) \rightarrow 2NO_2(g)$$
nitrogen monoxide + oxygen → nitrogen dioxide

More reactive metals than copper may produce dinitrogen monoxide (N_2O) or nitrogen (N_2).

3 WITH METAL OXIDES OR HYDROXIDES

Dilute nitric acid reacts with a metal oxide or hydroxide to produce a metal nitrate solution.

E.g.
$$CuO(s) + 2HNO_3(aq) \rightarrow Cu(NO_3)_2(aq) + H_2O(l)$$
copper(II) oxide + nitric acid → copper nitrate + water

$$NaOH(aq) + HNO_3(aq) \rightarrow NaNO_3(aq) + H_2O(l)$$
sodium hydroxide + nitric acid → sodium nitrate + water

4 WITH METAL CARBONATES

Dilute nitric acid reacts with metal carbonates in a similar way to other dilute acids producing carbon dioxide and water. With dilute nitric acid the salt produced is a nitrate. It is not necessary to heat the mixture.

E.g.
$$CaCO_3(s) + 2HNO_3(aq) \rightarrow Ca(NO_3)_2(aq) + H_2O(l) + CO_2(g)$$
calcium carbonate + nitric acid → calcium nitrate + water + carbon dioxide

$$Na_2CO_3(s) + 2HNO_3(aq) \rightarrow 2NaNO_3(aq) + H_2O(l) + CO_2(g)$$
sodium carbonate + nitric acid → sodium nitrate + water + carbon dioxide

25.10 Uses of Nitric Acid

Most of the nitric acid produced is immediately turned into ammonium nitrate (see Unit 26). This is used as a fertilizer.

Some other uses of nitric acid are in the manufacture of medicines and in explosives (e.g. TNT – trinitrotoluene).

25.11 Summary

Ammonia NH_3 is a compound of nitrogen and hydrogen. It can be prepared by heating a mixture of an ammonium compound and an alkali.

In industry ammonia is produced by the Haber process. Nitrogen and hydrogen are combined together to form ammonia. Iron is the catalyst for the reaction.

Ammonia and ammonium compounds are widely used as fertilizers.

Much of the ammonia is converted to nitric acid in a three stage process. Ammonia is first oxidized to nitrogen monoxide using platinum as a catalyst. Then the gases are cooled and nitrogen dioxide is formed from further reaction with the air. Finally nitrogen dioxide is dissolved in water in the presence of oxygen to form nitric acid.

Nitric acid has strong oxidizing properties, particularly when concentrated.

26 FEEDING THE WORLD

26.1 Introduction

One in ten of the world's population are facing starvation. There are more people alive today than have ever lived. Every hour we need an extra two and a half square miles of farmland to feed the world's increasing population.

Obviously, the problems of feeding all of the people of the world will be with us for a very long time. Scientists, especially chemists, have made and are making a significant contribution to solving these problems.

The solution of the problem is very complex. It is not just a matter of growing more food. Ways of improving the situation include:

1 Development and supply of new fertilizers to enable greater quantities of food to be grown on the same area of land.

2 Breeding of new varieties of plant and new strains of animal which are more suited to particular conditions.

3 Development of better agricultural pesticides, e.g. insecticides, so that less food is lost.

4 Improvement of storage conditions of food. It is estimated that half of the food produced in the world is eaten by rats.

5 Manufacture of new kinds of food. For example, cattle double their weight in 2 to 4 months while plants double their weight in 1 to 2 weeks. Yeast and bacteria double their weight in 20 minutes. A shallow lake the size of Essex could supply sufficient protein in the form of protein rich bacteria to supply the whole world's need for protein. Although we might not want to eat protein rich bacteria, it could be used for high protein animal feeds.

Synthetic meat products can be made from soya beans.

6 Prevention of loss of land by soil erosion.

7 Reclamation of land not used for agriculture from deserts etc.

8 Better education of farmers in underdeveloped countries so they know what to do to get the best from their land.

9 Efforts to reduce world population by encouraging people to have smaller families.

These are only some of the things we could do. You could perhaps think of others. In this unit we are going to consider some of these, especially the production of fertilizers.

26.2 Elements Needed for Plant Growth

In order to grow healthy plants various elements are needed in the soil. **Nitrogen**, **phosphorus** and **potassium** are needed in large quantities. Other elements such as magnesium, calcium and sulphur are also needed in quantity. Some elements are needed in very small amounts. These include iron, boron and copper and they are called **trace elements**.

As the soil is used from year to year, it is possible that the soil could become short of some elements. Fertilizers are then required to replace these vital elements.

26.3 Nitrogen Fertilizers

Nitrogen is required in large amounts by plants. It is absorbed through the roots in the form of nitrate solutions. These nitrates are required to build up proteins in the plant. Figure 26.1 explains the circulation of nitrogen in nature, called the **nitrogen cycle**.

Nitrogen in the air is fixed in the soil by three methods:

1 Lightning causes nitrogen and oxygen to react together forming nitrogen monoxide. This nitrogen monoxide finally forms nitrates in the soil.

2 Bacteria in root nodules of certain plants (called leguminous plants, e.g. clover, peas etc.) are able to absorb nitrogen directly from the air.

3 Certain bacteria in the soil are able to fix nitrogen directly from the air.

Nitrogen also enters the soil from the death and decay of plants and animals, from animal urine and faeces. Bacterial action converts proteins into ammonia and then, via nitrites, into nitrates.

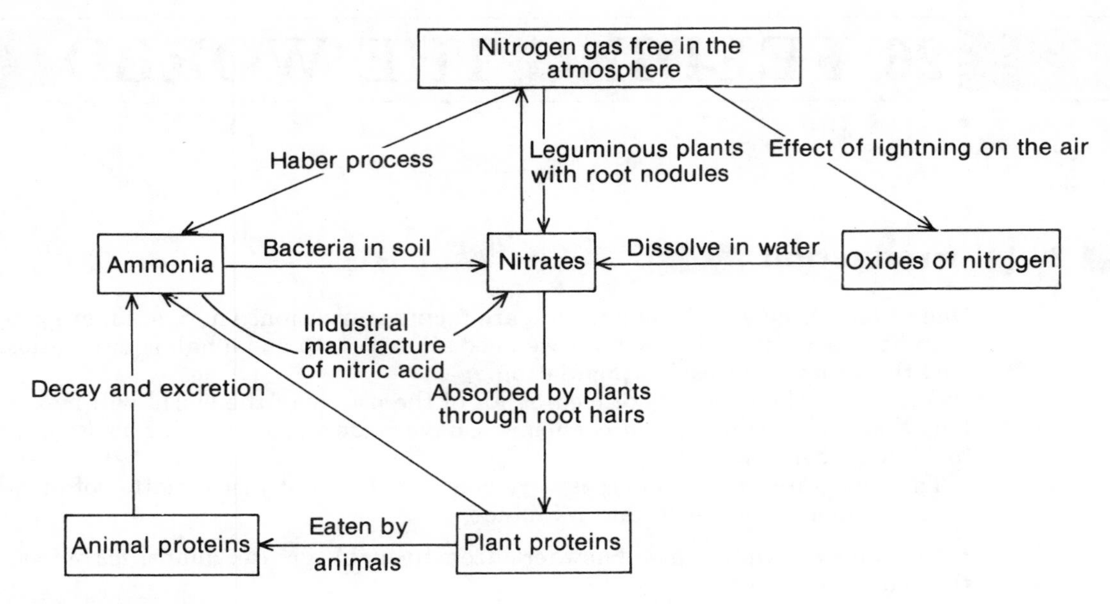

Fig 26.1 The nitrogen cycle

Because man intervenes in the nitrogen cycle by removing crops from the soil and not allowing them to decay, it becomes necessary to add nitrogen to the soil in the form of fertilizers.

Nitrogen can be supplied to the soil in the form of **manure** or **dried blood**. These natural forms of nitrogen often improve the quality of the soil. There are natural deposits of sodium nitrate in the desert areas of Chile.

Because of the insufficient supply of natural sources of nitrogen, it is necessary to supplement these with artificial fertilizers. These include **calcium nitrate**, **ammonium sulphate**, **ammonium nitrate** and **urea**. If a compound is very soluble in water (e.g. ammonium nitrate) it is readily washed out of the soil by rain, but before it is washed out its effects are rapid. Urea is soluble in cold water but it also reacts very slowly with water to produce ammonium compounds. It is therefore suitable as a long term fertilizer.

When considering which nitrogen fertilizer is most suitable in a particular situation, the following should be considered:

1 Percentage of nitrogen in the fertilizer (see Unit 29.9).
2 Cost of the fertilizer.
3 Solubility in water.

For example, Table 26.1 compares information concerning three fertilizers.

Table 26.1 Comparison of three fertilizers

Compound	*Formula*	*Mass of 1 mole*	*Price per kg*	*Solubility in water*
Ammonium nitrate	NH_4NO_3	80 g	50 p	Readily soluble
Calcium cyanamide	$CaCN_2$	80 g	60 p	Insoluble but reacting very slowly
Urea	$CO(NH_2)_2$	60 g	55 p	Soluble but reacting very slowly

1 Calculate the percentage of nitrogen in ammonium nitrate (see Unit 29.9).
2 Which fertilizer is most suitable for applying in spring so that it will continue to act throughout the summer and autumn?

Obviously ammonium nitrate is not suitable because it is too soluble. The choice between the other two depends on price and the percentage of nitrogen. On this basis, urea (which contains a greater percentage of nitrogen) would be chosen.

When nitrogen fertilizers get washed into streams and rivers they can cause serious water pollution problems (see Unit 19.11).

The production of fertilizers is a very large scale business. Ammonium nitrate is the most common nitrogen fertilizer used in Britain. Figure 26.2 summarizes the process used to make ammonium nitrate from ammonia and nitric acid.

In the final stage the solution of ammonium nitrate is evaporated. Solid ammonium nitrate is melted and sprayed down a tall tower. As the droplets fall they meet an upward flow of air. The fertilizer solidifies and forms small, hard pellets called **prills**. These are easy to handle and spread on to fields.

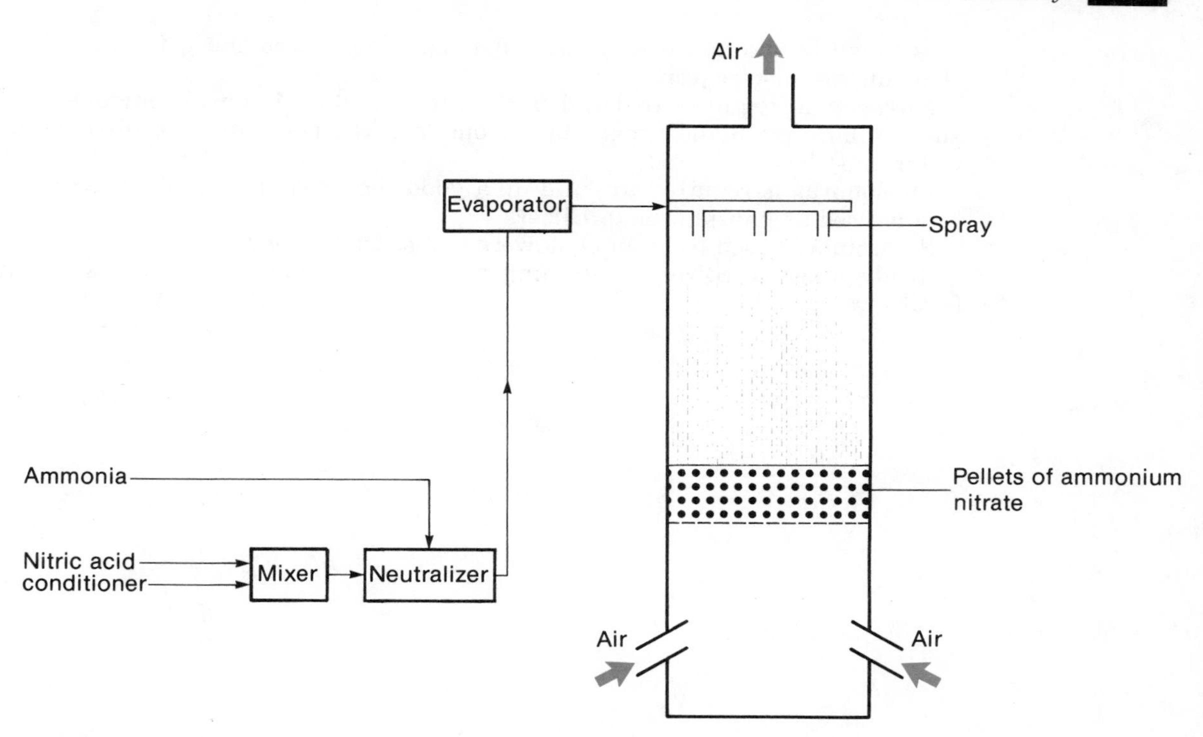

Fig 26.2 Flow diagram showing ammonium nitrate production

26.4 Phosphorus Fertilizers

Plants need phosphorus from the soil in order to produce a good root system. This is necessary before a healthy plant can develop.

Phosphorus can be supplied to the soil by **slag** or **bone meal**. Natural deposits of calcium phosphate $Ca_3(PO_4)_2$ are not very suitable because it is insoluble. However, if calcium phosphate is treated with concentrated sulphuric acid, **calcium superphosphate** is formed.

$$Ca_3(PO_4)_2(s) + 2H_2SO_4(l) \rightarrow Ca(H_2PO_4)_2(s) + 2CaSO_4(s)$$

calcium phosphate + sulphuric acid → calcium superphosphate + calcium sulphate

Calcium superphosphate contains soluble phosphates.

Ammonium phosphate $(NH_4)_3PO_4$ is a suitable phosphorus fertilizer and it also contains nitrogen.

26.5 Potassium Fertilizers

Plants need potassium for the production of flowers and seeds. Potassium can be added to the soil in the form of **wood ash** or by the addition of **potassium sulphate**.

26.6 Use of Fertilizers

Ready-mixed fertilizers are sometimes called **NPK fertilizers** because they supply nitrogen, phosphorus and potassium.

When fertilizers are used, they are sprinkled over the soil or sometimes injected into the soil (e.g. liquid ammonia). It is necessary to monitor the pH of the soil. Soils tend to become more acidic as soluble alkalis are washed out of the soil. The use of ammonium sulphate can make the soil more acidic. **Lime** (calcium hydroxide) can be used to neutralize the soil if it is acidic. However, lime and ammonium sulphate should not be used together as ammonia gas is released.

$$(NH_4)_2SO_4(s) + Ca(OH)_2(s) \rightarrow CaSO_4(s) + 2NH_3(g) + 2H_2O(l)$$

ammonium sulphate + calcium hydroxide → calcium sulphate + ammonia + water

26.7 Summary

Growing sufficient food to feed the population of the world is going to be an increasing problem. There are a number of ways of attempting to do it.

One of these is to increase the use of fertilizers.

For healthy plants the elements nitrogen, phosphorus and potassium are essential. Other elements are also required.

Nitrogen is required to build up healthy plants. Man-made nitrogen fertilizers include ammonium nitrate and urea. Apart from cost, the percentage of nitrogen and solubility in water must be considered.

Phosphorus is required to build up a good root system in a plant. Slag and bonemeal are widely used as phosphorus fertilizers.

Potassium is used to produce flowers and seeds in the plant.

Ready-mixed fertilizers containing nitrogen, phosphorus and potassium are called NPK fertilizers.

27 SULPHUR AND SULPHURIC ACID

27.1 Occurrence of Sulphur

Sulphur is found uncombined in volcanic regions of Italy, including Sicily. It is also found in underground deposits in Texas and Louisiana (USA).

Sulphur is also found in a wide range of compounds including:

1 Hydrogen sulphide (H_2S) sometimes present in petroleum and natural gas. Petroleum and natural gas from the North Sea do not contain sulphur.

2 Metallic sulphides, e.g. zinc blende ZnS, iron pyrites FeS_2.

3 Metallic sulphates, e.g. calcium sulphate $CaSO_4$ anhydrite.

At the beginning of this century almost all of the world's sulphur came from the volcanic deposits of Italy.

Now sulphur is obtained from either petroleum or natural gas or, to a lesser extent, from the Frasch process.

27.2 Extraction of Sulphur

Sulphur is extracted from natural gas in Lacq in Southern France. The natural gas is treated with an alkaline solution under pressure. The hydrogen sulphide and carbon dioxide (present only in small amounts) are dissolved because they are acid gases. The solution is then acidified to release the hydrogen sulphide and carbon dioxide gases. Partial combustion of the hydrogen sulphide with a limited supply of air in the presence of a catalyst produces sulphur.

$$2H_2S(g) + O_2(g) \rightarrow 2S(s) + 2H_2O(l)$$

$$\text{hydrogen sulphide} + \text{oxygen} \rightarrow \text{sulphur} + \text{water}$$

An ingenious method is used to extract sulphur from underground deposits in Texas and Louisiana. Mining cannot be carried out in the usual way because of layers of quicksand between the surface and the deposits and because of the poisonous gases within the sulphur. A method for obtaining the sulphur was devised by **Frasch**; it is called the **Frasch process** and is now widely used. A hole is drilled from the surface to the deposits using an oil drill. The pump, consisting of three tubes one inside another, is inserted into the hole (Fig. 27.1).

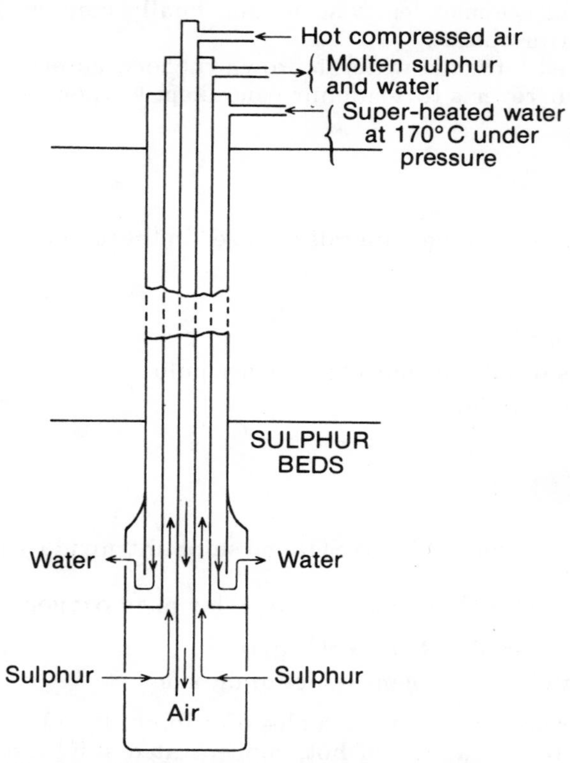

Fig 27.1 Extraction of sulphur by the Frasch process

Water, superheated to 170 °C under pressure, is pumped down the outer pipe and this melts the sulphur. Hot compressed air is pumped down the middle pipe and this forces the mixture of molten sulphur and water to the surface.

The sulphur obtained by this method is 99.5 per cent pure. It is sold in the form of cylinders and is called **roll sulphur**.

27.3 Allotropes of Sulphur

Sulphur can exist in different allotropic forms. The stable allotrope of sulphur below 96 °C is called **α-(octahedral or rhombic) sulphur**. Above 96 °C, the stable allotrope of sulphur is **β-(prismatic or monoclinic) sulphur**. This temperature (96 °C) is called the **transition temperature**. Both allotropes are stable at this temperature.

Both α- and β-sulphur are made up from regular but different arrangement of rings. Each ring contains eight sulphur atoms.

27.4 Preparation of α-Sulphur

A solution of sulphur in xylene or carbon disulphide (CS_2)* is prepared. The solution is allowed to evaporate at room temperature and pale yellow octahedral crystals are formed. The α-sulphur is formed because the temperature is below 96 °C throughout.

27.5 Preparation of β-Sulphur

Sulphur is heated until molten and then allowed to cool. Brown needle-like crystals are found beneath the crust which forms on the surface. The crystals form above 96 °C and so β-sulphur is formed. The β-sulphur changes to α-sulphur when kept at room temperature (or any temperature below 96 °C).

27.6 Action of Heat on Sulphur

When sulphur is heated, it melts at 119 °C to form an amber-coloured free-running liquid. The free-running liquid consists of S_8 rings.

As the temperature is raised, the liquid first turns red and then black. It also becomes viscous (like treacle). The S_8 rings open out to form long chains of sulphur atoms.

* Carbon disulphide is not now recommended for use in schools.

On further heating, the liquid becomes less viscous and finally boils at 444 °C. The long chains break down to form shorter chains.

When molten sulphur is poured into cold water, a brown rubbery material is formed called **plastic sulphur**. Plastic sulphur reverts to α-sulphur when kept at room temperature.

27.7 Uses of Sulphur

Most of the sulphur and sulphur containing minerals are used in the manufacture of sulphuric acid (see Unit 27.9).

Sulphur is also used for:

1 Hardening or vulcanizing rubber.
2 Producing calcium hydrogensulphite for bleaching paper pulp.
3 Making medicines containing sulphur.

27.8 Oxides of Sulphur

There are two oxides of sulphur – sulphur dioxide SO_2 and sulphur trioxide (sulphur(VI) oxide) SO_3.

Sulphur dioxide can be produced by burning sulphur in air or in oxygen.

$$S(s) + O_2(g) \rightarrow SO_2(g)$$
sulphur + oxygen → sulphur dioxide

This reaction is not suitable for preparing pure samples of sulphur dioxide in the laboratory.

Sulphur is usually prepared by the action of hot, concentrated sulphuric acid on copper turnings. (Dilute sulphuric acid does not react with copper.)

$$Cu(s) + 2H_2SO_4(l) \rightarrow CuSO_4(aq) + 2H_2O(l) + SO_2(g)$$
copper + sulphuric acid → copper(II) sulphate + water + sulphur dioxide

Sulphur dioxide can be dried with concentrated sulphuric acid and collected by downward delivery.

Most of the sulphur dioxide produced is used to make sulphuric acid. It is also used for bleaching wood pulp and as a preservative in foods.

Sulphur trioxide cannot be produced by burning sulphur in air or oxygen. It can, however, be produced by the catalytic oxidation of sulphur dioxide.

$$2SO_2(g) + O_2(g) \rightleftharpoons 2SO_3(g)$$
sulphur dioxide + oxygen ⇌ sulphur trioxide

This is the vital reaction in the industrial manufacture of sulphuric acid.

27.9 Industrial Manufacture of Sulphuric Acid (Contact Process)

1 PRODUCTION OF SULPHUR DIOXIDE

Sulphur dioxide is produced by burning sulphur or heating other sulphide minerals in air.

$$S(s) + O_2(g) \rightarrow SO_2(g)$$
sulphur + oxygen → sulphur dioxide

$$2ZnS(s) + 3O_2(g) \rightarrow 2ZnO(s) + 2SO_2(g)$$
zinc sulphide + oxygen → zinc oxide + sulphur dioxide

$$4FeS_2(s) + 11O_2(g) \rightarrow 2Fe_2O_3(s) + 8SO_2(g)$$
iron pyrites + oxygen → iron(III) oxide + sulphur dioxide

Sulphur dioxide is also produced by oxidation of hydrogen sulphide found in certain natural gas samples.

$$2H_2S(g) + 3O_2(g) \rightarrow 2H_2O(g) + 2SO_2(g)$$
hydrogen sulphide + oxygen → water + sulphur dioxide

2 PURIFICATION OF THE SULPHUR DIOXIDE

The sulphur dioxide produced by these methods contains certain impurities, e.g. arsenic compounds which if not removed would prevent the catalyst working ('poison the catalyst') in the next step.

The sulphur dioxide is passed through the electrostatic dust precipitators which remove charged particles, e.g. dust. The sulphur dioxide is then washed with water and dried.

3 CATALYTIC OXIDATION OF SULPHUR DIOXIDE

The important step in the manufacture of sulphuric acid is the reversible reaction:

$$2SO_2(g) + O_2(g) \rightleftharpoons 2SO_3(g) \qquad \Delta H = -385\ \text{kJ}$$

sulphur dioxide + oxygen ⇌ sulphur trioxide

Effect of temperature

As the forward reaction is exothermic (ΔH is negative), lowering the temperature should cause the equilibrium to move to the right, producing more sulphur trioxide.

Effect of pressure

Two moles of sulphur dioxide molecules react with 1 mole of oxygen molecules to produce 2 moles of sulphur trioxide molecules. This means there will be a reduction in the number of molecules, and in the volume, as the forward reaction proceeds.

An increase in pressure will cause the equilibrium to move to the right.

Effect of concentration

Removal of the sulphur trioxide produced would cause the equilibrium to move to produce more sulphur trioxide.

An increase in the oxygen concentration will cause the equilibrium to move to reduce this, producing more sulphur trioxide.

Industrial conditions for the Contact process

1 Atmospheric pressure. Although an increased pressure should improve the yield, about 98 per cent of the gases can be converted without increasing pressure.
2 Temperature of 450 °C.
3 Sulphur trioxide removed from the mixture.
4 Catalyst of vanadium(V) oxide (vanadium pentoxide) V_2O_5.

4 ABSORPTION OF SULPHUR TRIOXIDE

In theory, if sulphur trioxide is dissolved in water, sulphuric acid is produced. But this is not done in practice on a large scale because the reaction is too exothermic and boils the sulphuric acid produced.

The sulphur trioxide is dissolved first in concentrated sulphuric acid to form oleum (fuming sulphuric acid).

$$SO_3(g) + H_2SO_4(l) \rightarrow H_2S_2O_7(l)$$

sulphur trioxide + sulphuric acid → oleum

The oleum is then diluted with the correct amount of water to produce concentrated sulphuric acid.

$$H_2S_2O_7(l) + H_2O(l) \rightarrow 2H_2SO_4(l)$$

oleum + water → sulphuric acid

The acid produced is very pure because all the impurities have been removed during the production.

27.10 Properties of Sulphuric Acid

Concentrated sulphuric acid is a colourless oily liquid which does not show any acidic properties unless water is present. A great deal of heat is produced when concentrated sulphuric acid is diluted with water. It is therefore sensible to add the acid to water (rather than water to acid).

Concentrated sulphuric acid is hygroscopic. For this reason it is a good drying agent for most gases (e.g. SO_2).

The properties of sulphuric acid can be remembered under four headings:

1 As an acid – when dilute.
2 As a producer of other acids – when concentrated.
3 As a dehydrating agent – when concentrated.
4 As an oxidizing agent – when concentrated.

27.11 Sulphuric Acid as an Acid

Sulphuric acid is a dibasic mineral acid. It contains two replaceable hydrogens per molecule.
In the presence of water, sulphuric acid shows the usual acidic properties.
Dilute sulphuric acid turns blue litmus red.
Dilute sulphuric acid produces hydrogen gas with magnesium or zinc.

$$Mg(s) + H_2SO_4(aq) \rightarrow MgSO_4(aq) + H_2(g)$$
magnesium + sulphuric acid → magnesium sulphate + hydrogen

Dilute sulphuric acid produces carbon dioxide gas with a metal carbonate or hydrogencarbonate, e.g. sodium carbonate.

$$Na_2CO_3(s) + H_2SO_4(aq) \rightarrow Na_2SO_4(aq) + H_2O(l) + CO_2(g)$$
sodium carbonate + sulphuric acid → sodium sulphate + water + carbon dioxide

Dilute sulphuric acid produces salts (called sulphates) with metal oxides and hydroxides.

E.g. $$CuO(s) + H_2SO_4(aq) \rightarrow CuSO_4(aq) + H_2O(l)$$
copper(II) oxide + sulphuric acid → copper(II) sulphate + water

Because there are two replaceable hydrogens, it is possible to form acid salts.

E.g. $$2NaOH(aq) + H_2SO_4(aq) \rightarrow Na_2SO_4(aq) + 2H_2O(l)$$
sodium hydroxide + sulphuric acid → sodium sulphate + water

$$NaOH(aq) + H_2SO_4(aq) \rightarrow NaHSO_4(aq) + H_2O(l)$$
sodium hydroxide + sulphuric acid → sodium hydrogensulphate + water

Dilute sulphuric acid can be distinguished from other dilute acids because it is a sulphate and gives a positive sulphate test (see Unit 32.2).

27.12 Sulphuric Acid as a Producer of Other Acids

Concentrated sulphuric acid can be used to prepare nitric and hydrochloric acids.
If concentrated sulphuric acid is added to any metal nitrate (e.g. sodium nitrate) and the mixture is heated, nitric acid vapour is produced.

$$NaNO_3(s) + H_2SO_4(l) \rightarrow NaHSO_4(s) + HNO_3(g)$$
sodium nitrate + sulphuric acid → sodium hydrogensulphate + nitric acid

If concentrated sulphuric acid is added to any metal chloride (e.g. sodium chloride), hydrogen chloride gas is produced, which dissolves in water to form hydrochloric acid. Gentle heating may be necessary.

$$NaCl(s) + H_2SO_4(l) \rightarrow NaHSO_4(s) + HCl(g)$$
sodium chloride + sulphuric acid → sodium hydrogensulphate + hydrogen chloride

Both of these reactions take place because the nitric acid and hydrogen chloride have lower boiling points than sulphuric acid. (They are more volatile.) They escape from the reaction mixture in preference to sulphuric acid.

27.13 Sulphuric Acid as a Dehydrating Agent

This is a consequence of the affinity of concentrated sulphuric acid for water.

1 SUGAR $C_{12}H_{22}O_{11}$

Sugar (sucrose) is a carbohydrate and contains the constituent elements of water (hydrogen and oxygen).
If concentrated sulphuric acid is added to a sample of sugar, the sugar turns yellow, then brown, and finally black. The black solid residue is carbon, which is formed when the concentrated sulphuric acid has removed the hydrogen and oxygen. The reaction is very exothermic.

$$C_{12}H_{22}O_{11}(s) \rightarrow 12C(s) + 11H_2O(g)$$
sugar → carbon + water

Similar reactions take place when other carbohydrates are used. For this reason concentrated sulphuric acid has to be used carefully with carbon compounds.

2 COPPER(II) SULPHATE CRYSTALS $CuSO_4.5H_2O$

Copper(II) sulphate crystals contain water of crystallization. When concentrated sulphuric acid is added to blue copper(II) sulphate crystals, the crystals turn white because the water of crystallization has been removed by the concentrated sulphuric acid.

$$CuSO_4.5H_2O(s) \rightleftarrows CuSO_4(s) + 5H_2O(l)$$

copper sulphate crystals ⇄ anhydrous copper(II) sulphate + water

27.14 Sulphuric Acid as an Oxidizing Agent

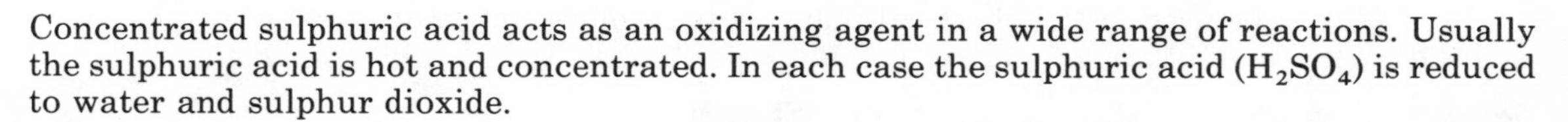

Concentrated sulphuric acid acts as an oxidizing agent in a wide range of reactions. Usually the sulphuric acid is hot and concentrated. In each case the sulphuric acid (H_2SO_4) is reduced to water and sulphur dioxide.

1 COPPER

$$Cu(s) + 2H_2SO_4(l) \rightarrow CuSO_4(aq) + 2H_2O(l) + SO_2(g)$$

copper + sulphuric acid → copper(II) sulphate + water + sulphur dioxide

2 CARBON AND SULPHUR

Hot concentrated sulphuric acid oxidizes carbon and sulphur to carbon dioxide and sulphur dioxide respectively.

$$C(s) + 2H_2SO_4(l) \rightarrow CO_2(g) + 2SO_2(g) + 2H_2O(g)$$

carbon + sulphuric acid → carbon dioxide + sulphur dioxide + water

$$S(s) + 2H_2SO_4(l) \rightarrow 3SO_2(g) + 2H_2O(g)$$

sulphur + sulphuric acid → sulphur dioxide + water

27.15 Uses of Sulphuric Acid

There are many uses of sulphuric acid. In the list that follows the major uses are given in order of importance.

1 One third of the sulphuric acid manufactured is used to make ammonium sulphate fertilizer and superphosphates (see Unit 26).

2 Sulphuric acid is used to make titanium dioxide, which is used in making paint pigments.

3 Sulphuric acid is used to make plastics and other chemicals.

4 Sulphuric acid is used to make soapless detergents from byproducts of oil refining (see Unit 20). Products include washing powders, washing-up liquids and shampoos.

5 Sulphuric acid is used to make man-made fibres such as rayon. It is also used for making dyes for textiles.

6 Sulphuric acid is used for removing the oxide coating from steel before giving the steel a coating to prevent rusting. This process is called **pickling**.

27.16 Summary

Sulphur is obtained from petroleum and natural gas, from underground deposits in the USA and from sulphur-containing minerals.

Sulphur can exist in different allotropic forms. The stable form of sulphur at room temperature is α-sulphur. Above 96 °C, β-sulphur is the stable allotrope. Plastic sulphur is an unstable form of sulphur consisting of chains of sulphur atoms.

Sulphur is used for hardening rubber and making chemicals to bleach paper and for medicines containing sulphur.

These are two oxides of sulphur – sulphur dioxide and sulphur trioxide. Sulphur dioxide can be produced directly from burning sulphur in air but sulphur trioxide cannot.

Sulphuric acid is manufactured by the Contact process. Sulphur dioxide and air are passed over a heated vanadium(V) oxide catalyst. Sulphur trioxide is produced which on dissolving produces sulphuric acid.

Apart from the usual acid properties, sulphuric acid is a maker of other acids, a dehydrating agent and a strong oxidizing agent.

Sulphuric acid is used in the production of fertilizers, paint pigments, plastics, soapless detergents and man-made fibres and in metallurgy for pickling steel.

28 SALT AND CHEMICALS FROM SALT

28.1 Introduction

The chemical name for **salt** is **sodium chloride**. It is a very important raw material for the chemical industry. In this unit we are going to consider where salt is available and how it can be used in the chemical industry to produce important products.

28.2 Occurrence of Salt

Salt is dissolved in all the seas. In Mediterranean countries, for example, solid salt is obtained by the evaporation of sea water. Shallow lakes of salt water are allowed to evaporate using the heat of the sun.

In Great Britain there are vast underground salt deposits in Cheshire. These were formed by the evaporation of seas millions of years ago. These deposits can be exploited in two ways:

1 Underground caverns can be excavated and solid rock salt can be mined. This is used for 'salting' roads in winter.

2 A hole can be drilled down to the deposits and water pumped down. The water dissolves the salt and the salt water, or **brine** as it is called, can be pumped back to the surface.

The availability of these salt deposits in Cheshire was an important reason for the development of the chemical industries in North Cheshire and Lancashire.

Removing salt from underground can cause problems of subsidence.

28.3 Electrolysis Processes Using Salt

In Unit 23.4 the extraction of sodium from molten sodium chloride was discussed. The byproduct of this process is chlorine, which is a very valuable byproduct.

Electrolysis of brine (sodium chloride solution) is an extremely important industry. There are two important alternative processes – the Diaphragm cell process and the Mercury cell process.

In both cases electrolysis of brine produces hydrogen and sodium hydroxide at the negative electrode and chlorine gas at the positive electrode. If the products are allowed to mix the chlorine gas reacts with the alkaline solution to form sodium chlorate(I) (sodium hypochlorite).

$$Cl_2(g) + 2NaOH(aq) \rightarrow NaOCl(aq) + NaCl(aq) + H_2O(l)$$

chlorine + sodium hydroxide → sodium chlorate(I) + sodium chloride + water

Therefore the products must be kept separate from one another. The two cells – the Diaphragm and Mercury cells – are two ways of doing this.

In the **Mercury cell** (sometimes called the Kellner–Solvay cell) purified, saturated brine passes between a flowing film of mercury (the cathode) and titanium plates (the anodes) (Fig. 28.1). The cell is sloping so that the mercury runs through the cell. Direct current of about five volts is used. During this electrolysis, the brine loses about 20 per cent of its sodium chloride.

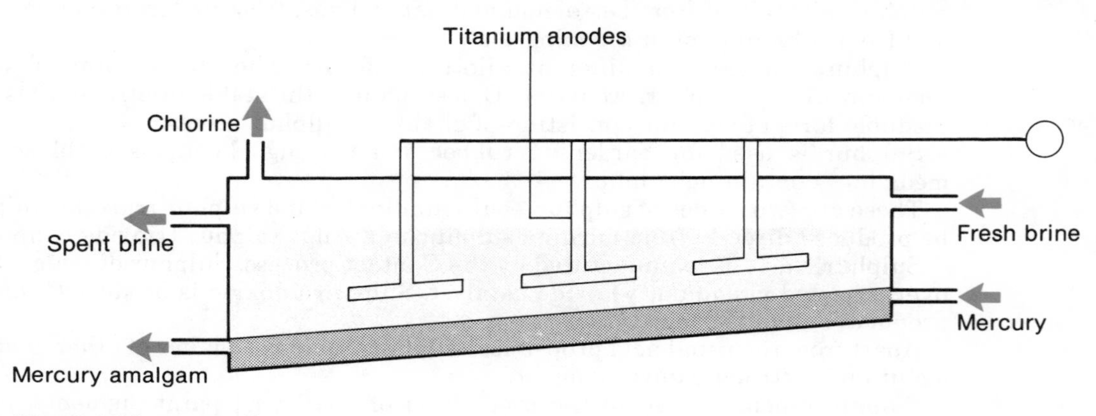

Fig 28.1 Mercury cell

The electrode reactions are:

anode $2Cl^-(aq) \rightarrow Cl_2(g) + 2e^-$
cathode $Na^+(aq) + e^- \rightarrow Na(s)$

At the cathode the sodium ions are discharged in preference to hydrogen ions, forming an amalgam with mercury ($Na_{amalgam}$).

When the sodium amalgam leaves the cell it passes through a tank of cold water and sodium hydroxide and hydrogen are produced.

$$2Na_{amalgam}(l) + 2H_2O(l) \rightarrow 2NaOH(aq) + H_2(g)$$
sodium amalgam + water → sodium hydroxide + hydrogen

The mercury is then recycled.

In the **Diaphragm cell** the electrolysis of purified, saturated brine takes place with a titanium anode and a steel cathode. The anode and cathode are in separate compartments separated by an asbestos diaphragm, which allows the brine to pass through but prevents the products, chlorine and sodium hydroxide solution, from coming into contact.

Figure 28.2 shows a simple representation of the cell. The level of liquid in the anode compartment is kept higher than the level in the cathode compartment to ensure that the flow of solution is from anode compartment to cathode compartment.

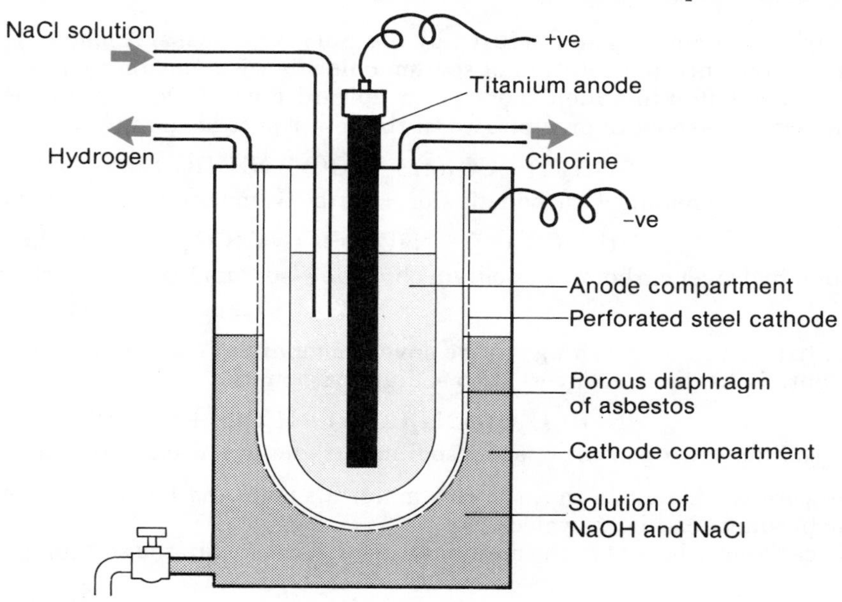

Fig 28.2 Simple representation of a Diaphragm cell

The electrode reactions are the same as for the mercury cell.

The solution leaving the cathode compartment contains approximately 12 per cent by mass of sodium hydroxide and 15 per cent sodium chloride. When this solution is evaporated to about one fifth of its volume the solution contains 50 per cent sodium hydroxide and less than 1 per cent sodium chloride.

Table 28.1 compares some of the factors which would affect the choice of which cell would be most suitable for a particular purpose.

Table 28.1

	Diaphragm cell	*Mercury cell*
Output (kilotonnes/year)		
chlorine	100	250
sodium hydroxide	113	282
Quantity of electricity required (kW/tonne Cl_2)	3000	3550
Cost of plant construction	Relatively simple and inexpensive	Expensive to build and provide mercury
Operating the cell	Frequent replacement of diaphragms necessary	Mercury potentially hazardous. Must all be recovered from effluent for economic and environmental reasons
Product quality	Product contains sodium chloride	High purity product

28.4 Uses of Chlorine and Sodium Hydroxide

About 30 per cent of the chlorine produced is used to make the polymer PVC (see Unit 20.10). This is used for electric wire insulation, floor coverings, clothing, etc. The chloroethene (vinyl chloride) produced during this process is extremely toxic and must be handled with the greatest of care.

About 20 per cent of the chlorine produced is used to make solvents. Tetrachloroethene and 1,1,1-trichloroethane are used as solvents in the dry cleaning of clothing.

Other uses include treatment of water (see Unit 19.1) and production of chemicals. Household bleach contains sodium chlorate(I) (sodium hypochlorite) and is produced by mixing chlorine and sodium hydroxide solution.

Sodium hydroxide has a wide range of uses. These include the manufacture of soap (see Unit 19.3), sodium compounds, rayon, paper pulp and organic chemicals, and purification of aluminium ores for aluminium extraction (see Unit 23.5).

Hydrogen is also produced in the electrolysis of brine. The uses of hydrogen are given in Unit 5.6.

28.5 Manufacture of Sodium Carbonate

Sodium carbonate can be manufactured by the **Solvay process**. Ammoniacal brine (made by saturating a concentrated solution of sodium chloride with ammonia) descends a large tower called the carbonator in which there is an upward flow of carbon dioxide under pressure. Sodium hydrogencarbonate precipitates in the lower part of the tower.

$$NH_3(g) + CO_2(g) + H_2O(l) \rightarrow NH_4HCO_3(aq)$$

ammonia + carbon dioxide + water → ammonium hydrogencarbonate

$$NH_4HCO_3(aq) + NaCl(aq) \rightarrow NaHCO_3(s) + NH_4Cl(aq)$$

ammonium hydrogencarbonate + sodium chloride → sodium hydrogencarbonate + ammonium chloride

After filtration and washing to remove ammonium compounds, the sodium hydrogencarbonate is heated to convert it to sodium carbonate.

$$2NaHCO_3(s) \rightarrow Na_2CO_3(s) + H_2O(g) + CO_2(g)$$

sodium hydrogencarbonate → sodium carbonate + water + carbon dioxide

This process is efficient in that the raw materials (salt and limestone) are inexpensive and certain byproducts can be recycled.

Sodium carbonate is used in the manufacture of glass, and in the softening of water (see Unit 19.4).

28.6 Summary

Salt (sodium chloride) is an important raw material for the chemical industry. Cheshire is an important source of salt.

Electrolysis of molten sodium chloride produces sodium and chlorine.

Electrolysis of brine (sodium chloride solution) produces sodium hydroxide, hydrogen and chlorine. There are two alternative ways of doing this – the Diaphragm cell and Mercury cell.

Sodium carbonate is produced by the Solvay process.

29 THE MOLE AND CHEMICAL CALCULATIONS

This section is the basis of most chemical calculations that appear on examination papers. Questions based on this section appear frequently and provide a useful discrimination between candidates. In the Scottish Standard grade syllabus, the 'mole' concept is restricted to that of gram formula weight and the definition of the mole as the Avogadro number of particles is deferred until the Higher grade.

29.1 Relative Atomic Mass

All atoms are too small to be weighed individually. It is possible, however, to compare the mass of one atom with the mass of another. This is done using a mass spectrometer. A magnesium atom has twice the mass of a carbon-12 atom and six times the mass of a helium atom.

The **relative atomic mass** of an atom is the number of times an atom is heavier than a hydrogen atom (or one twelfth of a carbon-12 atom). Relative atomic masses are not all whole numbers because of the existence of isotopes.

$$\text{relative atomic mass} = \frac{\text{mass of 1 atom of element}}{\text{mass of 1 atom of hydrogen}}$$

The relative atomic mass is simply a number and has no units. You are not expected to remember relative atomic masses. They are given on examination papers in one of the following ways:

1. $A_r(\text{Ca}) = 40$ or **2.** (Ca = 40, C = 12)

Throughout this book relative atomic masses will be shown as in **1**.

Relative atomic masses are sometimes just called atomic masses or atomic weights.

29.2 The Mole

As an alternative to comparing masses of individual atoms of different elements, it is possible to consider large numbers of atoms. For example:

1 atom of magnesium weighs twice as much as 1 atom of carbon-12
2 atoms of magnesium weigh twice as much as 2 atoms of carbon-12
100 atoms of magnesium weigh twice as much as 100 atoms of carbon-12

The mass of magnesium atoms will always be twice the mass of the carbon-12 atoms, providing the number of magnesium and carbon-12 atoms are the same.

We are used to collective terms to describe a number of objects, e.g. a dozen eggs, a gross of test tubes, etc. In chemistry, the term **mole** (abbreviation mol) is used in the same way. We speak of a mole of magnesium atoms, a mole of carbon dioxide or a mole of electrons.

A mole provides a quantity of material that can be used in the laboratory. A mole of carbon atoms (12 grams) is just a small handful.

A mole contains approximately 6×10^{23} particles (600 000 000 000 000 000 000 000). This number is called **Avogadro's** constant (L). A very large number is consequently difficult for us to appreciate. If one were to stand on a sandy beach and look along the beach in both directions, you would not see enough particles of sand to make 1 mole of grains of sand.

It is conveniently arranged that 1 mole of atoms of any element has a mass equal to the relative atomic mass (but with units of grams).

E.g. relative atomic mass of Na = 23 (i.e. $A_r(\text{Na}) = 23$)
$\therefore$ mass of 1 mole of sodium atoms = 23 g

The mole may be defined as the amount of substance which contains as many elementary units as there are atoms in 12 grams of carbon-12. These elementary units can be considered as:

atoms	e.g. Mg, C, He
molecules	e.g. CH_4, H_2O
ions	e.g. Na^+, Cl^-
specified formula units	e.g. H_2SO_4

There are other terms that might be seen in books or examination papers:

E.g. 1 gram-atom – mass of 1 mole of atoms
1 gram-molecule – mass of 1 mole of molecules
1 gram-ion – mass of 1 mole of ions
1 gram-formula – mass of 1 mole of formula unit
1 Faraday – 1 mole of electrons

The term '1 mole of chlorine' can be ambiguous. It could mean 1 mole of chlorine atoms (6×10^{23} atoms) or 1 mole of chlorine molecules (12×10^{23} atoms).

29.3 Volume of One Mole of Molecules of a Gas

There is no simple relationship that predicts the volume occupied by 1 mole of molecules in a solid or a liquid. However, 1 mole of molecules of any gas occupies 24 000 cm^3 (24 dm^3) at room temperature and pressure or 22 400 cm^3 (22.4 dm^3) at STP (standard temperature and pressure). This information is given on the examination paper if it is required.

29.4 Molar Solutions

When 1 mole of a substance is dissolved in water and the volume of solution made up to 1000 cm^3 (1 dm^3), the resulting solution is called a **molar (or M) solution**. If 2 moles of a substance are made up to 1000 cm^3 (or 1 mole made up to 500 cm^3) the solution is said to be a 2 M solution. E.g. 8 g of sodium hydroxide NaOH is dissolved in water to make 100 cm^3 of solution. What is the molarity of the solution? ($A_r(H) = 1$, $A_r(O) = 16$, $A_r(Na) = 23$)

mass of 1 mole of sodium hydroxide NaOH = 23 + 16 + 1 = 40 g
8 g of sodium hydroxide is $\frac{8}{40} = 0.2$ mole (see Unit 29.5)

0.2 mole of sodium hydroxide dissolved to make 100 cm^3 of solution
0.2×10 mole of sodium hydroxide dissolved to make 1000 cm^3 of solution

The solution produced is then 0.2×10 M, i.e. 2M sodium hydroxide.

29.5 Conversion of Mass (in grams) of Substance to Amount (in moles)

It is useful to convert, for example, 18 g of water to 1 mole of water molecules. This is because we know that there are 6×10^{23} water molecules in 1 mole of water but we have no idea how many particles there are in a given mass of water.

$$\text{number of moles} = \frac{\text{number of grams}}{\text{mass of 1 mole}}$$

E.g. How many moles of carbon dioxide molecules are present in 11 grams of carbon dioxide? ($A_r(C) = 12$, $A_r(O) = 16$.)

$$\text{mass of 1 mole of carbon dioxide } CO_2 = 12 + (2 \times 16)\text{ g} = 44\text{ g}$$

$$\text{number of moles of carbon dioxide} = \frac{11}{44} = 0.25 \text{ moles}$$

29.6 Conversion of Amount (in moles) to Mass (in grams) of Substance

This is the reverse of 29.5.

$$\text{number of grams} = \text{number of moles} \times \text{mass of 1 mole}$$

E.g. What is the mass of 2 moles of ethanol molecules (C_2H_5OH)? ($A_r(H) = 1$, $A_r(C) = 12$, $A_r(O) = 16$)

$$\text{mass of 1 mole of ethanol molecules} = (2 \times 12) + (5 \times 1) + 16 + 1\text{ g} = 46\text{ g}$$

$$\text{number of grams of ethanol} = 2 \times 46\text{ g} = 92\text{ g}$$

29.7 Finding the Formula of a Compound

In Unit 4 there is information to enable you to work out the formula of a compound. It should be remembered that each formula could be worked out following a suitable experiment involving weighing.

E.g. Magnesium oxide

A known mass of magnesium ribbon is burnt in a crucible in contact with air. The mass of magnesium oxide produced is found.

①	mass of crucible and lid	= 20.12 g
②	mass of crucible, lid and magnesium	= 20.36 g
	mass of magnesium	= 0.24 g i.e. ② − ①
③	mass of crucible, lid and magnesium oxide	= 20.52 g
	mass of magnesium oxide	0.40 g i.e. ③ − ①

0.24 g of magnesium combines with 0.16 g of oxygen (0.40 − 0.24) to form 0.40 g of magnesium oxide.

(This is the key statement that you must write down. It will help you and also will give you marks.)

Multiply through this statement by 100 to remove decimals and prevent arithmetical mistakes:

24 g of magnesium combines with 16 g of oxygen to form 40 g of magnesium oxide

Divide masses of magnesium and oxygen by the appropriate relative atomic masses to give the number of moles of atoms of magnesium and oxygen (see Unit 29.5) ($A_r(O) = 16$, $A_r(Mg) = 24$):

$$\frac{24}{24} \text{ moles of magnesium atoms combine with } \frac{16}{16} \text{ moles of oxygen atoms}$$

1 mole of magnesium atoms combine with 1 mole of oxygen atoms. Since 1 mole of magnesium atoms contains the same number of atoms as 1 mole of oxygen atoms (i.e. 6×10^{23}), the simplest formula of magnesium is MgO. Since these questions are very common, here are two further examples:

1 2.00 g of mercury combines with 0.71 g of chlorine to form 2.71 g of a mercury chloride. What is the simplest formula for the mercury chloride? ($A_r(Cl) = 35.5$, $A_r(Hg) = 200$.)

Key statement:
2.00 g of mercury combines with 0.71 g of chlorine to form 2.71 g of mercury chloride.

Multiply throughout by 100 to remove decimals:

200 g of mercury combines with 71 g of chlorine

Divide by the appropriate atomic masses:

$$\frac{200}{200} \text{ moles of mercury atoms combine with } \frac{71}{35.5} \text{ moles of chlorine atoms}$$

1 mole of mercury atoms combines with 2 moles of chlorine atoms

∴ simplest formula is $HgCl_2$

2 11.2 g of iron combines with 4.8 g of oxygen to form an iron oxide. What is the simplest formula for the iron oxide? ($A_r(O) = 16$, $A_r(Fe) = 56$.)

Key statement:
11.2 g of iron combines with 4.8 g of oxygen to form 16.0 g of iron oxide.

Multiply throughout by 10 to remove decimals:

112 g of iron combines with 48 g of oxygen

Divide by the appropriate relative atomic masses:

$$\frac{112}{56} \text{ moles of iron atoms combines with } \frac{48}{16} \text{ moles of oxygen atoms}$$

2 moles of iron atoms combine with 3 moles of oxygen atoms

∴ simplest formula is Fe_2O_3

29.8 Calculating the Simplest Formula from Percentages

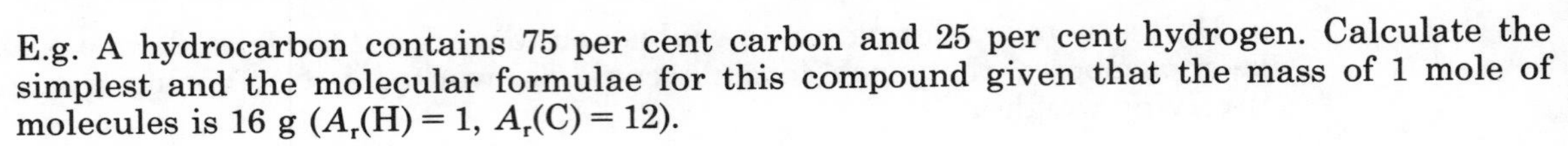

E.g. A hydrocarbon contains 75 per cent carbon and 25 per cent hydrogen. Calculate the simplest and the molecular formulae for this compound given that the mass of 1 mole of molecules is 16 g ($A_r(H) = 1$, $A_r(C) = 12$).

	C	H
percentage	75	25
relative atomic mass	12	1
divide percentage by relative atomic mass	6.25	25
divide by smallest, i.e. 6.25	1	4

Simplest formula (or empirical) formula = CH_4

This may not be the molecular formula. It could be C_2H_8, C_3H_{12} etc. – always four times as many hydrogens as carbons.

If the formula is CH_4, the mass of 1 mole of molecules is 16 g.

$\therefore$ molecular formula is CH_4

29.9 Calculating the Percentages of Elements in a Compound

E.g. Calculate the percentage of nitrogen in ammonium nitrate NH_4NO_3 ($A_r(H) = 1$, $A_r(N) = 14$, $A_r(O) = 16$).

$$\text{mass of 1 mol of ammonium nitrate } NH_4NO_3 = 14 + (4 \times 1) + 14 + (3 \times 16) = 80 \text{ g}$$

Each 80 g of ammonium nitrate contains 28 g of nitrogen (it contains two nitrogen atoms, i.e. 2×14):

$$\text{percentage of nitrogen} = \frac{28}{80} \times 100 = 35 \text{ per cent}$$

This type of calculation is useful for calculating the percentage of nitrogen in a fertilizer.

29.10 Calculations from Equations

This type of question appears frequently on GCSE examination papers. It is necessary to have a balanced equation. Usually this equation is supplied.

E.g. $Na_2CO_3(s) + 2HCl(aq) \rightarrow 2NaCl(aq) + H_2O(l) + CO_2(g)$

sodium carbonate + hydrochloric acid → sodium chloride + water + carbon dioxide

This equation gives the following information:

1 mole of sodium carbonate (Na_2CO_3) reacts with 2 moles of hydrochloric acid (HCl) to produce 2 moles of sodium chloride (NaCl), 1 mole of water (H_2O) and 1 mole of carbon dioxide (CO_2).

Using the relative atomic masses ($A_r(H) = 1$, $A_r(C) = 12$, $A_r(O) = 16$, $A_r(Na) = 23$, $A_r(Cl) = 35.5$) we can find the masses of the substances that react together and the masses of the substances produced.

$Na_2CO_3(aq)$	+ $2HCl(aq)$	→ $2NaCl(aq)$	+ $H_2O(l)$	+ $CO_2(g)$
$(2 \times 23) + 12 + (3 \times 16)$	$2(1 + 35.5)$	$2(23 + 35.5)$	$(2 \times 1) + 16$	$12 + (2 \times 16)$
106 g	73 g	117 g	18 g	44 g

At this stage it is worthwhile checking that the sum of the masses on the left-hand side (106 g + 73 g = 179 g) equals the sum on the right-hand side (117 g + 18 g + 44 g = 179 g). Silly arithmetical mistakes, frequently seen on examination papers, should be avoided if you do this.

The calculations now are just proportion sums.

1 Calculate the maximum mass of sodium chloride that could be produced from 5.3 g of sodium carbonate.

From the information following the equation:

106 g of sodium carbonate (1 mole) produces 117 g of sodium chloride (2 moles)

$$1 \text{ g of sodium carbonate produces } \frac{117}{106} \text{ g of sodium chloride}$$

$$5.3 \text{ g of sodium carbonate produces } \frac{117 \times 5.3}{106} \text{ g of sodium chloride} = \frac{11.7}{2} = 5.85 \text{ g}$$

2 Calculate the volume of 2 M hydrochloric acid which would exactly react with 5.3 g of sodium carbonate. Using the information above again:

106 g of sodium carbonate (1 mole) reacts with 73 g of hydrochloric acid (2 moles).

Since hydrochloric acid is in a dilute solution it is worth remembering that:

1 mole of hydrochloric acid dissolved and made up to 1000 cm^3 (1 dm^3) produces an M solution (see Unit 29.4).

2 moles of hydrochloric acid dissolved and made up to 1000 cm^3 produces a 2 M solution.

$\therefore$ 106 g of sodium carbonate reacts with 1000 cm^3 of 2 M hydrochloric acid

$$1 \text{ g of sodium carbonate reacts with } \frac{1000}{106} \text{ cm}^3 \text{ of 2 M hydrochloric acid}$$

$$5.3 \text{ g of sodium carbonate reacts with } \frac{1000}{106} \times 5.3 \text{ cm}^3 \text{ of 2 M hydrochloric acid}$$

$$= 50 \text{ cm}^3 \text{ of 2 M hydrochloric acid}$$

3 Calculate the volume of carbon dioxide (at room temperature and pressure) produced when 5.3 g of sodium carbonate reacts with excess hydrochloric acid. Using the information above again:

106 g of sodium carbonate (1 mole) produces 44 g of carbon dioxide (1 mole). However, 1 mole of any gas at room temperature and pressure occupies 24 000 cm^3 (24 dm^3) (see Unit 29.3).

$\therefore$ 106 g of sodium carbonate produces 24 000 cm^3 of carbon dioxide

$$1 \text{ g of sodium carbonate produces } \frac{24000}{106} \text{ cm}^3 \text{ of carbon dioxide}$$

$$5.3 \text{ g of sodium carbonate produces } \frac{24000 \times 5.3}{106} \text{ cm}^3 \text{ of carbon dioxide}$$

$$= 1200 \text{ cm}^3$$

(All measurements at room temperature and pressure.)

The chemical calculations using equations enable a chemist to calculate the quantities of materials required for a particular reaction and to calculate the quantities of products formed.

29.11 Summary

Chemical calculations appear regularly on all chemistry papers.

The relative atomic mass of an atom is the number of times an atom is heavier than a hydrogen atom (or one twelfth of a carbon-12 atom). Relative atomic masses are given to you on examination papers.

A mole is an amount of substance which contains 6×10^{23} particles. You can calculate the number of moles of particles using the formula:

$$\text{number of moles} = \frac{\text{number of grams}}{\text{mass of 1 mole}}$$

A molar (or M solution) is solution in which 1 mole of chemical is dissolved and made up to 1000 cm^3.

Calculations can be made using an equation. It is possible to calculate masses of reacting substances and products. It is also possible to calculate volumes of reacting solutions and volumes of gases required or produced.

30 QUANTITATIVE VOLUMETRIC CHEMISTRY

30.1 Introduction

In this section we are concerned with the volumes of standard solutions that react exactly together; with this information various calculations can be done.

In Unit 29, it was explained that the concentration of a solution can be expressed in terms of molarity. Alternatively the concentration of a solution may be written in terms of moles per cubic decimetre. A 0.1 M solution has a concentration of 0.1 moles/dm^3.

E.g. what is the concentration of a solution of hydrochloric acid containing 7.3 g of hydrogen chloride in 100 cm^3 of solution? ($A_r(H) = 1$, $A_r(Cl) = 35.5$.)

$$\text{mass of 1 mole of hydrogen chloride HCl} = 1 + 35.5\text{ g} = 36.5\text{ g}$$

7.3 g of hydrogen chloride in 100 cm^3 of solution has the same molarity as 73 g of hydrogen chloride in 1000 cm^3 of solution.

$$\text{the solution contains } \frac{73}{36.5} \text{ moles of hydrogen chloride}$$

$$\therefore \text{ the solution is 2 M (i.e. 2 mol/dm}^3\text{)}$$

Remember: 1 mole of a substance dissolved to make 1000 cm^3 of solution produces a molar (M) solution.

A solution of known concentration or known molarity is called a **standard solution**. In volumetric chemistry, a series of **titrations** are carried out. In each titration, a solution A is added in small measured quantities, from a burette, to a fixed volume of a solution B, measured with a pipette, in the presence of an indicator. At least one of the solutions must be a standard solution. The addition of a solution A is continued until the indicator just changes colour. At this stage, called the **end point**, the two substances in solution are present in quantities that exactly react.

In any titration, accuracy of measurement is very important.

30.2 Titration of Sodium Hydroxide with Standard Sulphuric Acid

A solution of sodium hydroxide (of unknown concentration) is going to be standardized (i.e. its concentration found) by titration with 0.1 M sulphuric acid.

Exactly 25 cm^3 of sodium hydroxide solution is added to a conical flask using a pipette. A couple of drops of screened methyl orange (indicator) are added and the solution turns green. (Litmus is not sufficiently sensitive.) 0.1 M sulphuric acid is put into the burette and the reading on the burette recorded. The sulphuric acid is added to the flask in small volumes. After each addition, the flask is swirled. The process is continued until the solution turns colourless. The final reading on the burette is recorded.

The procedure is repeated until consistent results are obtained. Sample results:

$$\text{volume of sodium hydroxide solution} = 25.00\text{ cm}^3$$
$$\text{volume of sulphuric acid} = 24.00\text{ cm}^3$$

(This result would be the average of the results obtained.)
Molarity of sulphuric acid = 0.1 M.

The reaction is represented by the equation:

$$2NaOH(aq) + H_2SO_4(aq) \rightarrow Na_2SO_4(aq) + 2H_2O(l)$$
$$\text{sodium hydroxide + sulphuric acid} \rightarrow \text{sodium sulphate + water}$$

If 1000 cm^3 of M sulphuric acid were used, this would contain 1 mole of sulphuric acid.

$$\therefore \text{ 24.00 cm}^3 \text{ 0.1 M sulphuric acid contains } 0.1 \times \frac{24}{1000} \text{ moles of sulphuric acid}$$
$$= 0.0024 \text{ moles sulphuric acid}$$

From the equation opposite:

1 mole of sulphuric acid exactly reacts with 2 moles sodium hydroxide
∴ 0.0024 moles of sulphuric acid exactly react with 0.0048 moles sodium hydroxide

Now, 0.0048 moles of sodium hydroxide is contained in 25.00 cm^3 of solution.

∴ 1000 cm^3 of sodium hydroxide solution would contain

$$\frac{0.0048 \times 1000}{25} = 0.192 \text{ moles}$$

molarity of sodium hydroxide = 0.192 M

30.3 Titration of Standard Sodium Carbonate with Hydrochloric Acid

A solution of hydrochloric acid (of unknown concentration) is going to be standardized by titration with 0.1 M sodium carbonate solution.

Exactly 25 cm^3 of sodium carbonate solution is added to a conical flask from a pipette. A couple of drops of screened methyl orange is added. Hydrochloric acid is added to the flask in small volumes and the volume of acid required to change the colour of the indicator is found. Again further titrations are carried out until consistent results are obtained.

Sample results:

volume of sodium carbonate solution = 25.00 cm^3
volume of hydrochloric acid = 23.50 cm^3 (average)
molarity of sodium carbonate solution = 0.1 M

The equation for the reaction is:

$$Na_2CO_3(aq) + 2HCl(aq) \rightarrow 2NaCl(aq) + CO_2(g) + H_2O(l)$$

sodium carbonate + hydrochloric acid → sodium chloride + carbon dioxide + water

1000 cm^3 of M sodium carbonate contains 1 mole of sodium carbonate.

∴ 25 cm^3 of 0.1 M sodium carbonate contain

$$\frac{0.1 \times 25}{1000} \text{ moles sodium carbonate} = 0.0025 \text{ moles sodium carbonate}$$

From the equation:
1 mole of sodium carbonate reacts with 2 moles of hydrochloric acid; 0.0025 moles of sodium carbonate reacts with 0.005 moles of hydrochloric acid. Now, 0.005 moles of hydrochloric acid is contained in 23.50 cm^3 of solution.

$$\therefore 1000 \text{ cm}^3 \text{ of solution would contain } \frac{0.005 \times 1000}{23.5} \text{ moles} = 0.213 \text{ moles}$$

molarity of hydrochloric acid = 0.213 M

N.B. If phenolphthalein had been used as indicator, the volume of acid required would have been 11.75 cm^3 (exactly half the volume required when screen methyl orange was used). This is because phenolphthalein is detecting the end point in the reaction.

$$Na_2CO_3(aq) + HCl(aq) \rightarrow NaHCO_3(aq) + NaCl(aq)$$

sodium carbonate + hydrochloric acid → sodium hydrogencarbonate + sodium chloride

30.4 Points to Remember when doing Volumetric Experiments

Simple volumetric experiments are frequently used in practical assessment. When doing this type of experiment accuracy is very important and the following points should be remembered:

1 Shake up all solutions thoroughly before use to make sure the solution is the same throughout.

2 Rinse out the conical flask with distilled water only.

3 The burette and pipette should be rinsed out with the solution that is to go into them. These rinsing solutions should then be discarded.

4 The last drop of solution in the pipette should not be blown or shaken out of the pipette.

5 Distilled water can be added to the conical flask during the titration.

30.5 Volumetric Analysis used to Determine an Equation

Suppose an exactly 0.1 M solution of a metal hydroxide $M(OH)_x$ is provided together with 0.2 M hydrochloric acid. It is possible to find the value of x and then the equation from a volumetric experiment.

E.g. 25.00 cm^3 of 0.1 M metal hydroxide exactly reacts with 25.00 cm^3 of 0.2 M hydrochloric acid.

25.00 cm^3 of 0.1 M metal hydroxide contains:

$$\frac{0.1 \times 25}{1000} \text{ moles of metal hydroxide} = \frac{1}{400} \text{ mole} = 0.0025 \text{ moles}$$

25.00 cm^3 of 0.2 M hydrochloric acid contains:

$$\frac{0.2 \times 25}{1000} \text{ moles of hydrochloric acid} = \frac{1}{200} \text{ moles} = 0.005 \text{ moles}$$

$\therefore$ 1 mole of $M(OH)_x$ would exactly react with 2 moles of hydrochloric acid

So the formula of the metal hydroxide is $M(OH)_2$ and the equation is:

$$M(OH)_2(aq) + 2HCl(aq) \rightarrow MCl_2(aq) + 2H_2O(l)$$

30.6 Titration without an Indicator

The end point in a titration can be found by following the pH or electrical conductivity during the reaction. For example, the titration of barium hydroxide solution with dilute sulphuric acid can be carried out using electrical conductivity to detect the end point. At the end point, the electrical conductivity is zero.

$$Ba(OH)_2(aq) + H_2SO_4(aq) \rightarrow BaSO_4(s) + 2H_2O(l)$$

barium hydroxide + sulphuric acid → barium sulphate + water

Figure 30.1 shows the apparatus required and the graph obtained. The end point is the minimum point on the graph.

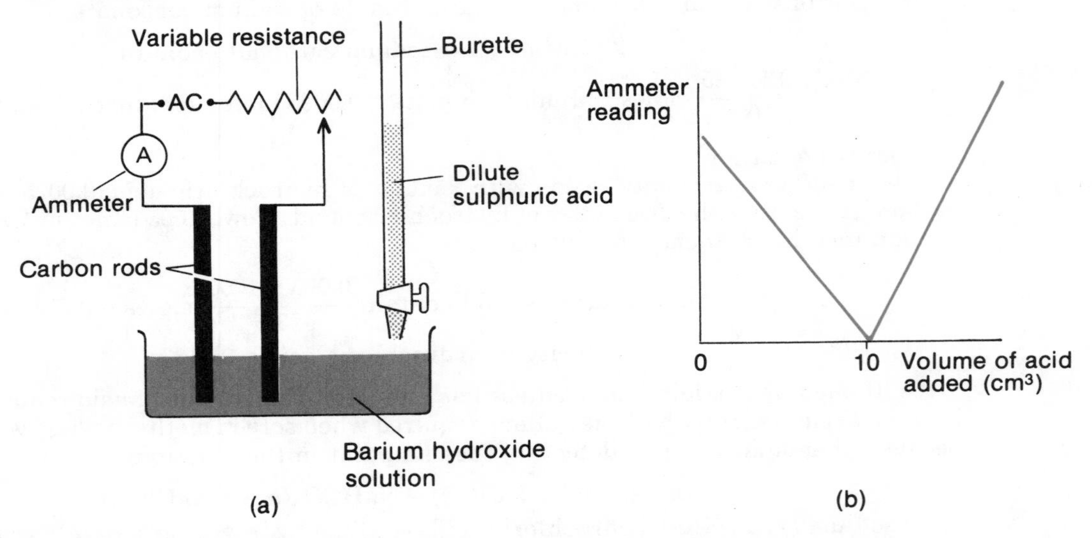

Fig 30.1 Neutralization of barium hydroxide solution with dilute sulphuric acid: (a) apparatus to follow the reaction by measuring conductivity changes; (b) graph of sample results

30.7 To Find the Number of Molecules of Water of Crystallization in a Sample of Sodium Carbonate

Hydrated sodium carbonate has a formula $Na_2CO_3 . yH_2O$. In this experiment the value of y is going to be calculated.

7.15 g of hydrated sodium carbonate was dissolved to make 250 cm^3 of solution. A 25 cm^3 sample of this solution was taken and screened methyl orange indicator added. 50.00 cm^3 of 0.1 M hydrochloric acid was added to change the colour of the indicator.

$$Na_2CO_3(aq) + 2HCl(aq) \rightarrow 2NaCl(aq) + H_2O(l) + CO_2(g)$$

sodium carbonate + hydrochloric acid → sodium chloride + water + carbon dioxide

50.00 cm^3 of 0.1 M hydrochloric acid contains:

$$\frac{0.1 \times 50}{1000} \text{ moles of hydrogen chloride} = 0.005 \text{ moles}$$

0.005 moles of hydrochloric acid react with 0.0025 moles of sodium carbonate.
Let the molarity of the sodium carbonate solution be zM.
1000 cm^3 of M sodium carbonate contain 1 mole of sodium carbonate.

$$25 \text{ cm}^3 \text{ of M sodium carbonate contain } \frac{1 \times 25}{1000} \text{ moles of sodium carbonate}$$

$$25 \text{ cm}^3 \text{ of } z\text{M sodium carbonate contain } \frac{1 \times 25 \times z}{1000} \text{ moles of sodium carbonate}$$

Now,

$$\frac{1 \times 25 \times z}{1000} = 0.0025$$

$$z = 0.1$$

sodium carbonate solution is 0.1 M

Concentration of Na_2CO_3 in sodium carbonate solution = 0.1 × 106
= 10.6 g/dm^3

Concentration of $Na_2CO_3.yH_2O = 7.15 \times 4 = 28.6$ g/dm^3

∴ 28.6 g of sodium carbonate crystals contain 10.6 g of sodium carbonate Na_2CO_3 and 18.0 g of water ($Na_2CO_3 = 106$, $H_2O = 18$)

0.1 moles of sodium carbonate combine with 1 mole of water.

∴ 1 mole of sodium carbonate combines with 10 moles of water

$$\therefore y = 10$$

Sodium carbonate crystals are $Na_2CO_3.10H_2O$.

30.8 Summary

Accurate experiments to find the volumes of solutions of known concentration which react together are called titrations. Many titrations involve reactions between acid and alkali and an indicator is used to find the end point, i.e. where the acid and alkali exactly react together.
This type of question always contains the use of an equation, as a central feature.

31 RADIOACTIVITY

31.1 Introduction

The nuclei of some heavier atoms are unstable and tend to split up with the emission of certain types of radiation and the formation of new elements. This decay is called **radioactive decay**.
It is possible for lighter atoms containing a large proportion of neutrons to undergo radioactive decay, e.g. ^3_1H(tritium) is a radioactive isotope of hydrogen.

31.2 Types of Radiation Emitted

The radiation emitted can be of three types:
α-particles. These are positively charged particles each of which is identical with the nucleus of a helium atom, i.e. two protons and two neutrons but no electrons. They are comparatively heavy and slow moving and have little penetrating power. For example, they are unable to penetrate a piece of paper. They are deflected by magnetic and electrostatic fields. Because of their low penetrating power they cannot be detected by the usual apparatus used in the laboratory.
β-rays. These are negatively charged particles. They are in fact electrons. They have greater penetrating power than α-particles and are deflected by magnetic and electric fields. They are detected by the apparatus used in the laboratory.
γ-rays. These are high energy electromagnetic waves with zero charge. They are very penetrating and are unaffected by electric and magnetic fields. Although they penetrate the apparatus, they are not recorded on laboratory apparatus.

31.3 Apparatus used for Radioactivity Measurements

In experiments in the laboratory, radioactivity measurements are made using a **Geiger counter** attached to a suitable counting tube. The sample is placed in the counting tube and the measurements are made on the Geiger counter.

Because radioactive decay is taking place in the atmosphere and surroundings, it is necessary to correct for **background radiation**. In an experiment, if the reading on the Geiger counter is 512 counts per second (c.p.s.) when the sample is in place but 10 c.p.s. in the absence of any radioactive sample, the corrected reading would be 502 c.p.s. (i.e. 512 − 10).

31.4 Half Life $t_{1/2}$

The rate of decay of a radioactive isotope is independent of temperature. The time taken for half the mass of a radioactive isotope to decay is called the **half life** and is a characteristic of the isotope. It is the time taken for the corrected reading on the Geiger counter to fall to half of its original value.

Half life of $^{214}_{84}$Po $= 1.5 \times 10^{-4}$ s
Half life of $^{226}_{88}$Ra $= 1620$ years

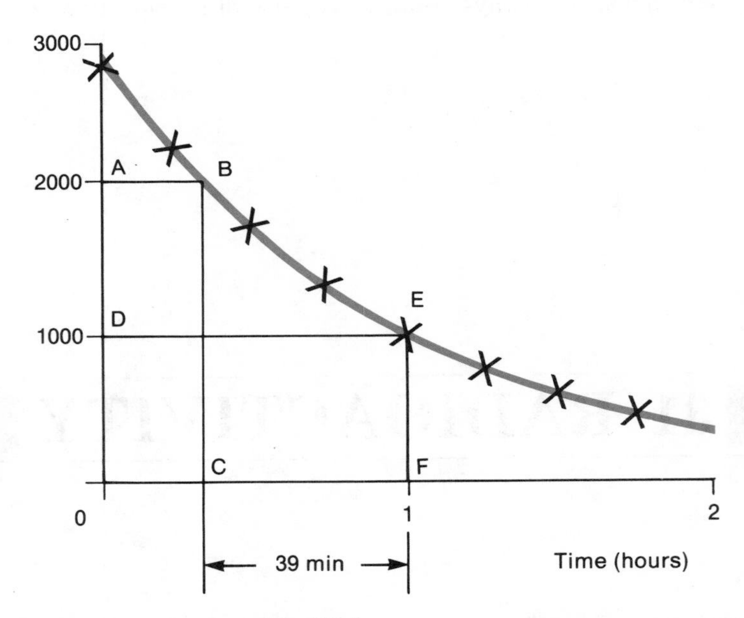

Fig 31.1 Radioactive decay and half life

A graph can be plotted as in Fig. 31.1 if regular readings are taken at intervals and then corrected. If a convenient reading is taken (say 2000) and the lines *AB* and *BC* drawn, then the lines *DE* and *EF* are drawn at a reading which is half of the original reading. The time difference between *BC* and *EF* is called the half life.

31.5 Changes that Accompany Radioactive Decay

Radioactive decay can change the number of protons and neutrons in the nucleus of an atom.

When an isotope loses an α-particle, it loses two protons and two neutrons. For example, α-decay of uranium-238:

$$^{238}_{92}U \rightarrow {}^{4}_{2}He + {}^{234}_{90}Th$$
$$\text{uranium-238} \rightarrow \alpha\text{-particle} + \text{thorium-234}$$

(N.B. In $^{238}_{92}U$, 238 is the mass number and 92 is the atomic number.)

The product of α-decay contains two fewer protons and, therefore, is two places to the left of uranium in the Periodic Table.

When an isotope undergoes β-decay, a neutron in the nucleus changes to a proton and an electron, which is emitted. The product, therefore, has one more proton than the starting isotope, and is therefore one place to the right in the Periodic Table. Thorium-234 undergoes β-decay:

$$^{234}_{90}Th \rightarrow {}^{234}_{91}Pa + \text{electron}$$
$$\text{thorium-234} \rightarrow \text{protactinium-234} + \text{electron}$$

The emission of γ-rays does not change the number of protons or neutrons in the isotope. Gamma radiation usually accompanies other types of emission.

31.6 Uses of Radioactivity

Radioactivity has a large number of uses in industry. These include:

1 Treating cancer by subjecting a patient to controlled amounts of γ-radiation from a cobalt-60 source.

2 Sterilizing instruments and equipment using γ-radiation.

3 Controlling the thickness of paper, rubber, metals and plastic accurately.

4 Controlling the filling of packets and containers.

5 Tracing the movement of a substance by following a radioactive isotope.

6 The energy produced by radioactive fission of uranium-235 is used within a nuclear power station to provide electricity.

7 Irradiation of food with γ-radiation from cobalt-60 to kill bacteria and ensure that the food remains fit to eat for longer.

The uses of a radioactive isotope depend upon its half life. If the half life is too short or too long it can have little practical application.

31.7 Disposal of Radioactive Waste

Radioactive waste, because of its very dangerous nature and its very long life, is extremely difficult to dispose of.

Used rods from power stations contain a whole range of radioactive materials including plutonium. At Sellafield, in Cumbria, much of this waste is treated.

Waste with only low levels of radioactivity can be pumped into the sea. Waste containing higher levels of radioactivity can be set in concrete and stored. Investigations are going on to find suitable sites for dumping radioactive waste inland.

Disposal of radioactive waste will always be a problem.

31.8 Summary

Radioactive materials can emit α-, β- or γ-radiation. The rate of this emission or decay does not depend upon temperature. The time taken for half of a radioactive sample to decay is called its half life. Half lives can vary between a fraction of a second and millions of years.

Emission of α- and β-particles changes the number of protons or neutrons in an atom, and therefore changes the isotope.

Radioactive isotopes have a wide range of uses. Apart from the production of atomic bombs and generation of electricity in a nuclear power station, there are many other industrial, medical and scientific uses.

32 QUALITATIVE ANALYSIS

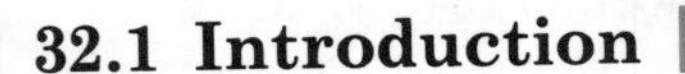

32.1 Introduction

Being able to identify simple inorganic substances by carrying out simple chemical reactions is an important skill that you should master in your GCSE Chemistry course. Often questions are set on this work on theory papers. This section will therefore be useful to you in preparing for theory and practical parts of your examination.

32.2 Tests for Anions (negative ions)

Ions marked * are not usually required but are included in case they are needed by you in practical situations.

CARBONATE (CO_3^{2-})

When dilute hydrochloric acid is added to a carbonate, carbon dioxide gas is produced. No heat is required. The carbon dioxide turns limewater milky.

E.g. $$Na_2CO_3(s) + 2HCl(aq) \rightarrow 2NaCl(aq) + H_2O(l) + CO_2(g)$$
sodium carbonate + hydrochloric acid → sodium chloride + water + carbon dioxide

HYDROGENCARBONATE (HCO_3^-)

Hydrogencarbonates behave in a similar way to carbonates with dilute hydrochloric acid.

When a solution of a hydrogencarbonate is heated, carbon dioxide is produced.

E.g. $$2NaHCO_3(aq) \rightarrow Na_2CO_3(aq) + H_2O(l) + CO_2(g)$$
sodium hydrogencarbonate → sodium carbonate + water + carbon dioxide

CHLORIDE (Cl^-)

When a chloride is treated with concentrated sulphuric acid, a colourless gas is produced. This gas (hydrogen chloride) forms steamy fumes in moist air and forms dense white fumes when mixed with ammonia gas.

$$NH_3(g) + HCl(g) \rightleftarrows NH_4Cl(s)$$
ammonia + hydrogen chloride ⇄ ammonium chloride

If a chloride is mixed with manganese(IV) oxide and concentrated sulphuric acid added, the chloride and acid react as above to give HCl. This reacts with the manganese(IV) oxide when the mixture is warmed and a greenish-yellow gas is produced. This gas is chlorine and it turns blue litmus red and then bleaches it.

$$MnO_2(s) + 4HCl(aq) \rightarrow MnCl_2(aq) + 2H_2O(l) + Cl_2(g)$$
manganese(IV) oxide + hydrochloric acid → manganese(II) chloride + water + chlorine

When a solution of a chloride is acidified with dilute nitric acid and silver nitrate solution added, a white precipitate of silver chloride is formed immediately. This precipitate turns purple in sunlight and dissolves completely in concentrated ammonia solution.

E.g. $$NaCl(aq) + AgNO_3(aq) \rightarrow AgCl(s) + NaNO_3(aq)$$
sodium chloride + silver nitrate → silver chloride + sodium nitrate

BROMIDE (Br^-)

When a solution of a bromide is acidified with dilute nitric acid and silver nitrate solution added, a creamish precipitate of silver bromide is formed immediately. This precipitate dissolves partially in concentrated ammonia solution.

E.g. $$NaBr(aq) + AgNO_3(aq) \rightarrow AgBr(s) + NaNO_3(aq)$$
sodium bromide + silver nitrate → silver bromide + sodium nitrate

IODIDE (I^-)

When a solution of an iodide is acidified with dilute nitric acid and silver nitrate solution added, a yellow precipitate of silver iodide is formed immediately. This precipitate is insoluble in ammonia solution.

E.g.
$$NaI(aq) + AgNO_3(aq) \rightarrow AgI(s) + NaNO_3(aq)$$
sodium iodide + silver nitrate → silver iodide + sodium nitrate

SULPHATE (SO_4^{2-})

When dilute hydrochloric acid and barium chloride solution are added to a solution of a sulphate, a white precipitate of barium sulphate is formed immediately.

E.g.
$$Na_2SO_4(aq) + BaCl_2(aq) \rightarrow BaSO_4(s) + 2NaCl(aq)$$
sodium sulphate + barium chloride → barium sulphate + sodium chloride

NITRATE (NO_3^-)

There are two tests that can be used to test for a nitrate in solution:

1 Sodium hydroxide solution is added to a suspected nitrate and aluminium powder is added. (Sometimes **Devarda's alloy** is used in place of aluminium. This is an alloy containing aluminium that reacts more slowly than pure aluminium.) The mixture is warmed and hydrogen is produced. If a suspected nitrate is added, it will be reduced to ammonia gas. This will turn red litmus paper blue.

$$3NO_3^-(aq) + 8Al(s) + 5OH^-(aq) + 2H_2O(l) \rightarrow 3NH_3(g) + 8AlO_2^-(aq)$$
nitrate ions + aluminium + hydroxide ions + water → ammonia + aluminate ions

2 An equal volume of iron(II) sulphate solution (acidified with dilute sulphuric acid) is added to a suspected nitrate in a test tube. Concentrated sulphuric acid is poured carefully down the inside of the test tube so that it forms a separate sulphuric acid layer below the aqueous layer. (This is because concentrated sulphuric acid is denser than water or aqueous solutions.) If a nitrate is present a brown ring forms at the junction of the two layers. The brown substance is $FeSO_4$. NO, produced by the reduction of nitrate to nitrogen monoxide by iron(II) ions.

$$NO_3^-(aq) + 4H^+(aq) + 3Fe^{2+}(aq) \rightarrow NO(g) + 3Fe^{3+}(aq) + 2H_2O(l)$$
nitrate ions + hydrogen ions + iron(II) ions → nitrogen monoxide + iron(III) ions + water

This is called the **brown ring test**. Nitrites and bromides can give similar results.

*NITRITE (NO_2^-)

When dilute hydrochloric acid is added to a nitrite, brown nitrogen dioxide gas is produced and the solution turns pale blue. No heat is required. The nitrogen dioxide turns blue litmus paper red but does not bleach it.

*SULPHIDE (S^{2-})

When dilute hydrochloric acid is added to a sulphide, colourless hydrogen sulphide gas is produced. The hydrogen sulphide smells of bad eggs and turns filter paper, soaked in lead nitrate solution, black.

$$Na_2S(s) + 2HCl(aq) \rightarrow 2NaCl(aq) + H_2S(g)$$
sodium sulphide + hydrochloric acid → sodium chloride + hydrogen sulphide

*SULPHITE (SO_3^{2-})

When dilute hydrochloric acid is added to a sulphite and the mixture is heated, colourless sulphur dioxide gas is produced. The sulphur dioxide has a pungent odour and turns potassium dichromate from orange to green. It does not change lead nitrate solution.

E.g.
$$Na_2SO_3(s) + 2HCl(aq) \rightarrow 2NaCl(aq) + SO_2(g) + H_2O(l)$$
sodium sulphite + hydrochloric acid → sodium chloride + sulphur dioxide + water

32.3 Tests for Cations (positive ions)

There are three tests that can be used to identify cations.

1 FLAME TESTS

A small quantity of the compound is taken and a couple of drops of concentrated hydrochloric acid are added. A clean piece of platinum wire is dipped into the mixture and put into a hot Bunsen burner flame. Certain cations colour the Bunsen flame. Common flame colours are shown in Table 32.1.

Table 32.1 Flame tests for identifying cations

Flame colour	*Cation*
Orange-yellow	Sodium Na^+
Lilac-pink	Potassium K^+
Brick red	Calcium Ca^{2+}
Pale green	Barium Ba^{2+}
Green	Copper(II) Cu^{2+}
Blue	Lead Pb^{2+}

2 WITH SODIUM HYDROXIDE SOLUTION

If a small quantity of the compound in solution is treated with sodium hydroxide solution an insoluble hydroxide may be precipitated. If a precipitate is formed it may re-dissolve in excess sodium hydroxide solution. A summary of the precipitation of metal hydroxides with sodium hydroxide solution is shown in Table 32.2.

Table 32.2 Precipitation of metal hydroxides with sodium hydroxide

	Addition of sodium hydroxide solution	
Cation	*A couple of drops*	*Excess*
Potassium K^+	No precipitate	No precipitate
Sodium Na^+	No precipitate	No precipitate
Calcium Ca^{2+}	White precipitate	Precipitate insoluble
Magnesium Mg^{2+}	White precipitate	Precipitate insoluble
Aluminium Al^{3+}	White precipitate	Precipitate soluble – colourless solution
Zinc Zn^{2+}	White precipitate	Precipitate soluble – colourless solution
Iron(II) Fe^{2+}	Green precipitate	Precipitate insoluble
Iron(III) Fe^{3+}	Red-brown precipitate	Precipitate insoluble
Lead Pb^{2+}	White precipitate	Precipitate soluble – colourless solution
Copper(II) Cu^{2+}	Blue precipitate	Precipitate insoluble
Silver Ag^+	Grey-brown precipitate	Precipitate insoluble

If no precipitate is formed, the solution is warmed. If the ammonium ion (NH_4^+) is present ammonia gas is produced, which turns red litmus blue.

E.g. $$NH_4Cl(aq) + NaOH(aq) \rightarrow NH_3(g) + NaCl(aq) + H_2O(g)$$
ammonium chloride + sodium hydroxide → ammonia + sodium chloride + water

3 WITH AQUEOUS AMMONIA SOLUTION (AMMONIUM HYDROXIDE)

If a small quantity of the compound in solution is treated with ammonia solution an insoluble hydroxide may be precipitated. If a precipitate is formed it may redissolve in excess ammonia solution. A summary is shown in Table 32.3.

Table 32.3 Precipitation of metal hydroxides with ammonia solution

	Addition of ammonia solution	
Cation	*A couple of drops*	*Excess*
Potassium	No precipitate	No precipitate
Sodium	No precipitate	No precipitate
Calcium	No precipitate	No precipitate
Magnesium	White precipitate	Precipitate insoluble
Aluminium	White precipitate	Precipitate insoluble
Zinc	White precipitate	Precipitate soluble – colourless solution
Iron(II)	Green precipitate	Precipitate insoluble
Iron(III)	Red-brown precipitate	Precipitate insoluble
Lead	White precipitate	Precipitate insoluble
Copper	Blue precipitate	Precipitate soluble – blue solution
Silver	Brown precipitate	Precipitate soluble

32.4 Testing for Gases

Table 32.4 summarizes the tests for common gases.

Table 32.4 Summary of properties of common gases

Gas	*Formula*	*Colour*	*Smell*	*Test with moist litmus*	*Test with lighted splint*	*Other tests*
Hydrogen	H_2	×	×	×	Squeaky pop splint extinguished	
Oxygen	O_2	×	×	×	Relights glowing splint	
Nitrogen	N_2	×	×	×	Extinguished	Forms compound with magnesium
Chlorine	Cl_2	Greenish-yellow	✓	Blue → red then bleaches	Extinguished	
Hydrogen chloride	HCl	×	✓	Blue → red	Extinguished	White fumes with ammonia
Carbon dioxide	CO_2	×	×	Little change	Extinguished	Turns limewater milky
Carbon monoxide	CO	×	×	×	Burns with blue flame	
Ammonia	NH_3	×	✓	Red → blue	Extinguished	White fumes with hydrogen chloride
Sulphur dioxide	SO_2	×	✓	Blue → red	Extinguished	Turns potassium dichromate green. No effect on lead nitrate
Hydrogen sulphide	H_2S	×	✓	Blue → slightly red	Burns producing sulphur dioxide	Lead nitrate paper turns black
Nitrogen dioxide	NO_2	Brown	✓	Blue → red	Extinguished	
Dinitrogen monoxide	N_2O	×	✓	×	Relights glowing splint	Quite soluble
Nitrogen monoxide	NO	×	—	—	—	Forms brown fumes in air

32.5 Summary

There are certain tests for anions that you should be able to do. Tests for carbonate (with dilute hydrochloric acid), sulphate (with barium chloride), chloride (with silver nitrate) and nitrate (by one of the two tests given) are especially important.

Most of the common cations you will meet can be identified using a flame test. Sodium hydroxide and ammonia solutions can be used to identify certain cations.

SELF TEST UNITS

Unit 1 States of Matter

1 When a substance changes from a solid to a liquid it is said to change its _______.

2 When a liquid changes to a gas it _______. This process is fastest at the _______ _______ of the liquid.

3 Sublimation is when a solid changes directly to a _______ without changing to a _______ first.

4 In a solid the particles are very close together and can only _______ about a fixed position.

5 The particles in a liquid are free to _______.

6 The ease with which a gas can be compressed is an indication that the particles are _______ _______.

7 The automatic mixing of two gases or two liquids is called _______. This is caused by the _______ _______ of the particles.

8 When smoke is viewed under a microscope, the particles of smoke appear to be jostled around in all directions. This movement is called _______ _______.

9 Pure substances have definite _______ and _______ _______.

10 The molecules in a dense gas move _______ than the molecules in a less dense gas.

Unit 2 Separation Techniques in Chemistry

1 When a mixture of sand and salt is added to water, the salt _______ and the sand sinks to the bottom because it is _______ in water.

2 When a mixture is filtered, the substance that is left in the filter paper is called the _______.

3 A pure solvent can be obtained from a solution by _______.

4 _______ _______ is a method of separating a mixture of liquids. This method depends on the liquids having different _______ _______.

5 Liquids that do not mix are called _______.

For questions 6–10 choose the best method of separation.

6 Oil and water.

7 Alcohol and water.

8 Nitrogen from liquid air.

9 Sodium chloride and ammonium chloride.

10 The dyes in ink.

Unit 3 Elements, Mixtures and Compounds

1 A pure substance that cannot be split up into anything simpler is called an _______.

2 There are about a _______ different elements.

3 Name two elements that are liquid at room temperature and pressure.

4 Most elements are _______ at room temperature and pressure.

5 When two or more elements react and join together a _______ is formed.

6 Which elements do the following contain:

copper sulphate
carbon dioxide
sodium hydrogencarbonate
ammonium
nitrate
a carbohydrate

7 What does the formula tell you about a compound?

8 Write down the formulae of the following compounds:

magnesium oxide
copper(II) hydroxide
lead(II) chloride
iron(III) sulphate
ammonium sulphate

9 The properties of a compound are _______ from the properties of the elements that it is made from.

10 When a compound is made directly from the elements in it, the process is called _______.

Unit 4 Chemical Equations

1 Write a word and symbol equation for the reaction of magnesium ribbon with dilute sulphuric acid.

2 What 'state symbol' is used to indicate a solution in water?

3 Balance the following equations:

$$Na + H_2O \rightarrow NaOH + H_2$$
$$CuCO_3 + HCl \rightarrow CuCl_2 + CO_2 + H_2O$$
$$AgNO_3 + ZnCl_2 \rightarrow AgCl + Zn(NO_3)_2$$

4 What advantage does a balanced symbol equation have over a word equation?

5 Write the simplest ionic equations for the following:
(a) the reaction between an acid and an alkali;
(b) the precipitation of silver chloride;
(c) the oxidation of iron(II) ions to iron(III) ions by chlorine.

Unit 5 Hydrogen

1 What is the normal test for hydrogen?

2 Name two metals that will produce hydrogen from cold water.

3 Which chemicals are normally used to produce hydrogen in the laboratory?

4 Name a metal that will produce hydrogen when it reacts with an alkali, e.g. sodium hydroxide.

5 Give two industrial uses of hydrogen.

6 In the normal laboratory preparation of hydrogen, copper sulphate is often added to act as a _______. The hydrogen can be collected over water because it is _______ in water. It could also be collected by _______ _______ because it is _______ _______ _______.

7 Complete the following equation and then write a fully balanced symbol, equation for the reaction:

copper(II) oxide + hydrogen → _______ + _______

8 What is produced when hydrogen burns in air?

9 How would you show that hydrogen diffuses faster than air?

10 In electrolysis, at which electrode is hydrogen usually produced?

Unit 6 Metals and Nonmetals

1 Give three physical properties that are common to most metals.

2 Name one metal that is a liquid at room temperature.

3 Name a nonmetal that conducts electricity.

4 What do all metals have in common when they react?

5 In order for a metal to displace _______ from a dilute acid the metal must be above _______ in the _______ _______.

6 What is an alloy?

7 Name two alloys and give their compositions and uses.

8 Metals can be distinguished from nonmetals by considering their oxides. Metal oxides are usually ______ at room temperature, whereas nonmetal oxides are usually ______ or ______. If a metal oxide dissolves in water an ______ solution is formed. If an oxide of a nonmetal dissolves in water an ______ solution is formed. Most metal oxides are ______ in water.

9 Carbon and silicon have some properties of both metals and nonmetals. What is this type of element called?

10 Give two uses of the element copper and say on which properties of the copper these uses depend.

Unit 7 Atomic Structure and Bonding

1 Atoms are made up from three main particles: ______, ______ and ______. The small centre of the atom is called the ______ and contains ______ and ______. The ______ go round the centre in distinct energy levels or orbits.

2 The atomic number of an element is the number of ______ in an atom of the element.

3 An atom is electrically neutral because it has an equal number of ______ and ______.

4 What is an isotope?

5 Carbon has two stable isotopes, ${}^{12}_{6}C$ and ${}^{13}_{6}C$. In terms of protons, neutrons and electrons, describe atoms of these two isotopes.

(MEG specimen P. Science)

Use the following information to answer questions 6–10:

	Atomic number	*Mass number*
magnesium	12	24
calcium	20	40
fluorine	9	19
chlorine	17	35

6 What is the electron arrangement (configuration) of a calcium atom?

7 How many electrons will a magnesium *ion* have and how will they be arranged?

8 How many neutrons does an atom of fluorine have?

9 What do the electron arrangements of chlorine and fluorine have in common?

10 What is the charge on a chloride ion and how does this come about?

11 When an atom reacts it tries to obtain a ______ outer shell of ______. If the shell is nearly empty the atom ______ ______ and so becomes a positive ion. If the shell is nearly full the atom ______ ______ and becomes a negative ion.

12 Some atoms complete their shells by sharing electrons. This is called ______ bonding.

13 Draw a diagram to show how the electrons are arranged in a molecule of water.

14 What are the main properties of an ionic compound?

15 Give an example of:
(a) a giant structure of atoms; **(b)** a giant structure of ions; **(c)** a giant structure of molecules.

16 Why are metals good conductors of electricity?

17 What is meant by the term 'allotropy'?

18 Name the two allotropes of carbon.

Unit 8 The Reactivity Series of Metals

1 Name three metals that react with cold water.

2 Name a metal that reacts with dilute acid but does not react with cold water.

3 How is the stability of a metal compound related to the position of the metal in the reactivity series?

4 Which metal carbonates are *not* decomposed by heating?

5 Write a balanced symbol equation for the effect of heat on copper carbonate.

6 Which gas is always given off when a metal nitrate is heated?

7 If a piece of magnesium ribbon is dropped into a test tube containing copper(II) sulphate solution, the blue colour fades, the test tube becomes warm and a brown solid sinks to the bottom:

(a) Complete the equation for the reaction

$$Mg + CuSO_4 \rightarrow \_\_\_\_\_\_\_\_.$$

(b) Explain why the above reaction takes place.

(c) What does the fact that the test tube becomes warm tell you about the reaction?

(MEG specimen Science)

8 Write a balanced equation for the action of heat on silver nitrate.

9 Why do you think that gold has been known and used for many thousands of years whereas sodium has been known for only about two hundred years?

10 Sodium, zinc, lead, iron, aluminium, calcium, magnesium. Choose a metal from this list which:

(a) reacts very slowly with cold water but burns in steam;

(b) reacts violently with cold water to form a hydroxide type MOH;

(c) is used as a protective shield from radiation;

(d) forms an oxide whose formula is of the type M_2O_3;

(e) cannot reduce lead(II) nitrate to lead. (MEG specimen)

Unit 9 Chemical Families and the Periodic Table

1 Elements are grouped together because of their similar properties. Give three properties that lithium, sodium and potassium have in common.

2 Which is the most reactive of sodium, lithium and potassium?

3 Explain in terms of electron arrangement why fluorine is more reactive than chlorine.

4 Describe what you would see if a piece of potassium were put into a trough of water.

5 Why are the noble gases so unreactive?

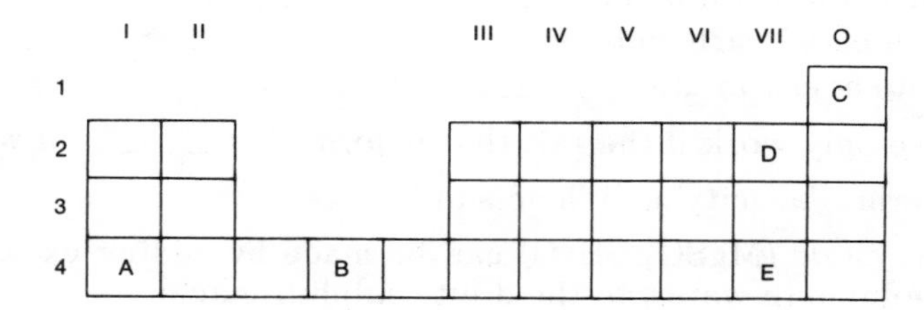

Use the skeleton Periodic Table to answer the following questions. Write down the letters for:

6 Two elements in the same group.

7 An alkali metal.

8 A noble gas.

9 A transition metal.

10 What type of bonding would you expect in a compound of A and D?

Unit 10 Oxidation and Reduction

1 Oxidation can be defined as the gaining of _______ or the loss of _______

2 Oxidation and _______ always go together.

3 A substance that brings about oxidation is called the _______ _______ and it is always _______ in the reaction.

4 In each of the following reactions say which substance is oxidized:

magnesium + oxygen → magnesium oxide

copper oxide + hydrogen → copper + water

methane + chlorine → carbon + hydrogen chloride

5 Define oxidation in terms of electron transfer.

6 Why is chlorine a good oxidizing agent?

7 Name three reducing agents.

8 Indicate in each of the following reactions which substance is reduced and which substance is oxidized.

$$\text{magnesium} + \text{copper oxide} \rightarrow \text{magnesium oxide} + \text{copper}$$
$$Cl_2 + 2I^- \rightarrow 2Cl^- + I_2$$
$$CO_2 + C \rightarrow 2CO$$

9 Which process always happens at the cathode in electrolysis?

10 Under what conditions does copper react with sulphuric acid and why is this an oxidation reaction?

Unit 11 Acids, Bases and Salts

1 An acid is a substance which turns litmus paper _______.

2 Before anything can act as an acid, _______ must be present. This allows the acid to _______ and so form a solution which contains a high concentration of _______ _______.

3 The strength of an acid is measured on the _______ scale. A strong acid has a _______ number of _______. A weak acid has a _______ number between _______ and _______.

4 A substance that is neither acid or alkali is said to be _______. An example of a _______ substance is water.

5 Explain why when hydrogen chloride gas is dissolved in water the temperature of the water increases and the resulting solution has acidic properties.

6 Explain why sulphuric acid is a strong acid whereas ethanoic acid is a weak acid.

7 Dilute acids have four general reactions. They are:

dilute acid + a fairly reactive metal → _______ + _______
dilute acid + a metal oxide → _______ + _______
dilute acid + a metal carbonate → _______ + _______ + _______ _______
dilute acid + an alkali → _______ + _______

8 The four reactions above only work if the salt that is formed is _______ in water.

9 What is meant by the terms 'basicity' and 'acid salt'?

10 Magnesium sulphate crystals ($MgSO_4.7H_2O$) can be made by adding excess magnesium oxide (MgO), which is insoluble in water, to the dilute sulphuric acid.

(a) Why was the magnesium oxide in excess?

(b) The following apparatus could be used to separate the excess magnesium oxide from the solution. Say what each of the letters on the diagram refer to.

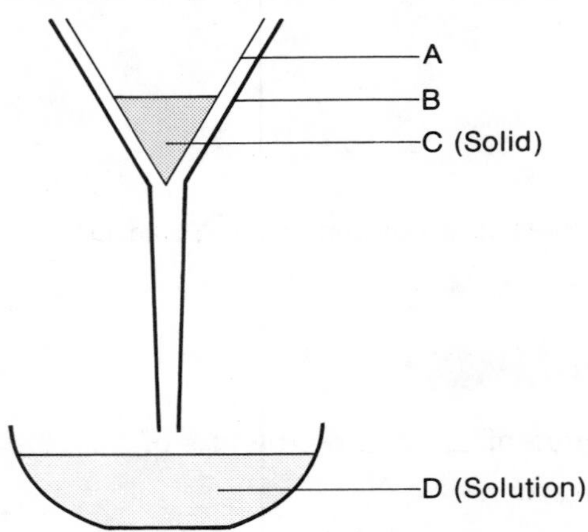

(c) Given the relative atomic masses: Ar(H) = 1, Ar(Mg) = 24, Ar(O) = 16, Ar(S) = 32, calculate the relative formula mass of:

magnesium oxide MgO;
magnesium sulphate crystals $MgSO_4.7H_2O$.

(d) Use your answer from **c** to calculate the maximum mass of magnesium sulphate crystals that could be obtained from 2.0 g of magnesium oxide.

(e) Describe how you could obtain pure dry crystals of magnesium sulphate from magnesium sulphate solution.

(LEAG specimen)

11 Name three insoluble salts.

12 How would you prepare an insoluble salt? Choose an example and write a word and symbol equation for the reaction.

Unit 12 The Effect of Electricity on Chemicals

1 What is an electrolyte?

2 Name three classes of substances that are electrolytes.

3 Solid sodium chloride will not conduct electricity. This is because the ions are not _______ _______ _______. When sodium chloride is in aqueous solution or when it is _______ it will conduct because of the ions can now _______.

4 In any electrolysis the positive ions are attracted to the negative electrode, which is called the _______, and the negative ions are attracted to the positive electrode, which is called _______.

5 In the electrolysis of molten lead(II) bromide, _______ is produced at the negative electrode and _______ is produced at the positive electrode.

6 In the electrolysis of an aqueous solution, there will always be a few _______ ions and _______ ions from the water.

7 Write down the equations for the electrode reactions in the electrolysis of aqueous sulphuric acid.

8 Metals _______ _______ in the reactivity series are *not* usually produced in the electrolysis of an aqueous solution; _______ is produced instead.

9 Give three industrial uses of electrolysis.

10 If a current of 5 A is passed through molten calcium bromide $CaBr_2$ for 3 minutes 13 seconds, calculate;

(a) the number of coulombs of electricity passed;
(b) the number of Faradays passed (1 F = 96 500 coulombs);
(c) the mass of calcium deposited at the negative electrode;
(d) the mass of bromine given off at the positive electrode.

(Ar(Ca) = 40, Ar(Br) = 80)

Unit 13 Rates of Reaction

1 Explain what has to happen in order that two substances might react together.

2 What is meant by the term 'activation energy'?

3 Name three observable changes that you could use to follow the rate of a reaction.

4 What is a catalyst? Give two examples of industrial processes that use a catalyst. Say what is reacting, what is produced and what the catalyst is.

5 An experiment was carried out to investigate the rate of reaction between magnesium and dilute sulphuric acid. 0.07 g of magnesium ribbon were reacted with *excess* dilute sulphuric acid.

The volume of gas produced was noted every 5 seconds. The following results were obtained.

Time (s)	*Volume (cm^3)*	*Time (s)*	*Volume (cm^3)*
0	0	25	63
5	18	30	67
10	34	35	69
15	47	40	70
20	57	45	70

(a) On a piece of graph paper, plot these results with the volume as the vertical axis and time as the horizontal axis. Draw a smooth curve through the points.
(b) When is the reaction fastest?
(c) How long does it take for the 0.07 g of magnesium to react completely?
(d) At what time was 0.02 g of magnesium *left unreacted*?

(e) On your graph, sketch a curve that might have been obtained if 0.07 g of magnesium powder had been used instead of the magnesium ribbon.

(f) Give two other changes that would alter the rate of this reaction.

(g) Copy and complete the equation for the reaction.

$$Mg + H_2SO_4 \rightarrow \_\_\_\_\_\_$$

(h) How may the gas produced be identified?

(MEG specimen Science)

Unit 14 Reversible Reactions and Equilibrium

1 What is meant by a 'reversible reaction'?

2 For a reversible reaction in a closed vessel and under fixed conditions, a chemical _______ will be set up. When this happens the reaction will appear to _______. In fact this is not the case; the rate of the forward reaction is _______ _______ the rate of the reverse reaction.

3 What happens to reversible reaction if the conditions are altered?

4 What effect does a catalyst have on an equilibrium reaction?

5 Explain why a high pressure favours the formation of ammonia in the following reversible reaction.

$$N_2 + 3H_2 \rightleftharpoons 2NH_3$$

Unit 15 Energy Changes in Chemistry

1 What is meant by the term 'exothermic reaction'?

2 Give an example of an exothermic reaction.

3 Draw an energy level diagram for the combustion of methane. The heat of combustion of methane $CH_4 = -890$ kJ/mol.

4 Two grams of a liquid fuel were burnt using the apparatus shown in Fig. 15.4. The temperature of 100 g of water in the can rose by 50 °C. Assuming that all the heat from the burning fuel was transferred to the water, calculate (specific heat capacity of water = 4.2 J/g/°C):

(a) the heat produced by the 2 g of fuel; **(b)** the heat produced by 1 g of fuel;

(c) the heat produced by 1 mol of the fuel (relative molecular mass of the fuel = 50).

5 If simple cells are set up between the following pairs of metals:

(a) copper and zinc; **(b)** copper and copper; **(c)** copper and magnesium;

which one would produce the greatest potential difference and which one would not produce any potential difference?

Unit 16 Social, Economic and Environmental Considerations

Normally questions on this section will be included in questions on most other sections of the syllabus. Below are a few general considerations on the topic.

1 What are the main things you would consider when deciding where to site a new oil refinery?

2 Why are several of the United Kingdom's aluminium extraction plants situated in Scotland?

3 What is meant by 'eutrophication'?

4 What gas is the main cause of 'acid rain'?

5 Why is recycling an important process?

Unit 17 Oxygen and the Air

1 What is the approximate percentage of oxygen in the air?

2 Oxygen is obtained industrially by the _______ _______ of liquid air. The liquid air can be separated in this way because it is a _______ and the different components have different _______ _______.

3 One of the easiest ways to make oxygen in the laboratory is by the decomposition of a solution of _______ _______. A catalyst of _______ _______ is normally used. The oxygen can be collected _______ _______ because it is only slightly _______ _______ _______.

4 List three large scale uses of oxygen.

5 What is photosynthesis?

6 Although respiration and burning use up oxygen, the approximate percentage of oxygen in the air stays the same. Why is this?

7 Name three gases that pollute the air.

8 Why is there an increasing demand for 'lead free' petrol?

9 What are the main effects of acid rain?

10 How could car exhaust gases be 'cleaned up'?

Unit 18 Chalk, Limestone and Marble

1 Chalk, limestone and marble are all forms of _______ _______.

2 When a piece of marble was heated strongly, _______ _______ was given off. The shiny surface of the marble disappeared and the hard marble turned _______.

After the solid had cooled down, a few drops of water were added to it. The solid seemed to expand and steam was given off showing that the reaction produced _______. When the solid was tested with Universal Indicator paper, the paper turned _______ showing that the substance was an _______.

3 What is the chemical name for limewater?

4 Describe what you would see when carbon dioxide is bubbled through limewater until there is no further change. Write down equations for any reactions taking place.

5 Which substance causes 'temporary hard water'?

6 How are stalactites and stalagmites formed?

7 Which substances are normally used to prepare carbon dioxide in the laboratory? Why is sulphuric acid not used?

8 Give two uses of carbon dioxide.

9 For what purpose is limestone used in the extraction of iron?

10 Name two other large scale uses of limestone.

Unit 19 Water

1 What is meant by saying that 'water is a good solvent'?

2 In general the solubility of a solid _______ with temperature but gases are _______ soluble in hot water.

3 The solubility of a substance is the _______ of the solute that will dissolve in _______ of water at a given _______.

4 What is a saturated solution and how would you recognize one?

5 What is meant by hard water?

6 What is the difference between 'temporary hard water' and 'permanent hard water'?

7 Give two ways that 'permanent hard water' can be softened.

8 What is a detergent?

9 Give one advantage of a soapless detergent compared with soap.

10 Name two metals that react with cold water.

11 How does magnesium react with water?

12 What is 'water of crystallization'?

13 When copper(II) sulphate crystals are heated gently, they soon lose their _______ colour and turn _______. As they do this they give off _______. The form of copper sulphate that is left is called _______, which means _______ _______. If a few drops of water are added to the solid it becomes very _______ and turns _______ again. This reaction can be used to test a liquid to see if it contains water. Another substance that can be used to test for water in a similar way is _______ _______. The _______ crystals turn _______ when the water is driven off.

14 What is meant by the terms 'efflorescent' and 'deliquescent'? Give an example for each case.

15 What does 'biodegradable' mean?

16 Why is it harmful to release large amounts of warm water from power stations into rivers?

17 How would you show that a liquid was pure water?

18 What effect do high concentrations of nitrates have on ponds and streams?

Unit 20 Chemicals from Petroleum

1 Petroleum is a complex mixture of ______. These are compounds of ______ and ______.

2 Petroleum is a fossil fuel and was formed ______ of years ago from dead ______ and ______.

3 What are the main characteristics of an homologous series?

4 Write down the names, formulae and structural formulae of the first three alkanes.

5 What is the difference between a saturated and an unsaturated hydrocarbon?

6 What is an isomer? Write down and name three isomers which have the molecular formula C_5H_{12}.

7 Give an example of an addition reaction with an alkene.

8 What is meant by 'cracking' and why is it so important?

9 A ______ is a large molecule made by linking many similar units together. There are two types of ______: addition ______ and ______ ______.
In addition ______ there is only one simple unit or ______ which joins together. Nothing else is formed. These units are often alkenes.
In ______ ______ two different types of unit each with reactive end groups link together with the elimination of a small molecule, often ______.
Two examples of addition ______ are polythene and ______.

10 Give two advantages and two disadvantages of addition polymers.

Unit 21 Ethanol and Ethanoic Acid

1 Write out the structural formula for ethanol.

2 What is fermentation?

3 How can ethanol be obtained from a mixture of ethanol and water?

4 What is the starting material in the industrial production of ethanol?

5 Apart from alcoholic drinks, what is ethanol used for?

6 Write the equation for the burning of ethanol.

7 Ethanol can be oxidized by the air or by a good oxidizing agent such as ______ to form ______ ______. This is a weak monobasic acid.

8 An ester is the compound formed when an organic ______ reacts with an ______. An example of an ester is ______ ______.

9 Name two examples of condensation polymers.

10 What is a carbohydrate? Give two examples of carbohydrates and say why carbohydrates are important foods.

Unit 22 Fuels

1 What is a fuel?

2 What characteristics do you think that a good fuel should have?

3 Give two examples of fossil fuels.

4 What are the main products of the destructive distillation of coal?

5 What is the main gas present in the North Sea gas?

Unit 23 Extraction of Metals

1 A rock that contains a metal is called an ______.

2 In the extraction of any metal from one of its compounds, the compound has to be ______.

3 Metals at the top of the reactivity series are more ______ to obtain from their compounds.

4 Aluminium is extracted from ______, which is a form of aluminium oxide. Aluminium oxide has a very high ______ ______ and so for this electrolysis it is dissolved in molten ______. The anode is made of large blocks of ______ which have to be frequently replaced because the oxygen produced at the anode reacts with it to form ______ ______.

5 The main chemicals added to the blast furnace are ______, ______ and ______. The reducing agent in the blast furnace is ______ ______. The molten iron produced runs to the bottom of the furnace and is then covered with a layer of ______. This contains several impurities but is mainly ______ ______.

6 Most of the iron produced by the blast furnace is converted into ______.

7 Which metal is used to reduce titanium chloride in order to obtain the titanium?

8 From which compound is sodium extracted?

9 Give the common name for an ore of iron.

10 Why is the recycling of copper so important?

Unit 24 Corrosion of Metals

1 Why are sodium and potassium stored in oil?

2 Although aluminium is quite high in the reactivity series it does not appear to corrode very quickly and is a very useful construction metal. How can you account for its apparent lack of reactivity?

3 What are the conditions needed for iron to rust?

4 Give three ways that we try to prevent iron from rusting.

5 What is meant by 'sacrificial corrosion'? Give an example of when it is used.

6 The thickness of the layer of aluminium oxide on the surface of the aluminium can be increased by electrolysis. This process is called ______.

7 The green colour on a copper roof is a form of ______ ______.

Unit 25 Ammonia and Nitric Acid

1 Ammonia is a compound of ______ and ______.

2 The industrial manufacture of ammonia is a synthesis reaction, i.e. the compound is made from ______ ______.

3 The industrial process is called the ______ process.

4 The catalyst in this reaction is ______.

5 Two uses of ammonia are in the manufacture of ______ and of ______.

6 The normal laboratory preparation of ammonia is to heat an ______ ______ with an ______.

Two suitable chemicals are ______ ______ and ______ ______.

7 Ammonia is collected by ______ ______ because it is ______ than air. It cannot be collected over water because it is ______ ______.

8 Ammonia can be identified by its strong smell and by its effect on Universal Indicator. It is the only common ______ gas. A good test for ammonia is using the stopper from the concentrated hydrochloric acid bottle. In the presence of ammonia, dense ______ fumes of ______ ______ are formed.

9 What would you see if ammonia solution was added a drop at a time until it was in excess, to a solution of copper(II) sulphate?

10 Write the word equation for the reaction of ammonia with copper(II) oxide. How is the ammonia reacting?

11 Ammonia is converted to nitric acid by catalytic ______. The catalyst in this reaction is ______.

12 Nitric acid is produced in the laboratory by heating a mixture of potassium nitrate and ______ ______ ______. The apparatus is usually all ______ because the freshly prepared nitric acid attacks rubber and cork.

13 Hot concentrated nitric acid is a good _______ _______. It changes copper to _______ _______, carbon to _______ _______ and iron(II) to _______ _______.

14 Which of the main reactions of a dilute acid does not work very well with nitric acid?

15 Give two industrial uses of nitric acid.

16 All metal nitrates decompose on heating and give off _______. Some nitrates give off a brown gas as well. This is _______ _______.

17 All nitrates are _______ in water.

Unit 26 Feeding the World

1 In what form do most plants take in nitrogen?

2 Iron, boron and copper are required by plants in very small quantities. These are called _______ elements.

3 Name a natural fertilizer that contains nitrogen.

4 Which group of plants have bacteria living in their roots which enable them to make use of atmospheric nitrogen?

5 Calculate the percentage of nitrogen in ammonium sulphate $(NH_4)_2SO_4$.

6 Give an example of a fertilizer that contains phosphorus.

7 Why do plants need potassium?

8 What is meant by a NPK fertilizer?

9 Why must lime and an ammonium salt not be added to the soil at the same time?

10 Farmers use large amounts of nitrogen-containing fertilizers. Some of these fertilizers are washed off the farm land by rain into streams and lakes. Also present in streams and lakes are phosphates from domestic detergents. The dissolved fertilizers and phosphates increase the amount of chemicals needed by plants in the water. The surface of the water then becomes covered with algae. Because of this the plants below the surface die. When the algae decay, the amount of dissolved oxygen in the water is lowered.

(a) Give the chemical name of a fertilizer which contains nitrogen.
(b) Why are fertilizers used on farms?
(c) Name an element other than nitrogen which is needed for plant growth.
(d) How do phosphates get into the water system?
(e) What would be the effect of lowering the amount of dissolved oxygen in streams and lakes?
(MEG specimen)

Unit 27 Sulphur and Sulphuric Acid

1 There are various forms of sulphur. These include: α-sulphur, β-sulphur and plastic sulphur. Which of these forms:
(a) does not consist of an arrangement of S_8 rings?
(b) is most stable at room temperature?
(c) is formed when molten sulphur is cooled rapidly by pouring into cold water?
(d) is formed when a solution of sulphur evaporates at 40 °C?

2 Sulphur is extracted from underground deposits by the _______ process. A hole is drilled down to the deposits and a pump is placed in the hole. Super-heated _______ at 170 °C is pumped down the _______ tube to _______ the sulphur. Hot compressed _______ is pumped down the middle pipe to force the sulphur to the surface.

3 A lot of sulphur is now extracted from natural gas. The compound in the gas that contains the sulphur is _______ _______. In order to obtain the sulphur, the _______ _______ is oxidized using a catalyst of _______.

4 Give two uses of sulphur.

5 Sulphur dioxide can be prepared in the laboratory by reacting copper with _______ _______ _______. The gas is usually collected by _______ _______. Sulphur dioxide is very soluble in water as can be demonstrated by the _______ experiment. When sulphur dioxide dissolves in water _______ _______ is formed. This is a weak dibasic acid. The salts of this acid are _______ and _______.

6 Sulphur dioxide is a good _______ agent, and when it acts in this way it is always oxidized to _______ _______.

7 Sulphuric acid is manufactured by the _______ process. The main stage is the _______ of sulphur dioxide to _______ _______. A catalyst of _______ is used. The product of the reaction is not dissolved directly in water because the reaction is too _______. Instead it is dissolved in _______ _______ _______ to form _______. This can then be carefully reacted with water to produce the sulphuric acid.

8 When concentrated sulphuric acid is added to copper sulphate crystals, the crystal lose their _______ colour and turn _______. This is because the sulphuric acid is removing _______ from the copper sulphate crystals. The sulphuric acid is acting as a _______ agent.

9 If concentrated sulphuric acid is added to sodium chloride crystals a steamy gas is given off. This is _______ _______.

10 Concentrated sulphuric acid reacts with sugar by first removing _______ from the carbohydrate material and then oxidizing the carbon produced to _______ _______.

Unit 28 Salt and Chemicals from Salt

1 A strong solution of salt is called _______. This solution can be electrolysed using a Diaphragm cell or a Mercury cathode cell. In both cases the final three products are _______ gas, _______ gas and _______ _______ solution.

2 Write the equation for the reaction at the anode in the electrolysis of salt solution.

3 Give three uses of chlorine.

4 What is the main chemical in 'household bleach'?

5 Give two uses of sodium hydroxide.

6 What is manufactured by the 'Solvay process'?

Unit 29 The Mole and Chemical Calculations

1 Write down the numbers of moles of atoms in the following (see Fig. 9.2 on page 28 for relative atomic masses):

(a) 60 g of carbon;
(b) 64 g of oxygen;
(c) 100 g of calcium;
(d) 64 g of sulphur;
(e) 39 g of potassium;
(f) 12 g of magnesium;
(g) 3 g of aluminium;
(h) 4 g of iron;
(i) 1 g of mercury;
(j) 0.1 g of hydrogen.

2 What is the mass of the following:
(a) 6 moles of nitrogen atoms; **(b)** 5 moles of lead atoms;
(c) 3 moles of aluminium atoms; **(d)** $\frac{1}{2}$ mole of magnesium atoms;
(e) 0.01 moles of iron atoms?

3 Calculate the formula of an oxide of sulphur where 2 g of sulphur join with 3 g of oxygen.

4 Calculate the formula of a compound containing 28 per cent iron, 24 per cent sulphur and 48 per cent oxygen by mass.

5 What is the percentage of water of crystallization in copper sulphate $CuSO_4.5H_2O$?

6 The formula of potassium hydrogencarbonate is $KHCO_3$.
(a) What is the mass of 1 mole of potassium hydrogencarbonate?
(b) What is the percentage of oxygen in potassium hydrogencarbonate?

7 How many times heavier is one atom of magnesium than one atom of carbon?

8 What are the masses of the following:
(a) 10 moles of water H_2O; **(b)** 0.5 moles of ammonium nitrate NH_4NO_3;
(c) 2 moles of ethanol C_2H_5OH; **(d)** 0.01 moles of lead nitrate $Pb(NO_3)_2$.

9
$$CaCO_3 + 2HCl \rightarrow CaCl_2 + CO_2 + H_2O$$
(a) What mass of calcium chloride would be produced if 10 g of calcium carbonate were reacted with excess acid?

(b) What volume of carbon dioxide would be produced if 5 g of calcium carbonate were reacted with excess acid? (1 mole of a gas has a volume of about 24 dm^3)

10
$$2C_2H_6 + 7O_2 \rightarrow 4CO_2 + 6H_2O$$
(a) What volume of oxygen is needed to react with 20 cm^3 of ethane?

(b) What volume of carbon dioxide would be produced by the complete combustion of 20 cm^3 of ethane?

Unit 30 Quantitative Volumetric Chemistry

1 Which piece of apparatus is usually used to measure exactly 25 cm^3 of liquid into a flask?

2 Which piece of apparatus is usually used to add the acid in a titration?

3 What is a molar solution?

4 What mass of sulphuric acid (H_2SO_4) is needed to make 25 cm^3 of 2 M acid?

5

$$HCl + NaOH \rightarrow NaCl + H_2O$$

25 cm^3 of 0.1 M sodium hydroxide were titrated with 0.5 M hydrochloric acid.

(a) What mass of sodium hydroxide is needed to make 1 dm^3 of 0.1 M sodium hydroxide solution?

(b) How many moles of sodium hydroxide are present in 25 cm^3 of 0.1 M solution?

(c) How many moles of hydrochloric acid will react with this amount of sodium hydroxide?

(d) What volume of 0.5 M acid would this be?

Unit 31 Radioactivity

1 Radioactivity originates in the _______ of an atom.

2 What are the three types of radiation?

3 Which form of radiation is the most penetrating?

4 What is meant by the half life of an isotope?

5 Give two uses of radioactivity.

6 Why is the disposal of radioactive waste such a problem?

Unit 32 Qualitative Analysis

1 What are the flame test colours for:
(a) sodium compounds; **(b)** potassium compounds; **(c)** calcium compounds.

2 What are the colours of:
(a) copper(II) hydroxide; **(b)** iron(II) hydroxide; **(c)** iron(III) hydroxide.

3 Name two metal hydroxides that are insoluble in water, but that will dissolve in excess sodium hydroxide solution.

4 Name a substance that gives off carbon dioxide when it is reacted with a dilute acid.

5 What sort of substances produce a white precipitate if solutions of silver nitrate and nitric acid are added?

6 A white solid gives a lilac flame test and is soluble in water. When barium chloride and hydrochloric acid are added to the solution, a white precipitate is formed. What is the original white solid?

7 What is the usual test for chlorine gas?

8 Which gas relights a glowing splint?

9 Which is the lightest of all gases?

10 Is carbon dioxide 'heavier' or 'lighter' than air?

Unit Test Answers

UNIT 1 STATES OF MATTER

1 State.
2 Evaporates; boiling point.
3 Gas; liquid.
4 Vibrate.
5 Move.
6 Widely spaced.
7 Diffusion; random motion.
8 Brownian motion.
9 Melting; boiling points.
10 Faster.

UNIT 2 SEPARATION TECHNIQUES IN CHEMISTRY

1 Dissolves; insoluble.
2 Residue.
3 Distillation.
4 Fractional distillation; boiling points.
5 Immiscible.
6 Separating funnel.
7 Fractional distillation.
8 Fractional distillation.
9 Sublimation.
10 Chromatography.

UNIT 3 ELEMENTS, MIXTURES AND COMPOUNDS

1 Element.
2 100.
3 Mercury; bromine.
4 Solid.
5 Compound.
6 Copper, sulphur and oxygen
carbon and oxygen
sodium, hydrogen, carbon and oxygen
nitrogen, hydrogen and oxygen
carbon, hydrogen and oxygen.
7 It tells you which elements it contains and how much of each element.
8 MgO, $PbCl_2$, $(NH_4)_2SO_4$, $Cu(OH)_2$, $Fe_2(SO_4)_3$.
9 Different.
10 Synthesis.

UNIT 4 CHEMICAL EQUATIONS

1 Magnesium + dilute sulphuric acid → magnesium sulphate + hydrogen.

$$Mg + H_2SO_4 \rightarrow MgSO_4 + H_2$$

2 (aq).
3 $2Na + 2H_2O \rightarrow 2NaOH + H_2$
$CuCO_3 + 2HCl \rightarrow CuCl_2 + CO_2 + H_2O$
$2AgNO_3 + ZnCl_2 \rightarrow 2AgCl + Zn(NO_3)_2$.
4 It tells you how much of each substance is reacting.
5(a) $H^+ + OH^- \rightarrow H_2O$ **(b)** $Ag^+ + Cl^- \rightarrow AgCl$ **(c)** $2Fe^{2+} + Cl_2 \rightarrow 2Fe^{3+} + 2Cl^-$.

UNIT 5 HYDROGEN

1 It 'pops' with a lighted splint.
2 Potassium, sodium, lithium, calcium.
3 Zinc and dilute sulphuric acid.
4 Aluminium, zinc.

5 Manufacture of margarine; manufacture of ammonia.
6 Catalyst; insoluble; upward delivery; lighter than air.
7 Copper(II) oxide + hydrogen → copper + water.

$$CuO + H_2 \rightarrow Cu + H_2O.$$

8 Water.
9 Fill a porous pot with hydrogen and connect it to a manometer tube. The pressure inside the pot will go down as the hydrogen diffuses out faster than the air diffuses in.
10 Cathode.

UNIT 6 METALS AND NONMETALS

1 Shiny, high density, high melting point, bend, good conductors of heat and electricity, strong.
2 Mercury.
3 Carbon.
4 They lose electrons and so form positive ions.
5 Hydrogen; hydrogen; reactivity series.
6 It is a metal made by mixing two or more metals together.
7 Solder, steel, duralumin, brass and bronze (see Table 6.3).
8 Solids; gases; liquids; alkaline; acidic; insoluble.
9 Metalloid.
10 Coins – hard wearing; wires – very good conductor; pipes – easily shaped; and saucepans – good conductor of heat.

UNIT 7 ATOMIC STRUCTURE AND BONDING

1 Protons; neutrons; electrons; nucleus; protons; neutrons; electrons.
2 Protons.
3 Protons; electrons.
4 Isotopes are atoms of the same element with different numbers of neutrons.
5 Carbon 12 has 6 protons, 6 neutrons and 6 electrons;
carbon 13 has 6 protons, 7 neutrons and 6 electrons.
6 2.8.8.2.
7 10; 2.8.
8 10.
9 Seven electrons in their outer shell.
10 When a chlorine atom forms an ion it gains one electron. It will then have a 1− charge.
11 Full; electrons; loses electrons; gains electrons.
12 Covalent.

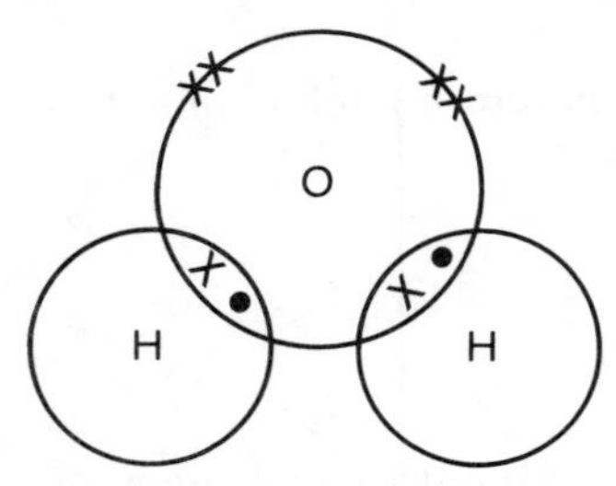

13 A molecule of water

14 High melting point crystalline solid, usually soluble in water, conducts electricity when molten or in aqueous solution.
15(a) A metal, diamond or graphite.
(b) Sodium chloride, magnesium oxide or any other ionic compound.
(c) Sulphur crystals, iodine crystals or a polymer.
16 A metal has the metal ions in a 'sea' of electrons; electrons can easily move through the structure.
17 Allotropy is when an element can exist in two or more forms in the same state.
18 Diamond and graphite.

UNIT 8 THE REACTIVITY SERIES OF METALS

1 Potassium, sodium, lithium, calcium.

2 Magnesium, zinc, iron.
3 The higher up the reactivity series the metal is, the more stable its compounds are.
4 Potassium and sodium.
5 $CuCO_3 \rightarrow CuO + CO_2$.
6 Oxygen.
7(a) $Mg + CuSO_4 \rightarrow MgSO_4 + Cu$.
(b) The reaction takes place because magnesium is higher up the reactivity series than copper.
(c) The reaction produces heat, i.e. it is exothermic.
8 $2AgNO_3 \rightarrow 2Ag + 2NO_2 + O_2$.
9 Gold is very unreactive and can be found uncombined in the ground. Also gold is easy to obtain from its compounds. Sodium is very reactive. It is never found uncombined in the ground and it is very difficult to extract from its compounds.
10(a) Magnesium. **(b)** Sodium. **(c)** Lead. **(d)** Aluminium or iron. **(e)** Lead or iron.

UNIT 9 CHEMICAL FAMILIES AND THE PERIODIC TABLE

1 React with water, soft, metals, stored in oil, float in water, low melting points for metals.
2 Potassium.
3 Both want to gain one electron; in the case of fluorine this extra electron is in the second shell and so is closer to the positive attraction of the nucleus (electron arrangements are fluorine 2.7; chlorine 2.8.7).
4 A small piece of potassium would float on the surface of the water; it would melt into a ball and skim across the water; it would hiss and usually the hydrogen given off would burn with a lilac flame; the potassium would disappear into the water.
5 Noble gases are unreactive because they already have full shells of electrons, i.e. a stable electron arrangement.
6 D and E.
7 A.
8 C.
9 B.
10 Ionic.

UNIT 10 OXIDATION AND REDUCTION

1 Oxygen; hydrogen.
2 Reduction.
3 Oxidizing agent; reduced.
4 Magnesium; hydrogen; methane.
5 Oxidation is the loss of electrons.
6 Because it likes to gain electrons.
7 Hydrogen, carbon, carbon monoxide, sulphur dioxide, hydrogen sulphide, metals.
8 Magnesium is oxidized: copper oxide is reduced; iodide ions are oxidized: chlorine is reduced; carbon is oxidized: carbon dioxide is reduced.
9 Reduction.
10 The acid must be hot and concentrated; the copper loses electrons.

UNIT 11 ACIDS, BASES AND SALTS

1 Red.
2 Water; ionize; hydrogen ions.
3 pH; pH; 1; pH; 3; 6.
4 Neutral; neutral.
5 The hydrogen chloride changes from a covalent compound to hydrogen ions and chloride ions; energy is given out as the ions are formed; the hydrogen ions give the solution its acidic properties.
6 Sulphuric acid completely ionizes in water but ethanoic acid only partially ionizes in water.
7 Acid + metal → salt + hydrogen
acid + metal oxide → salt + water
acid + metal carbonate → salt + water + carbon dioxide
acid + alkali → salt + water.
8 Soluble.
9 Basicity is the number of replaceable hydrogen atoms in one molecule of the acid. An acid salt is a salt where only part of the replaceable hydrogen of an acid has been replaced.

10(a) To make sure that all the acid was used up.

(b) A: filter paper; B: filter funnel; C: residue (magnesium oxide); D: filtrate (magnesium sulphate solution).

(c) $MgO = 40$; $MgSO_4.7H_2O = 246$.

(d) 12.3 g.

(e) Evaporate about half the solution, leave it to cool and to evaporate further at room temperature, filter the crystals, rinse with a little water and leave to dry.

11 Any carbonates except sodium, potassium or ammonium; lead, silver or mercury (I) chlorides; lead or barium sulphates.

12 An insoluble salt is prepared by mixing together two soluble salts each containing 'half' the required insoluble salt. e.g.
silver nitrate + sodium chloride → silver chloride + sodium nitrate

$$AgNO_3 + NaCl \rightarrow AgCl + NaNO_3$$

UNIT 12 THE EFFECT OF ELECTRICITY ON CHEMICALS

1 An electrolyte is a compound that conducts electricity when it is molten or in aqueous solution and is decomposed by the electricity.

2 Acids, bases and salts.

3 Free to move; molten; move.

4 Cathode; anode.

5 Lead; bromine.

6 Hydrogen; hydroxide.

7 Cathode equation: $H^+ + e^- \rightarrow H$, $2H \rightarrow H_2$
anode equation: $2OH^- - 2e^- \rightarrow H_2O + O$, $2O \rightarrow O_2$.

8 Above hydrogen; hydrogen.

9 Extraction of aluminium, purifying copper, anodizing aluminium, manufacture of chlorine and sodium hydroxide, electroplating.

10(a) 965 C. **(b)** 0.01 F. **(c)** 0.2 g calcium. **(d)** 0.8 g bromine.

UNIT 13 RATES OF REACTION

1 The particles of the reactants must collide with enough energy.

2 The 'activation energy' is the energy needed to be reached by a collision for a reaction to take place.

3 Colour, mass, volume of gas evolved, pH, cloudiness.

4 A catalyst is a chemical that can alter the rate of a reaction. The catalyst is not used up in the reaction and the reaction is not changed in any other way: e.g. iron in the Haber process (ammonia from hydrogen and nitrogen); vanadium(V) oxide in the Contact process (sulphuric acid from sulphur dioxide); platinum in the conversion of ammonia to nitric acid.

5(a)

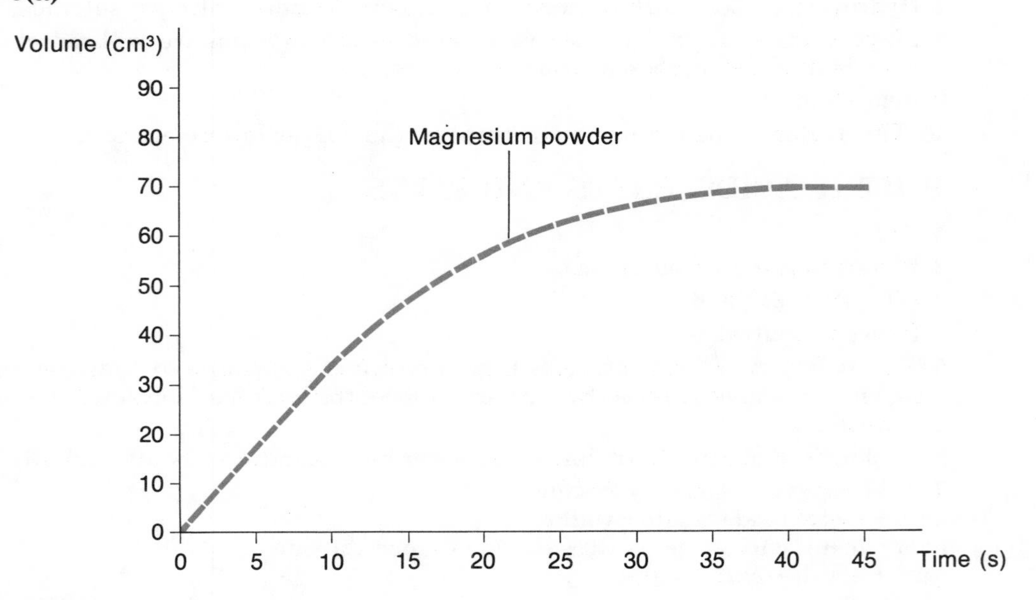

(b) At the start.

(c) 40 s.

(d) When 0.05 g *had* reacted, 50 cm^3 of gas had been given off; from the graph this is after 18 s.

(e) See the graph on previous page.

(f) The concentration of the acid, the temperature, a catalyst.

(g) $Mg + H_2SO_4 \rightarrow MgSO + H_2$.

(h) 'Pops' with a lighted splint.

UNIT 14 REVERSIBLE REACTIONS AND EQUILIBRIUM

1 A reversible reaction is a reaction that can go in either direction depending on the conditions.

2 Equilibrium; stop; equal to.

3 The reaction moves in a direction to oppose any change.

4 A catalyst speeds up the establishment of the equilibrium; it does *not* alter the position of the equilibrium.

5 As the reaction proceeds to the right there is a decrease in volume (4 moles to 2 moles) and so a decrease in pressure; high pressure would make the reaction want to reduce the pressure, i.e. move to the right.

UNIT 15 ENERGY CHANGES IN CHEMISTRY

1 An exothermic reaction is one that gives out heat.

2 Burning is one of many examples of exothermic reactions.

3

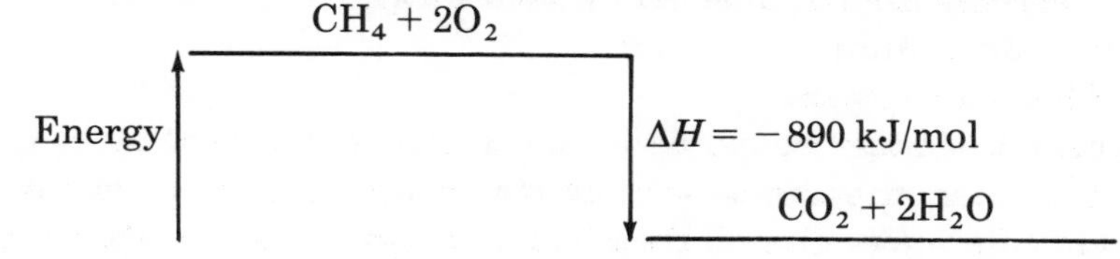

4(a) 21 kJ. **(b)** 10.5 kJ. **(c)** 525 kJ.

5 Copper and magnesium; copper and copper.

UNIT 16 SOCIAL, ECONOMIC AND ENVIRONMENTAL CONSIDERATIONS

1 Near the coast, good transport system, possible pollution consequences, local danger, effect on the environment.

2 Because of relatively cheap hydroelectric power.

3 The removal of oxygen from water by excess nitrogen compounds causing rapid bacterial and plant growth.

4 Sulphur dioxide.

5 Because the mineral resources of the earth are finite, i.e. they will run out.

UNIT 17 OXYGEN AND THE AIR

1 20 per cent.

2 Fractional distillation; mixture; boiling points.

3 Hydrogen peroxide; manganese(IV) oxide; over water; soluble in water.

4 Steel making, hospitals, space rockets, welding and cutting metals.

5 The process whereby green plants make carbohydrate material from water and carbon dioxide in the presence of sunlight.

6 Oxygen is produced by photosynthesis.

7 Carbon monoxide, sulphur dioxide, hydrogen sulphide, 'lead fumes'.

8 'Lead fumes' from petrol are thought to cause brain damage especially in children.

9 Damage to stonework, killing fir trees, killing pond life.

10 Using a converter containing a platinum catalyst.

UNIT 18 CHALK, LIMESTONE AND MARBLE

1 Calcium carbonate.

2 Carbon dioxide; powdery; heat; blue; alkali.

3 Calcium hydroxide solution.

4 The clear solution of limewater soon turns cloudy; more carbon dioxide causes the cloudiness to disappear.

$Ca(OH)_2 + CO_2 \rightarrow CaCO_3 + H_2O$

$CaCO_3 + CO_2 + H_2O \rightleftharpoons Ca(HCO_3)_2$

5 Calcium hydrogencarbonate.

6 A weak solution of calcium hydrogencarbonate, formed by the action of rain water on limestone, falls from the roof of caves and the reverse reaction takes place reforming the calcium carbonate.

7 Calcium carbonate (marble chips) and dilute hydrochloric acid. Sulphuric acid is not used because the calcium sulphate that would be formed is only slightly soluble in water.

8 Fizzy drinks, fire extinguishers, dry ice.

9 To produce carbon dioxide and to react with impurities to form the slag.

10 Making cement, making sodium carbonate.

UNIT 19 WATER

1 Water is very good at dissolving things.

2 Increases; less.

3 Mass; 100 g; temperature.

4 A saturated solution will not dissolve any more solute at that temperature; it is recognized by excess solid remaining undissolved even after prolonged stirring.

5 Water that will not lather easily by boiling; permanent hard water cannot.

7 Ion exchange column, adding sodium carbonate, distillation.

8 A soapless cleaning agent.

9 Soapless detergents are not affected by hard water.

10 Potassium, sodium, lithium, calcium.

11 It burns if heated in steam.

12 A fixed amount of water bound up in the crystal structure of many crystals.

13 Blue; white; water; anhydrous; without water; hot; blue; calcium chloride; purple; blue.

14 Efflorescent substances give off their water of crystallization when exposed to the air, e.g. sodium carbonate; deliquescent substances absorb water from the air and will eventually dissolve in the water, e.g. sodium hydroxide.

15 Can be broken down by natural substances and bacteria.

16 Oxygen is less soluble in hot water.

17 Measure its boiling point (100 °C).

18 Increased nitrates in the water cause rapid bacteria and plant growth; this in turn removes a lot of the dissolved oxygen from the water so killing other plant and animal life.

UNIT 20 CHEMICALS FROM PETROLEUM

1 Hydrocarbons; carbon; hydrogen.

2 Millions; plants; animals.

3 It can be represented by a general formula; its members have similar name endings, have similar reactions and show a gradation of physical properties.

4 Methane: CH_4 ethane: C_2H_6 propane: C_3H_8.

```
     H              H  H              H  H  H
     |              |  |              |  |  |
  H--C--H        H--C--C--H        H--C--C--C--H
     |              |  |              |  |  |
     H              H  H              H  H  H
```

5 Unsaturated compounds contain a multiple bond between two carbon atoms (alkenes contain a double bond); saturated compounds contain only single bonds.

6 Isomers are compounds that have the same molecular formula but different structural formulae.

pentane

C_5H_{12}

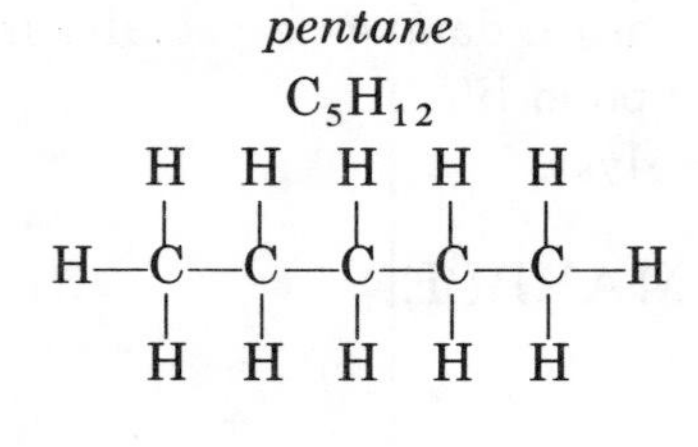

2-methyl butane

C_5H_{12}

```
     H       H       H  H
     |       |       |  |
  H--C-------C-------C--C--H
     |       |       |  |
     H    H--C--H    H  H
             |
             H
```

2,2-dimethyl propane

C_5H_{12}

```
             H
             |
     H    H--C--H    H
     |       |       |
  H--C-------C-------C--H
     |       |       |
     H    H--C--H    H
             |
             H
```

7 The reactions with hydrogen or bromine are the best examples:

ethene + hydrogen → ethane

ethene + bromine → 1,2-dibromoethane.

8 Cracking is the breaking down of large hydrocarbon molecules into smaller more useful ones.

9 Polymer; polymers; polymers; condensation polymers; polymers; monomer; condensation polymers; water; polymers; PVC, polystyrene, PTFE, perspex.

10 Polymers are light and strong and do not corrode; but the non-biodegradable nature is also a disadvantage especially in the case of litter; another disadvantage is that they often burn easily and give off poisonous fumes.

UNIT 21 ETHANOL AND ETHANOIC ACID

1 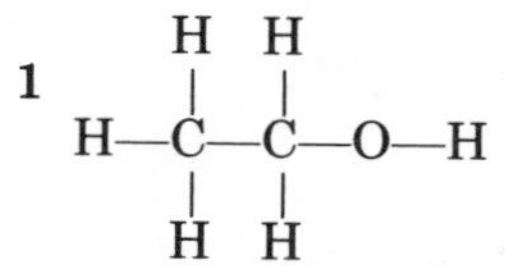

2 The process whereby enzymes in yeast convert sugar to alcohol.

3 Fractional distillation.

4 Ethene obtained from petroleum.

5 It is a good solvent.

6 Ethanol + oxygen → carbon dioxide + water

$$C_2H_5OH + 3O_2 \rightarrow 2CO_2 + 3H_2O$$

7 Potassium dichromate(VI); ethanoic acid.

8 Acid; alcohol; ethyl ethanoate.

9 Nylon, terylene or other polyesters.

10 A compound of carbon, hydrogen and oxygen; starch, sucrose, glucose; they provide energy.

UNIT 22 FUELS

1 A substance that is 'burnt' to provide energy.

2 Cheap, burns easily, no ash or smoke, easy to light, easily stored.

3 Oil, coal, natural gas.

4 Coke, coal gas, ammonia, tar.

5 Methane.

UNIT 23 EXTRACTION OF METALS

1 Ore.

2 Reduced.

3 Difficult.

4 Bauxite; melting point; cryolite; carbon; carbon dioxide.

5 Coke; limestone; iron ore; carbon monoxide; slag; calcium silicate.

6 Steel.

7 Sodium.

8 Sodium chloride.

9 Haematite, magnetite (limonite, pyrite).

10 Because there is a limited amount of copper in the ground and if it is not recycled it will soon run out.

UNIT 24 CORROSION OF METALS

1 To prevent them reacting with water or oxygen in the air.

2 It has a protective oxide coating.

3 Water, oxygen and a trace of carbon dioxide.

4 Painting, greasing, electroplating, galvanizing, plastic coating, tin plating, sacrificial protection.

5 Placing a more reactive metal than iron in contact with the iron so that it corrodes away instead of the iron; magnesium is usually used.

6 Anodizing.

7 Copper carbonate.

UNIT 25 AMMONIA AND NITRIC ACID

1 Nitrogen; hydrogen.

2 Its elements.

3 Haber.

4 Iron.
5 Fertilizers; nitric acid.
6 Ammonium salt; alkali; ammonium chloride; calcium hydroxide.
7 Upward delivery; 'lighter'; very soluble.
8 Alkaline; white; ammonium chloride.
9 A light blue precipitate of copper hydroxide would form in the blue copper sulphate solution; on adding excess ammonia solution the precipitate would redissolve to form a royal-blue solution.
10 Ammonia + copper oxide $\rightarrow$ copper + nitrogen + water.
The ammonia is acting as a reducing agent.
11 Oxidation; platinum.
12 Concentrated sulphuric acid; glass.
13 Oxidizing agent; copper ions; carbon dioxide; iron(III) ions.
14 Acid + a fairly reactive metal to give hydrogen.
15 Making fertilizers, dyes, drugs and explosives.
16 Oxygen; nitrogen dioxide.
17 Soluble.

UNIT 26 FEEDING THE WORLD

1 Nitrates in solution from the soil.
2 Trace.
3 Urea, manure.
4 Leguminous plants.
5 21 per cent.
6 Calcium superphosphate.
7 To help form flowers and seeds.
8 A fertilizer that contains nitrogen, phosphorus and potassium.
9 The ammonium salt would react with the lime and give off ammonia to the air, so losing the nitrogen content.
10(a) Any ammonium compound, sodium nitrate, nitrochalk.
(b) To replace food taken out of the soil by other plants, so allowing more and better plants to be grown.
(c) Carbon, oxygen, hydrogen.
(d) From domestic detergents
(e) Fish, plants and other aquatic animals would die.

UNIT 27 SULPHUR AND SULPHURIC ACID

1(a) Plastic sulphur. **(b)** α-sulphur. **(c)** Plastic sulphur. **(d)** α-sulphur.
2 Frasch; water; outer; melt; air.
3 Hydrogen sulphide; hydrogen sulphide; bauxite.
4 Vulcanizing rubber, making bleach for paper and in medicines.
5 Concentrated sulphuric acid; downward delivery; fountain; sulphurous acid; sulphites; hydrogensulphites.
6 Reducing; sulphate ions.
7 Contact; oxidation; sulphur trioxide; vanadium(V) oxide; violent; concentrated sulphuric acid; oleum.
8 Blue; white; water; dehydrating.
9 Hydrogen chloride.
10 Water; carbon dioxide.

UNIT 28 SALT AND CHEMICALS FROM SALT

1 Brine; hydrogen; chlorine; sodium hydroxide.
2 $Cl^- - e^- \rightarrow Cl$
$2Cl \rightarrow Cl_2$.
3 Bleach, disinfectant, making plastics, drugs, insecticides.
4 Sodium chlorate(I) (sodium hypochlorite).
5 Making soap, paper, rayon; purification of aluminium ore; fat solvent.
6 Sodium carbonate and sodium hydrogencarbonate.

UNIT 29 THE MOLE AND CHEMICAL CALCULATIONS

1(a) 5; **(b)** 4; **(c)** 2.5; **(d)** 2; **(e)** 1;
(f) 0.5; **(g)** 0.11; **(h)** 0.0714; **(i)** 0.005; **(j)** 0.1.

2(a) 84 g; **(b)** 1035 g; **(c)** 81 g; **(d)** 12 g; **(e)** 0.56 g.

3 SO_3.

4 $Fe_2(SO_4)_3$.

5 36 per cent.

6(a) 100 g; **(b)** 48 per cent.

7 Twice as heavy.

8(a) 180 g; **(b)** 40 g; **(c)** 92 g; **(d)** 3.31 g.

9(a) 11.1 g; **(b)** 1.2 dm^3.

10(a) 70 cm^3; **(b)** 40 cm^3.

UNIT 30 QUANTITATIVE VOLUMETRIC CHEMISTRY

1 A pipette.

2 A burette.

3 A solution where 1 dm^3 of solution contains 1 mole of solute.

4 4.9 g.

5(a) 4.0 g. **(b)** 0.0025 moles. **(c)** 0.0025 moles. **(d)** 5 cm^3.

UNIT 31 RADIOACTIVITY

1 Nucleus.

2 Alpha, beta and gamma.

3 Gamma.

4 The time taken for half of the radioactive nuclei to decay.

5 Treating cancer, sterilizing instruments, controlling the thickness of paper and metals, following the course of a process, atomic energy, preserving food.

6 Because of its dangerous nature and long life.

UNIT 32 QUALITATIVE ANALYSIS

1(a) Yellow/orange. **(b)** Lilac. **(c)** Brick red.

2(a) Light blue. **(b)** Pale green. **(c)** Orange.

3 Zinc hydroxide, aluminium hydroxide, and lead hydroxide.

4 A metal carbonate or hydrogencarbonate.

5 A solution containing chloride ions.

6 Potassium sulphate.

7 It bleaches moist indicator paper.

8 Oxygen.

9 Hydrogen.

10 'Heavier'.

TYPES OF EXAMINATION QUESTIONS

In this section the various types of question used on GCSE papers will be considered. Hints will be given to help you do each type of question.

Objective Questions

These are sometimes called **multiple choice questions**. They are used by some examination groups for testing candidates of all abilities. At first sight these questions may seem comparatively easy because it is simply a matter of choosing the correct answer from the possible answers given. In practice candidates do not always do as well as they expect, because the questions are specially designed, written and tested before use and because the candidates are limited in the amount of time available.

Because these questions can be easily and accurately marked by hand or by machine and because the standard of the examination can be judged by pre-testing, multiple choice questions have an important part to play in examining.

There are three types of multiple choice questions used for GCSE examinations.

TYPE 1 SIMPLE MULTIPLE CHOICE

In this type of question you are given an incomplete statement or a question (called the stem) together with four or five possible answers (or responses) and you have to choose the only one that correctly fits the stem. The wrong answers are called distractors.

E.g. What colour is hydrated copper(II) sulphate?

A Blue
B Green
C White
D Yellow
E Black

A is the correct answer (or key). B, C, D and E are distractors. This question is entirely a recall of information that a candidate should have learnt. If he or she confuses hydrated copper(II) sulphate with anhydrous copper(II) sulphate, the answer C might be given.

Many questions like the one above appear on GCSE papers. They test factual learning. Providing you have mastered the units in this book these questions should not be too difficult.

Other questions of this type require, in addition to factual learning, understanding and/or the application of principles.

E.g. Which one of the italicized substances is an element?

A A *green substance* which is separated into two substances by chromatography.
B A *black liquid* which boils over a range of temperature.
C A *black solid* which burns in oxygen completely to form a single colourless gas.
D A *white substance* which turns yellow on heating but white again on cooling.
E A *colourless liquid* which turns to a colourless solid on cooling.

Any candidate who has just learnt the definition of the term 'element' but does not understand it is unable to answer this question. The correct response is C because the black solid burns completely to form only one product. There is no need to try to identify these substances.

Sometimes a number of these items may be linked to the same experimental situation.

TYPE 2 CLASSIFICATION

This type of question is used when a number of similar type 1 multiple choice questions are being set with identical responses. The five lettered responses are given first followed by a series of questions.

E.g. For each of the questions 1–4 choose the one process labelled A, B, C, D or E with which it is chiefly associated.

A Cracking
B Polymerization
C Oxidation
D Hydrogenation
E Neutralization

1 Decane vapour is changed to ethene when passed over heated china.

2 Margarine is produced by passing hydrogen through a heated oil in the presence of a catalyst.

3 Sodium carbonate is added to ethanoic acid (acetic acid) until no further carbon dioxide is evolved.

4 Ethene is completely burnt in excess oxygen.

The correct responses to these questions are:
1 A **2** D **3** E **4** C

In this type of question each response may be used once, more than once or not at all.

TYPE 3 MULTIPLE COMPLETION

There is some doubt about whether this type of question is suitable for candidates of all abilities. However, they are being used by some groups in a simple form at present.

In this type of question there is a stem together with three responses (numbered 1, 2 and 3). One or more than one of these responses is (are) correct. The candidate chooses:

A if 1, 2 and 3 are correct;
B if 1 and 2 only are correct;
C if 2 and 3 only are correct;
D if 1 only is correct;
E if 3 only is correct.

These directions can be summarized in a table.

A	*B*	*C*	*D*	*E*
1, 2, 3 correct	1, 2 only	2, 3 only	1 only	3 only

E.g. A transition metal such as manganese:

1 Conducts electricity
2 Forms a wide range of coloured compounds
3 Is placed in Group I of the Periodic Table

Manganese is a metal and therefore conducts electricity – response 1 is correct. Transition metals form coloured compounds – response 2 is correct. Transition metals are placed in a block between groups II and III, not in group I which contains alkali metals – response 3 is incorrect. You should then record the answer B – responses 1 and 2 only are correct. You should derive some satisfaction that the combination of responses you have made is one of the ones given. If you thought responses 1 and 3 were correct you would have a problem. You should then look again at response 2, perhaps this was correct also and the correct answer should be A.

Questions of this type are more involved, although not necessarily more difficult. They usually appear on the end of the paper when you are getting tired and perhaps running out of time.

POINTS TO BE REMEMBERED WHEN ATTEMPTING A MULTIPLE CHOICE TEST

1 Read through each question carefully. Often a candidate gives a wrong answer because the question has not been read and understood. Often some of the distractors are designed to appeal to the candidate who has not completely read and understood the question.

2 Do not spend too much time on the early items or on any single item. Invariably the questions get longer and more involved as the test progresses, and you will need more time to tackle the questions at the end of the test.

3 Because there are a large number of questions in each test, perhaps 50, and the test is designed to cover the whole syllabus, it is unwise to study only parts of the syllabus. You should attempt to study as much of the syllabus as possible.

4 Do not be afraid to guess. Guessing sensibly can help you and you do not lose marks for a wrong answer. If there are five possible responses and you know that three are wrong but you cannot decide between the other two, then guess. You have increased your chances of success by ruling out incorrect responses. Never leave any question unattempted at the end of the test.

5 Make sure you use all of the information given in the questions. It would not be given unless it was required.

6 Research has shown that in multiple choice tests repeated checking of your answers does not improve the final mark obtained. You are at least as likely to change a correct answer to an incorrect one. The amount of time allowed is not intended to give you 'checking time'.

On pages 147–154 you will find a sample multiple choice test of the type used for GCSE. There are 50 questions and the test should take no longer than 1 hour.

Short Answer Questions

Short answer questions are very common on GCSE papers. They have the same advantage to the examiner as multiple choice questions because with a large number of questions virtually the whole syllabus can be covered. With short answer questions, the candidate often has to come up with an answer rather than just select one of the given answers.

Short answer questions can be used for candidates of all ability but are especially suitable for candidates achieving grades C–G.

There are many different types of short answer question. Sometimes the answer required is a single word, several words or a sentence. Sometimes, however, the candidate may be required to complete a table or a diagram.

On pages 155–157 there is a selection of short answer questions.

Structured Questions

Structured questions are perhaps the most common questions on GCSE papers. A structured question consists of some information given in the question followed by a number of questions based upon the information given. Usually there are spaces for you to fill in the answers. Bear in mind the amount of space you are given. If you are given three lines the examiner is expecting more than one or two words. Also the number of marks for each part is given, and if two marks, for example, are available usually more than a single word would be required.

On pages 158–168 there is a selection of GCSE structured questions.

Comprehension Questions

Comprehension questions appear on GCSE papers and provide a convenient way of introducing the candidate to up-to-date information especially the social, economic and environmental aspects of chemistry.

It is most important to read the information thoroughly before attempting to answer the question. Candidates usually score good marks on this type of question. However, there is a temptation to spend too much time on the comprehension questions.

On pages 172–174 there are some sample comprehension questions.

Essay Questions

Essay questions are suitable only for candidates attempting to achieve grades A, B or C. Apart from chemical knowledge and understanding, the candidate needs to be able to select, organize and present the information in an acceptable form.

Usually to help the candidate the question is broken down into parts and each part has a separate mark allocation. This should help you to know how much time to spend on each part. Just as you would with an English essay, you should plan your essay and there is nothing wrong with putting your essay plan down on the answer paper and then neatly crossing it out. It does at least show the examiner that there was planning!

Do not just 'waffle'. In chemistry credit is given only for specific points in your answer and so there is no benefit to just writing to produce a long essay. If you do this you are just wasting time.

On pages 175–176 there are some sample essay questions. On page 176 you will find sample answers or sample answer plans.

SAMPLE QUESTIONS AND ANSWERS

Multiple Choice Questions

These questions are designed for all abilities of candidate. This test should take about 1 hour.

1 Magnesium (atomic number 12) forms Mg^{2+} ions. The number of electrons in a magnesium ion is

A 4
B 6
C 10
D 12
E 14

2 Which one of the following substances dissolves in water to form a solution with a pH greater than 7?

A Sulphur dioxide
B Ammonia
C Chlorine
D Copper(II) oxide
E Hydrogen chloride

3 Which one of the following reactions would result in the formation of an element?

A Burning carbon in excess air
B Heating copper(II) sulphate crystals
C Neutralizing an acid with a base
D Reducing lead(II) oxide with hydrogen
E Heating lead(II) carbonate

4 Dehydration of methanoic acid (formic acid) HCOOH produces a gas. This gas is

A carbon monoxide
B carbon dioxide
C hydrogen
D carbon
E water

5 An indicator is used during a neutralization reaction in order to

A detect the acid and the alkali
B show when exactly reacting quantities of acid and alkali are present
C speed up the rate of reaction between the acid and the alkali
D measure the amount of heat liberated
E show whether the reaction is reversible

6 Crystals of sodium carbonate decahydrate (washing soda) are efflorescent. When these crystals are exposed to air the crystals

A lose mass and remain solid
B gain mass and remain solid
C gain mass and become liquid
D gain mass, change to liquid and evolve bubbles of gas
E remain unchanged

7 When a piece of copper is added to silver nitrate solution, silver is displaced. Iron reacts slowly with warm, dilute hydrochloric acid to produce hydrogen but silver and copper do not react. The metals in order of reactivity, with the most reactive first, are

A copper, silver, iron
B iron, silver, copper
C iron, copper, silver
D copper, iron, silver
E silver, iron, copper

8 Crude oil is separated into fractions with different boiling points by fractional distillation. In which one of the following are the fractions arranged in the correct order of increasing boiling point?

A Petrol, bitumen, diesel oil, paraffin
B Bitumen, diesel oil, paraffin, petrol
C Diesel oil, petrol, bitumen, paraffin
D Paraffin, diesel oil, bitumen, petrol
E Petrol, paraffin, diesel oil, bitumen

9 2.38 g of tin, when treated with concentrated nitric acid and heated, produced 3.02 g of anhydrous tin oxide. What is the formula of this oxide? ($A_r(O) = 16$, $A_r(Sn) = 119$)

A SnO
B SnO_2
C Sn_2O
D Sn_2O_3
E SnO_3

10 $2NH_3(g) + 3CuO(s) \rightarrow N_2(g) + 3H_2O(g) + 3Cu(s)$
What is the volume of the gaseous product formed when 80 cm^3 of ammonia is passed over heated copper(II) oxide?
(All volumes measured at room temperature and pressure.)

A 20 cm^3
B 40 cm^3
C 80 cm^3
D 120 cm^3
E 160 cm^3

QUESTIONS 11–15 REFER TO THE FOLLOWING TYPES OF CHEMICAL REACTION:

A Hydrolysis
B Polymerization
C Dehydration
D Neutralization
E Precipitation

Select the term from the list above which best describes the reaction represented by each of the following equations:

11 $(C_6H_{10}O_5)_n(aq) + n\ H_2O(l) \rightarrow n\ C_6H_{12}O_6(aq)$

12 $OH^-(aq) + H^+(aq) \rightarrow H_2O(l)$

13 $C_2H_5OH(g) \rightarrow C_2H_4(g) + H_2O(g)$

14 $MgO(s) + 2HNO_3(aq) \rightarrow Mg(NO_3)_2(aq) + H_2O(l)$

15 $MgSO_4(aq) + Na_2CO_3(aq) \rightarrow MgCO_3(s) + Na_2SO_4(aq)$

16 The diagram represents apparatus which can be used to pass a measured quantity of air, to and fro, over heated copper wire until no further reaction occurs.

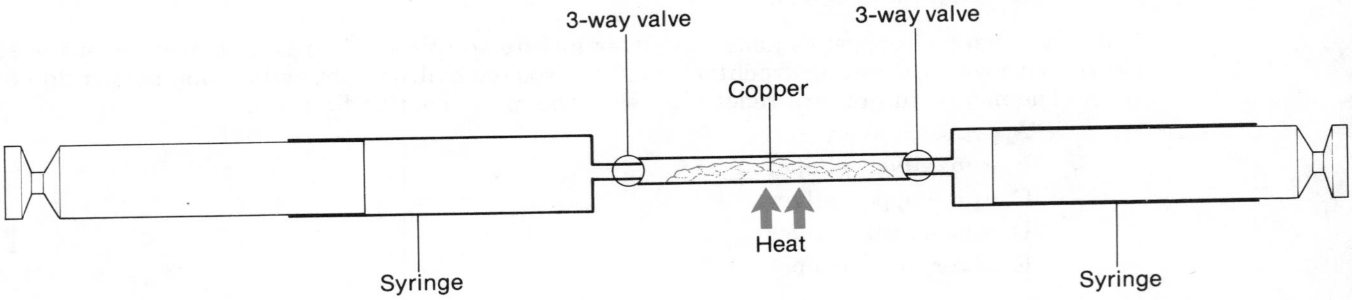

3-way valve
3-way valve
Copper
Heat
Syringe
Syringe

In one experiment it was found that 16 cm^3 of gas remained after the apparatus had cooled. How many cubic centimetres of air were there in the apparatus at the beginning of the experiment?

A 20
B 40
C 60
D 80
E 100

17 Three liquids were available:

(i) a solution of dry hydrogen chloride dissolved in dry methylbenzene (toluene)
(ii) pure ethanol
(iii) pure water

Which one of the following would have a pH less than 7?

A (i)
B (ii)
C (i) and (ii) mixed together
D (i) and (iii) mixed together
E (ii) and (iii) mixed together

18 A metal M forms a hydroxide $M(OH)_3$. The relative formula mass of this hydroxide is 78. What is the relative atomic mass of the metal M? ($A_r(H) = 1$, $A_r(O) = 16$)

A 27
B 30
C 59
D 61
E 75

19 A metallic element X forms an ion X^{3+}, what will be the formula of the sulphate of X?

A $X\ SO_4$
B $X\ (SO_4)_3$
C $X_2\ (SO_4)_3$
D $X_3\ SO_4$
E $X_3\ (SO_4)_2$

20 An ion with a single positive charge becomes an atom by

A losing an electron
B losing a neutron
C gaining a proton
D gaining a neutron
E gaining an electron

21 If a mixture of fine pollen grains in water is examined closely it will be seen that the pollen grains are always in motion.
This motion is most likely to be due to

A the convection currents in the water
B the chemical reaction between the pollen and the water
C the attraction and repulsion between charged particles
D the collisions between pollen grains and water molecules
E the diffusion of pollen grains

22 The products of the hydrolysis of starch with hydrochloric acid can be identified using

A iodine
B sodium carbonate
C chromatography
D distillation
E enzymes

QUESTIONS 23–6 REFER TO THE EXPERIMENT DESCRIBED BELOW

The apparatus shown in the diagram was used to produce a sample of ethene from liquid paraffin.

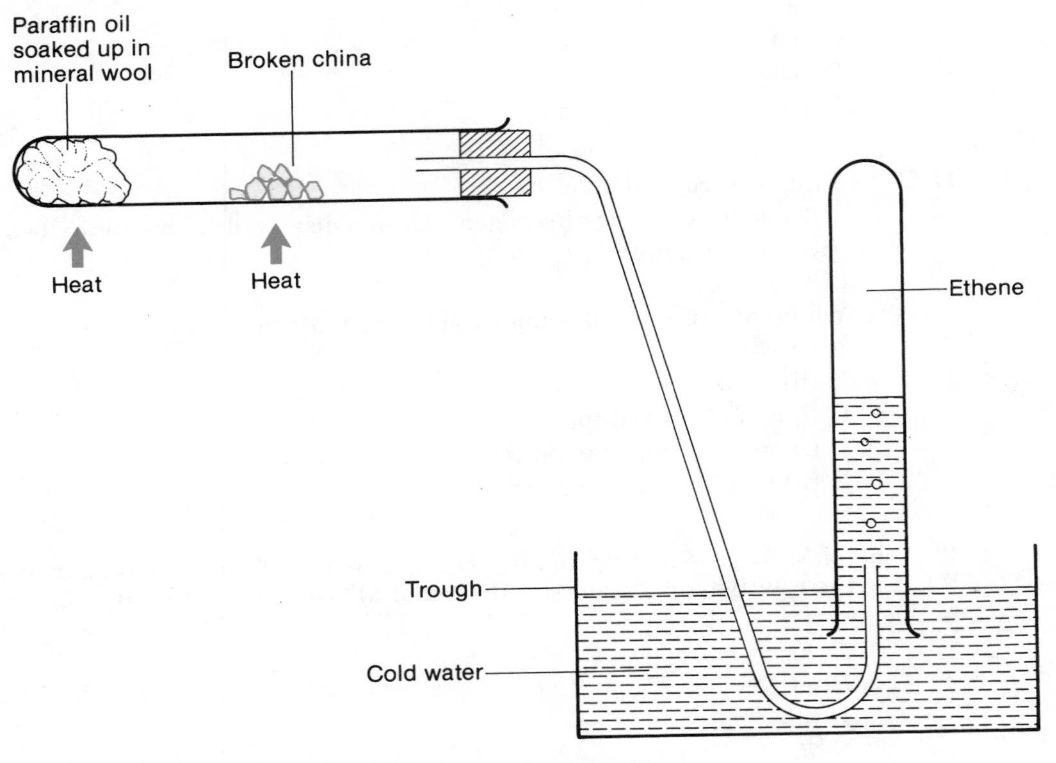

23 What is the reason for using the mineral wool?

A To act as a catalyst for the reaction
B To prevent heat escaping from the test tube
C To prevent water from the trough entering the test tube
D To hold the liquid paraffin at the end of the tube
E To lower the boiling point of the liquid paraffin

☐

24 The process taking place during this experiment is

A cracking
B polymerization
C combustion
D distillation
E saponification

☐

25 The liquid found floating on the water at the end of the experiment

A is produced by reaction between ethene and water
B comes from the broken china
C is unreacted liquid paraffin
D is produced when ethene comes in contact with air
E is produced by reaction between ethene and substances in the water

☐

26 Ethene and paraffin can be distinguished *chemically* by the fact that ethene, but not paraffin

A will burn in air
B will decolourize bromine water
C is a hydrocarbon
D can be obtained from crude oil
E is a gas at room temperature

☐

QUESTIONS 27–30 REFER TO THE CURVES A–E EACH CURVE SHOWS THE CHANGE OF PROPERTY WITH TIME

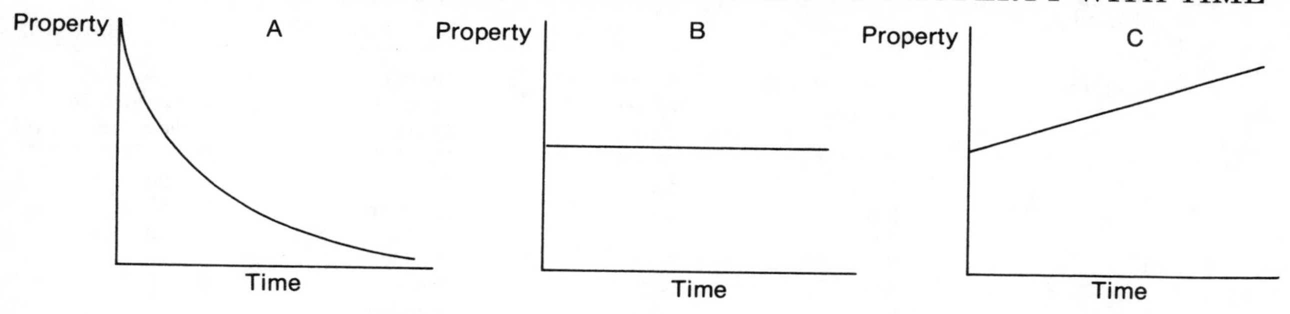

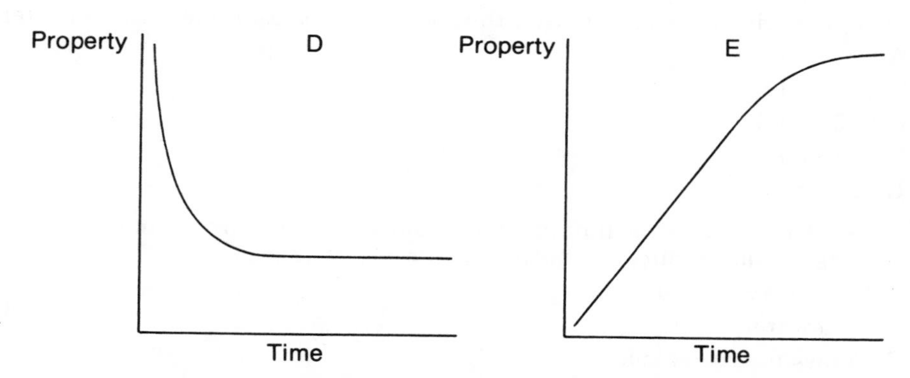

Select the curve which best shows the change of the property in italics with time for each of the following:

27 *The volume of carbon dioxide evolved* during the reaction between calcium carbonate and hydrochloric acid ☐

28 *The mass of a catalyst* during the decomposition of ammonia ☐

29 *The mass of lead(II) oxide* when heated in a stream of hydrogen ☐

30 *The mass of a copper cathode* during the electrolysis of copper(II) sulphate solution with copper electrodes and a constant current ☐

QUESTIONS 31–2 REFER TO THE FOLLOWING EXPERIMENT

Dry air is passed over heated zinc in the apparatus shown in the figure. The gases remaining escape at X.

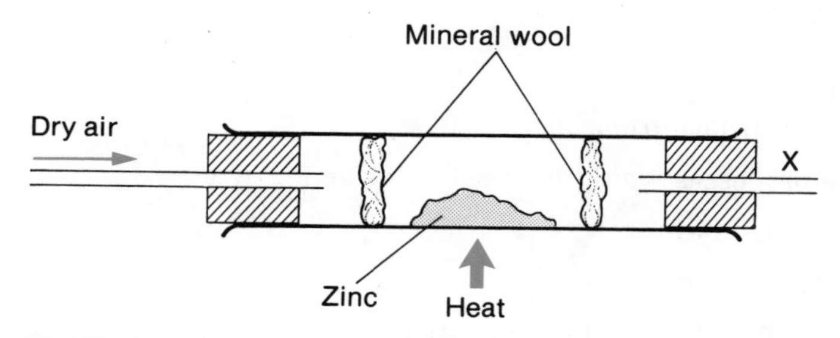

31 The gas leaving the apparatus at X is mainly

A argon
B carbon dioxide
C nitrogen
D oxygen
E water vapour ☐

32 The main chemical compound present in the apparatus at the end of the experiment is

A zinc
B zinc oxide
C zinc nitrate
D zinc nitride
E zinc hydroxide ☐

33 The table below gives the results of six experiments involving the reaction between zinc and hydrochloric acid. In all experiments 0.2 g of zinc was used together with the same volume of acid.

Experiment number	*Concentration of acid*	*Temperature of acid (°C)*	*State of division of zinc*	*Time for the reaction to be completed (s)*
1	1 M	25	Foil	190
2	2 M	25	Powder	85
3	2 M	35	Foil	62
4	2 M	50	Powder	15
5	2 M	35	Powder	15
6	3 M	50	Powder	11

Which set of experiments suggests that the speed of a reaction increases with temperature?

A 1, 3 and 4
B 2, 3 and 4
C 2, 3 and 5
D 2, 4 and 5
E 2, 5 and 6

34 Which one of the following liquids, when added to a strongly alkaline solution in the correct amounts, could produce a solution with a pH of 7?

A Distilled water
B Limewater
C Potassium hydroxide
D Sulphuric acid
E Universal Indicator

35 An atom of an element with an atomic number 3 and a mass number 7 contains

A 3 electrons
B 3 neutrons
C 4 electrons
D 4 protons
E 7 neutrons

QUESTIONS 36–40 REFER TO THE FOLLOWING PROCESSES

A Chromatography
B Crystallization
C Distillation
D Filtration
E Fractional distillation

Which of the above processes would be most suitable for

36 removing sand from river water?

37 obtaining pure water from sea water?

38 separating a mixture of petrol and diesel oil?

39 investigating the number of colourings added to a fruit drink?

40 obtaining ammonium sulphate from an aqueous solution of ammonium sulphate?

41 Magnesium reacts with hydrochloric acid according to the following word equation:

magnesium + hydrochloric acid → magnesium chloride + hydrogen

In an experiment magnesium was reacted with *excess* hydrochloric acid. Which one of the results A–E, plotted on the vertical axis, would produce the graph shown in the figure?

A The concentration of the hydrochloric acid
B The concentration of magnesium chloride in solution
C The mass of hydrogen formed
D The mass of magnesium remaining
E The volume of hydrogen formed

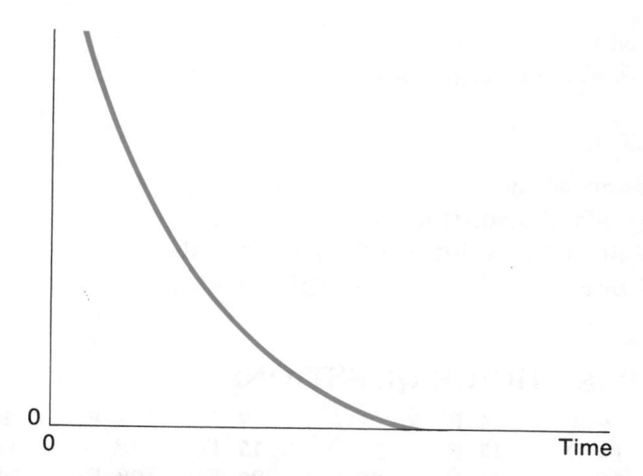

QUESTIONS 42–5 REFER TO THE TABLE BELOW

The table gives some properties of elements A–E. These letters are not the chemical symbols for these elements.

Element	A	B	C	D	E
Electrical conductivity in the solid state	Nil	Good	Good	Nil	Good
Melting point °C	112	63	1535	−210	−39
Boiling point °C	444	760		−195	357
Appearance of oxide	Colourless gas	White solid	Black solid	Colourless gas	Red solid
pH of solution of oxide in water	3	14	Insoluble	3	Insoluble

Which one of the elements A–E

42 forms an alkaline oxide solution? ☐

43 is a liquid at room temperature (20 °C)? ☐

44 could be part of the air? ☐

45 would best form the wire of a heating coil for an electric fire? ☐

For questions 46 to 50 one or more of the given answers are correct. Decide which one or ones are correct, then write your answer using the following code.

A If 1, 2 and 3 are correct
B If 1 and 2 are correct
C If 2 and 3 are correct
D If 1 only is correct
E If 3 only is correct

Example: Which of the following gases are elements?

1 H_2
2 O_2
3 CO_2

1 and 2 are correct, therefore the answer is B.

46 Chemical processes which always make molecules smaller include

1 reduction
2 depolymerization
3 cracking ☐

47 Reactions producing carbon dioxide include

1 reaction of a metal carbonate with an acid
2 burning ethanol in excess air
3 fermentation of sugar ☐

48 A substance whose structure, at room temperature, consists of a regular arrangement of molecules

1 produces a regular X-ray diffraction pattern
2 conducts electricity when molten
3 has a high melting point ☐

49 Hard water can be softened by
1 passing through an ion exchange column
2 distillation
3 adding sodium carbonate ☐

50 Crystals of lead can be produced by
1 splitting open crystals of lead(II) nitrate
2 adding a piece of silver to a solution of lead(II) nitrate
3 adding a piece of zinc to a solution of lead(II) nitrate. ☐

ANSWERS TO MULTIPLE CHOICE QUESTIONS

1 C;	**2** B;	**3** D;	**4** A;	**5** B;	**6** A;	**7** C;	**8** E;	**9** B;	**10** B;
11 A;	**12** D;	**13** C;	**14** D;	**15** E;	**16** A;	**17** D;	**18** A;	**19** C;	**20** E;
21 D;	**22** C;	**23** D;	**24** A;	**25** C;	**26** B;	**27** E;	**28** B;	**29** D;	**30** C;
31 C;	**32** B;	**33** D;	**34** D;	**35** A;	**36** D;	**37** C;	**38** E;	**39** A;	**40** B;
41 D;	**42** B;	**43** E;	**44** D;	**45** C;	**46** C;	**47** A;	**48** D;	**49** A;	**50** E

REASONING BEHIND THE ANSWERS

The answer may not help you to see how to do similar questions in the future. Here briefly is an explanation of how the answer is obtained. This is especially useful where the question tests application of your knowledge to different situations. Where the question is recall reference is made to the appropriate unit in the text.

1 A magnesium atom contains 12 electrons and loses two electrons when forming a 2+ ion.

2 See Unit 6.

3 Reduction is removing oxygen. If oxygen is removed from lead(II) oxide only lead remains.

4 Dehydration is the removal of water. If two hydrogen atoms and one oxygen atom are removed from HCOOH only CO (carbon monoxide) remains.

5 See Unit 11.

6 Efflorescence is the loss of water to the atmosphere and therefore there would be a loss of mass.

7 Only iron is reactive enough to displace hydrogen from hydrochloric acid. Copper replaces silver and is therefore more reactive than silver.

8 See Unit 20.

9 2.38 g of tin combine with 0.64 g of oxygen. 0.02 moles of tin atoms combine with 0.04 moles of oxygen atoms.

10 From the equation, 80 cm^3 of NH_3 produce half the volume of nitrogen (40 cm^3). However, candidates often forget that the water has condensed and give an answer of 160 cm^3.

11 Hydrolysis is the splitting up of a compound with water.

12 This is the ionic equation for neutralization.

13 This is removal of water – dehydration.

14 Acid + base, therefore it is neutralization (see Unit 11).

15 $MgCO_3$(s) tells you that the solid is precipitated.

16 One fifth of the 'air' removed during the experiment.

17 Water must be present before acid properties are shown.

18 There are three OH^-; mass $3 \times 17 = 51$. Subtract this from 78.

19 The ions are X^{3+} and SO_4^{2-}.

20 Positive ions are formed when atoms lose electrons. Atoms are formed when positive ions gain electrons.

21 Brownian motion (see Unit 1).

22 Iodine is used as a test for starch and iodine is a frequent wrong answer here. Starch is split up into sugars by hydrolysis and these can be separated and identified by chromatography.

23 Otherwise the liquid will run along the tube.

24 This is breaking down large molecules in the liquid into smaller molecules in the gas.

25 This liquid is produced by distillation. The liquid paraffin boils. Some of it does not react when passed over the heated china and condenses in contact with the cold water.

26 See Unit 20. This is frequently not known by candidates.

27 There is no carbon dioxide at the start. The volume increases during the reaction until one or both of the reactants are used up. No further gas is produced.

28 The mass of catalyst remains unchanged.

29 The mass of lead oxide decreases during the reaction until it reaches a minimum.

30 The mass of the cathode increases as copper is deposited and the increase is steady because of the constant current.

31 Oxygen is removed; mainly nitrogen remains.

32 Reaction between zinc and oxygen forms zinc oxide.

33 2, 4 and 5 – all the other conditions apart from temperature are the same – i.e. concentration of acid and state of division. In these three experiments only temperature is different.

34 Sulphuric acid is the only strong acid solution present. When a strong acid and a strong alkali are mixed in the right amounts a neutral solution of pH 7 results.

35 Atomic number 3, mass number 7 – 3 protons, 3 electrons and 4 neutrons.

36 This is removing an insoluble solid from a liquid (see Unit 2).

37 This is separating a solvent from a solution (see Unit 2).

38 This is separating liquids with different boiling points (see Unit 2).

39 See Unit 2.

40 This is obtaining crystals from an aqueous solution.

41 Excess acid and so no magnesium will remain at the end of experiment. Mass decreases to zero. It is not answer **a** because the concentration of acid is not zero at the end of the experiment.

42 The pH 14 is alkaline.

43 The melting point is below 20 °C and boiling point above 20 °C.

44 This is the only gas – melting point and boiling point both below 20 °C.

45 This is a good conductor with a high melting point.

46 Addition of hydrogen can be defined as reduction.

47 All produce carbon dioxide.

48 No ions are present, therefore it is not 2 or 3.

49 See Unit 19.

50 Candidates forget that lead(II) nitrate is a compound and does not contain uncombined lead.

Short Answer Questions

The questions in this section are suitable for GCSE pupils of all abilities.

1 The following list consists of common gases.

Hydrogen, carbon dioxide, sulphur dioxide, oxygen, helium

Choose your answers to the following questions from the list of gases above. Each gas should only be used once.

Choose the gas which is:

(a) a noble gas

(b) used in fire extinguishers;

(c) produced when fuels containing sulphur are burnt;

(d) used in the manufacture of margarine;

(e) used in turning impure iron into steel.

2(a) What do the members of each of the following groups have in common?

(i) Methane, ethane, propane and butane.

(ii) Crystallization, distillation, filtration, chromatography.

(iii) Nitrogen, hydrogen, ammonia, finely divided iron, heat, high pressure.
(b) In each of the lists below the one in **bold** type is the 'odd one out'.
In each case give a reason why.
(i) Lithium, **calcium**, sodium, potassium.
(ii) Coke, iron ore, limestone, **hydrogen**, oxygen.
(iii) Thermometer, metre rule, balance, **gas jar**, stop clock.

(MEG Chemistry 1986 II Q2)

3 This bar chart shows the average amount of sulphur dioxide in town air. In 1981 the average was 120 micrograms/metre^3. Add this information to the chart in the figure.

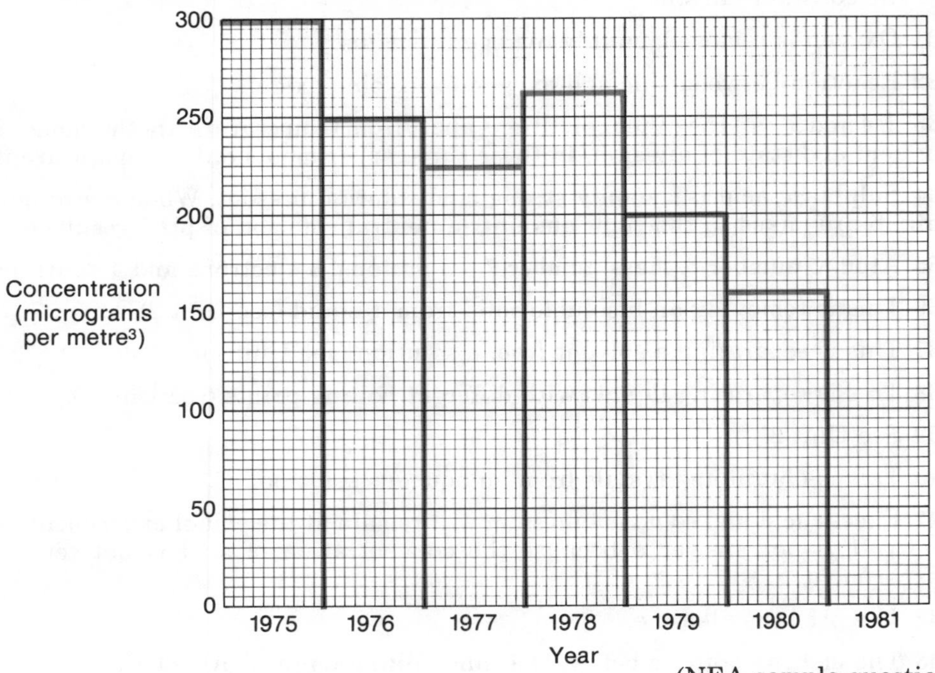

(NEA sample questions Syll. A I QA3)

4 Choose your answers from this list of substances:

air
ammonia
brass
carbon
carbon dioxide
copper
dilute hydrochloric acid
oxygen
sugar
sulphur
silver
sodium
sea water

(a) Name:
(i) **two** metallic elements;
(ii) **two** nonmetallic elements;
(iii) **two** compounds;
(iv) **two** mixtures.
(b) State which **one** of the substances is *used*:
(i) in the manufacture of fertilizers;
(ii) in the production of sulphuric acid;
(iii) in everyday coinage;
(iv) to produce chlorine and sodium hydroxide when it is electrolysed.

(MEG specimen papers II Q1)

5 Complete the following table:

Particle	*Atomic number*	*Mass number*	*Number of*		
			Protons	*Neutrons*	*Electrons*
Carbon atom	6	12	—	—	—
Sodium	11	—	—	12	—
Sodium ion Na^+	—	—	—	—	—

6 Complete the following table which shows various methods of preparing salts.

	Names of reacting substances	*Name of salt prepared*	*Names of other substances produced*
a	Sodium hydroxide + Nitric acid	**A**	**B**
b	Zinc + **C**	Zinc sulphate	**D**
c	Sodium sulphate + **E**	Barium sulphate	**F**
d	**G** + **H**	Iron(III) chloride	—
e	**I** + **J**	Lead(II) nitrate	Carbon dioxide and water

7 This question is about the states of matter

(a) Two of the states of matter are *liquid* and *gas*. What is the third state of matter?

(b) A liquid changes slowly into a vapour at room temperature. What word is used to describe this change?

(c) Liquids and gases consist of particles. What happens to these particles when liquids and gases are heated?

(d) Why can gases be compressed easily?

(e) The smell of perfume *rapidly* fills a room when the bottle is opened. Why is this?

8 This question is about metals

(a) Europium is a metal. Suggest three properties which europium might have because it is classed as a metal.

(b) Rubidium is similar to potassium. What two substances would you expect to be formed when rubidium reacts with water?

(LEAG Joint O/CSE 1986 Syll. A, Paper II. Qs 23–31)

ANSWERS TO SHORT ANSWER QUESTIONS

1(a) Helium. **(b)** Carbon dioxide. **(c)** Sulphur dioxide. **(d)** Hydrogen. **(e)** Oxygen.

2(a)(i) They are hydrocarbons (or compounds of carbon and hydrogen only) – more exactly they all belong to the alkane family.

(ii) They are methods of separating mixtures.

(iii) They are all associated with the Haber process.

(b)(i) The others are alkali metals – calcium is not.

(ii) The others are associated with the extraction of iron from iron ore.

(iii) All are pieces of measuring apparatus except the gas jar.

3

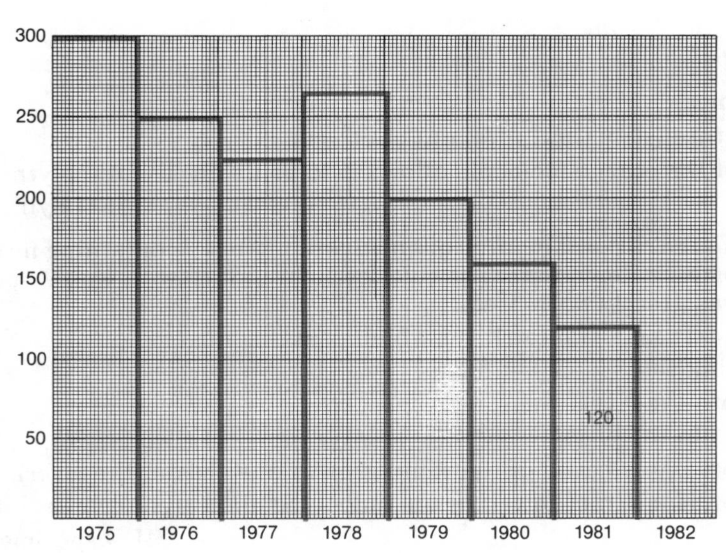

4(a)(i) Copper, silver, sodium (any two).
(ii) Oxygen, sulphur, carbon (any two).
(iii) Ammonia, carbon dioxide, sugar (any two).
(iv) Dilute hydrochloric acid, brass, air, sea water (any two).
(b)(i) Ammonia.
(ii) Sulphur.
(iii) Copper.
(iv) Sea water.

5 Carbon atom: 6p, 6n, 6e.
Sodium atom: mass no. 23, 11p, 11e.
Sodium ion (formed when a sodium atom loses an electron): at. no. 11, mass no. 23, 11p, 12n, 10e.

6(a) A – sodium nitrate; B – water.
(b) C – dilute sulphuric acid; D – hydrogen.
(c) E – barium chloride (or nitrate); F – hydrochloric (or nitric) acid.
(d) G and H – iron and chlorine.
(e) I and J – lead(II) carbonate and dilute nitric acid.

7(a) Solid.
(b) Evaporation.
(c) The particles move faster.
(d) The particles are widely spaced.
(e) The particles move quickly in all directions to fill all of the available space – this is diffusion.

8(a) E.g. high density; high melting and boiling points; good conductor of heat and electricity; shiny; ductile; malleable (**physical properties**).
Burns in oxygen to form a neutral or alkaline oxide (**chemical property**).
(b) Rubidium hydroxide and hydrogen.

Structured Questions

In this section questions that are unstarred are suitable for candidates of all abilities. Questions that are starred are intended for candidates aiming at grades A, B and C.

1(a) Name and give the formula of the main compound in natural gas.
(b) A compound found in the oil fraction kerosene is decane $C_{10}H_{22}$.
(i) Complete the word equation for burning decane completely in air.

decane + oxygen → ______________

(ii) What conditions might cause a poisonous gas to be formed when burning decane?
(iii) Name the poisonous gas referred to in **(b)(ii)**.
(c) Polythene [poly(ethene)] is made by joining together many ethene molecules.
(i) What is this process called?
(ii) The formula of polythene is:

$$\left(\begin{array}{cc} H & H \\ | & | \\ C - & C \\ | & | \\ H & H \end{array}\right)_n$$

What does the 'n' mean?
(iii) Draw the structure (structural formula) of ethene.
(iv) What feature of the ethene molecule allows it to be changed into polythene?

(SEG Syll. A, Paper I Q16)

2 River water was found to contain large quantities of nitrogen-containing fertilizers and also phosphates from domestic detergents. The river became choked with weeds and the amount of dissolved oxygen was greatly decreased.
(a) Give the chemical name of a fertilizer which contains nitrogen.
(b) How do you think the:
(i) fertilizers got into the river?
(ii) phosphates got into the river?
(c) What is the most likely cause of the rapid growth of weeds in the river?
(d) What would be the effect of lowering the amount of dissolved oxygen in the river?

(MEG specimen papers II Q8)

3 The table below includes the structural formulae of some monomers and the polymers that can be made from them.

Monomer	*Polymer*
$F_2C{=}CF_2$ (F, F / C=C / F, F) Tetrafluoroethene	$\left[-CF_2-CF_2-\right]_n$ Polytetrafluoroethene
$H(Cl)C{=}CH_2$ (H, Cl / C=C / H, H) Chloroethene	Polychloroethene
$CH_3(H)C{=}CH_2$ (CH_3, H / C=C / H, H) Propene	$\left[-CH(CH_3)-CH_2-\right]_n$ Polypropene

(a) Which one of the monomers is a hydrocarbon? (1 mark)

(b) The molecular formula of propene is C_3H_6. Write the molecular formula of chloroethene. (1 mark)

(c) Write in the table the structural formula of polychloroethene. (1 mark)

(d) What similarity in structure exists in the three monomers? (1 mark)

(e) What colour change would be observed if propene gas were bubbled through a solution of bromine? (1 mark)

(f) Ethene (C_2H_4) can be polymerized under different conditions to form either low or high density polyethene. Both forms of polyethene can be easily moulded using heat and/or vacuum methods.

Low density polyethene is cheaper to produce but has a lower melting point and is less strong than high density polyethene.

(i) What is the usual industrial source of ethene?

(ii) In which form of polyethene are the chains of atoms more closely packed? Give a reason for your answer. (3 marks)

(g) Washing up bowls used to be made of steel coated with paint. They are now often made of polyethene.

(i) Give *two* reasons why polyethene is more suitable than painted steel for this purpose.

(ii) Give a reason why high density polyethene is more suitable than low density polyethene for this purpose. (3 marks)

(h) Addition polymers such as polyethene and polypropene are very difficult to dispose of.

(i) Why are these polymers difficult to dispose of?

(ii) One possible method of disposal is burning. Name two possible products of combustion of polyethene and polypropene. (3 marks)

(i) The table below contains information about four materials in household refuse.

Material	*Added to water*
Polymers	Floats on water
Iron	Sinks in water
Aluminium	Sinks in water
Paper	Floats on water

Paper, however, sinks in water when it is thoroughly wetted.

Assuming household refuse is a mixture only of polymers, paper, iron and aluminium, how could:

(i) iron be removed from the refuse?

(ii) polymers be removed from the refuse? (3 marks)

(j) It is impossible at present to separate pure polyethene from household refuse. Usually the mixture of polymers is melted and made into cheaper blocks for lining walls.

(i) What properties of addition polymers are important in making and using these blocks?

(ii) What would be the economic advantage of being able to separate pure polyethene from household refuse? (3 marks)

(Total 21 marks)

(LEAG Syll. B Paper II Q4)

4 The table below shows how four metals (W, X, Y and Z) react with dilute hydrochloric acid. (W, X, Y and Z are *not* the chemical symbols)

Metal	*Reaction with dilute hydrochloric acid*	
W	Slow production of hydrogen.	
X	Rapid production of hydrogen.	Solution gets warm.
Y	No reaction.	
Z	Very fast production of hydrogen.	Solution gets hot.

(a) Describe a suitable test for hydrogen. (2 marks)

(b) Arrange the four metals in order of reactivity with the most reactive metal first. (2 marks)

(c) When a piece of W is added to copper(II) sulphate solution, a reaction takes place very slowly. A brown deposit forms on the metal and the solution turns colourless.

(i) What is the brown deposit formed?

(ii) What type of reaction is taking place? (2 marks)

(LEAG Syll. B Paper II Q5)

5 In a discussion on the use of rock salt, two pupils suggested that

ICE MELTS MORE QUICKLY IF ROCK SALT IS ADDED TO IT.

To test their prediction, they did the following experiment:

Step 1 Two funnels were filled with lumps of ice.
2 Rock salt was added to one of the funnels.
3 Water from the melting ice was collected in measuring cylinders.
4 The volume of water was recorded every minute.

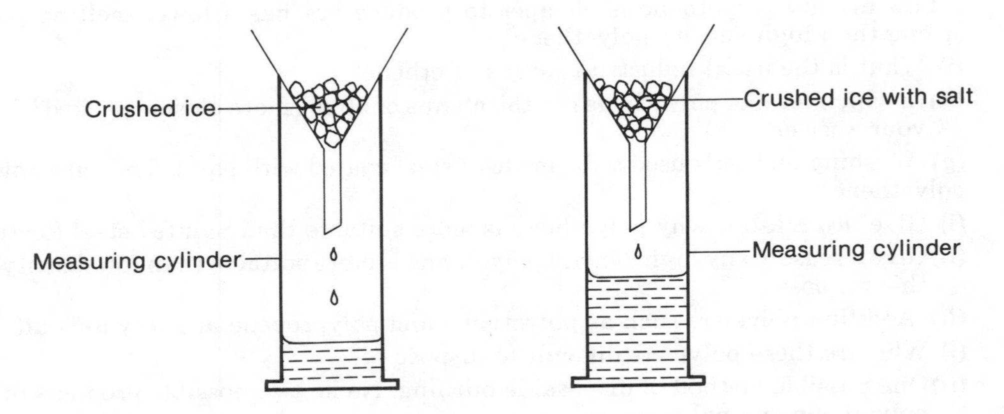

(a) State one precaution which should have been taken in Step 1 to make this a fair test. (1 mark)

(b) The volume of water collected from the ice *without* added salt is shown in the table below.

Time (minutes)	1	2	3	4	5	6
Volume of water collected (ml)	2	4	7	9	11	14

Draw a line graph to show how much water was collected as the ice melted and label this line A. (2 marks)

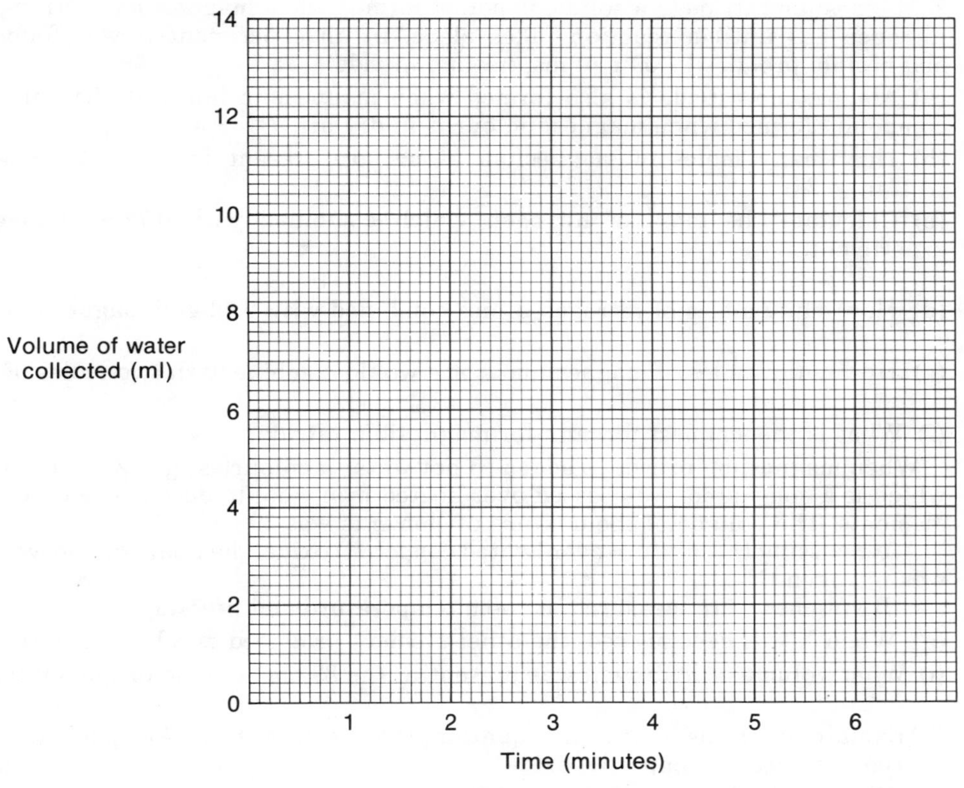

(c) The volume of water collected in the second measuring cylinder proved that the pupils' prediction was correct. Draw in another line on your graph to show this result and label this line B. (1 mark)

(d) Why does ice melt more quickly when rock salt is added? (1 mark)

(e) Give one *disadvantage* of using salt on icy roads. (1 mark)

(Total 6 marks)

(NEA Syll. A Paper I QC3)

6 The substance calcium carbonate occurs in nature as marble.

(a) Name *two* other forms of calcium carbonate which occur in nature. (2 marks)

(b) Write down the chemical formula of calcium carbonate. (1 mark)

(c) Carbon dioxide can be detected using a solution of another calcium compound.

(i) Write down the common name of the calcium compound which is used to test for carbon dioxide. (1 mark)

(ii) What is the chemical name for this calcium compound? (1 mark)

(d) Describe what you see when carbon dioxide is first passed into the test solution. (1 mark)

(e) Describe what you see when carbon dioxide is passed into the test solution for some time. (1 mark)

(f) What is the name of the solid formed when calcium carbonate is strongly heated? (1 mark)

(g) When water is added dropwise to the cold solid from **(f)**:

(i) Describe what you observe. (2 marks)

(ii) Will the resulting solution be acidic or alkaline? (1 mark)

(h) Water from limestone areas is sometimes called hard water. Explain what is meant by the term hard water. (1 mark)

(i) Describe how the reaction between rain water and limestone forms temporary hard water. (2 marks)

(j) Write a word equation for this reaction. (1 mark)

(k) Write a symbol equation for this reaction. (1 mark)

(l) Name one substance which would give rise to permanent hardness in water. (1 mark)

(m) State a method for removing temporary hardness from water. (1 mark)

You are given three bottles of water labelled A, B and C. You are told that the bottles contain, separately, distilled water, temporary hard water and permanent hard water.

(n) Describe how you would test the liquids so as to label the bottles correctly, i.e. Distilled Water; Temporary Hard Water; Permanent Hard Water. (3 marks)

(Total 21 marks)

(NI Paper II Q1)

7 Magnesium nitride is a solid compound formed when magnesium is strongly heated in dry nitrogen gas until a reaction takes place. In an experiment it was found that 0.72 g of magnesium produced 1.00 g of magnesium nitride.

(a) Magnesium nitride is also formed when magnesium burns in dry air. Name the other compound formed when magnesium burns in dry air. (1 line, 1 mark)

(b) How many moles of magnesium atoms are present in 0.72 g of magnesium ribbon? (A_r(Mg) = 24) (1 line, 1 mark)

(c) Calculate the mass of nitrogen which combined with 0.72 g of magnesium in this experiment. (1 line, 1 mark)

(d) How many moles of nitrogen atoms combined with 0.72 g of magnesium? (A_r(N) = 14). (1 line, 1 mark)

(e) How many moles of magnesium atoms would combine with two moles of nitrogen atoms? (1 line, 1 mark)

(f) What is the simplest formula for magnesium nitride? (1 line, 1 mark)

When magnesium nitride is added to hot water a colourless gas X is produced. 40 cm^3 of X, which is a compound, was passed over heated iron wool to decompose it completely into two elements. The volume of the mixture of elements was 80 cm^3.

After passing the mixture over heated copper(II) oxide, the final volume was 20 cm^3. This gas was nitrogen (N_2).

(All volumes were measured at room temperature and pressure.)

(g) When X was decomposed, the other element produced was hydrogen (H_2).

(i) What volume of hydrogen was present in the 89 cm^3 sample of the mixture of element? (1 line, 1 mark)

(ii) Explain, in words or with an equation, the reaction that takes place when the mixture is passed over heated copper(II) oxide. (2 lines, 2 marks)

(h) What is the formula of the gas X? (1 line, 1 mark)

(Southern Regional Examinations Board R(B))

8 The letters A, B, C, D, E, F and G represent seven elements. (These letters are not the actual symbols for these elements.) You are not expected to try to identify any of these elements.

	Atomic number	*Melting point °C*	*Boiling point °C*	*Element heated in air*	*pH of oxide residue*	*Reaction of element with water*
A	82	327	1744	Slow reaction	7	None
B	19	64	760	Burns on heating	12	Violent reaction
C	12	650	1110	Burns on heating	10	Reacts very slowly with cold water but rapidly with steam
D	3	180	1330	Burns on heating	11	Reacts steadily with cold water
E	80	−39	357	Slow reaction over several days	7	None
F	11	98	890	Burns on heating	11	Fast reaction
G	16	119	445	Burns on heating	5	None

Use the information above to answer the following questions. In each question give the letter or letters required and a brief reason for your answer.

(a) Which element is liquid at room temperature (20 °C)? (1 line, 1 mark)

(b) Which element is most reactive? (1 line, 1 mark)

(c) Give one element that forms a neutral oxide. (1 line, 1 mark)

(d) Which two elements are adjacent (next to each other) in a period of the Periodic Table? (1 line, 1 mark)

(e) Which three elements are in the same group of the Periodic Table? (1 line, 1 mark)

(f) Which element is a nonmetal? (1 line, 1 mark)

(g) What is the structure of element G at room temperature? (1 line, 1 mark)

If samples of A are put into solutions of the nitrates of C and E, a reaction takes place with the nitrate of E but not with the nitrate of C.

(h) Using all the information in the question, arrange the four elements A, C, D and E in order of reactivity, with the most reactive first. (3 lines, 2 marks)

(Southern Regional Examinations Board R(B))

9 The following table shows part of the Periodic Table with only a few symbols included.

GROUP / PERIOD	I	II											III	IV	V	VI	VII	O
1																		He
2																	F	
3	Na												Al	Si		S		
4		Ca				Cr											Br	
5																		
6																		

(a) Using only the elements in the table write down the symbol for each of the following:

(i) A metal stored in oil. (1 line, 1 mark)

(ii) A noble or inert gas. (1 line, 1 mark)

(iii) A transition metal. (1 line, 1 mark)

(iv) A halogen which is a gas at room temperature and pressure. (1 line, 1 mark)

(v) The element with the largest number of protons in the nucleus of each atom. (1 line, 1 mark)

(vi) The element which would contain most atoms in a 100 g sample. (1 line, 1 mark)

(vii) An element extracted from sand when a mixture of dry sand and magnesium powder is heated. (1 line, 1 mark)

(b) When Mendeleef devised the Periodic Table many of the elements that are now known had not been discovered. The following account refers to such an element. It is represented by the symbol X but this is not the usual symbol for the element.

X is an element with a melting point of 30 °C and a boiling point of 2440 °C. It conducts electricity at room temperature.

It burns in oxygen to form an oxide which has a pH of 7. The oxide is a colourless solid with a formula X_2O_3.

X forms a compound with fluorine which is a high melting point solid that conducts electricity when molten. The similar compound with bromine has a formula XBr_3 but is a low melting point solid.

The approximate relative atomic mass of X is 70.

(i) Is X a solid, liquid or gas at room temperature (20 °C)? (1 line, 1 mark)

(ii) Give a reason why X is regarded as a metal. (2 lines, 1 mark)

(iii) Predict the formula for the fluoride of X. (1 line, 1 mark)

(iv) What are the products of the electrolysis of the molten fluoride of X? (2 lines, 2 marks)

(v) Give the symbol for the element in the above table which most closely resembles X. (1 line, 1 mark)

(vi) In which group and period of the Periodic Table will X be placed? (1 line, 1 mark)

(East Anglian Examinations Board)

10(a) Some powdered starch was mixed thoroughly with dry copper(II) oxide and the mixture was heated in a dry test tube. Droplets of a colourless liquid collected on the cooler part of the test tube and carbon dioxide was produced.

The residue in the test tube at the end of the experiment includes a brown solid.

(i) The colourless liquid produced is water. Describe a test that could be used to show this. (1 line, 1 mark)

(ii) Identify the brown solid present in the residue. (1 line, 1 mark)

(iii) What is the reason for adding copper(II) oxide to the starch before heating? (2 lines, 1 mark)

(iv) What can be concluded about starch as a result of this experiment? (1 line, 1 mark)

(b) An aqueous solution of starch was divided into four portions. These were used for four separate experiments.

(i) A few drops of iodine solution were added. A dark blue-black colouration was formed.

(ii) Starch solution and dilute hydrochloric acid were mixed and boiled. A pale yellow solution was produced when iodine solution was added.

(iii) Saliva was added to starch solution. A pale yellow solution was again produced when iodine solution was added.

(iv) Yeast was added to starch solution and the mixture kept in a warm place. The solution started to froth and a colourless gas was produced. Ethanol (boiling point 78 °C) was present in the solution at the end of the experiment.

1 What is meant by the term 'aqueous solution'? (1 line, 1 mark)

2 What can be concluded from experiments **(ii)** and **(iii)**?

3 At the end of experiment **(ii)** a reducing sugar is present in the solution.

a Which reagent could be used to test for a reducing sugar? (1 line, 1 mark)

b What colour change would be observed when this reagent is heated with the solution from experiment **(ii)**? (1 line, 1 mark)

4 What name is given to the process occurring in experiment **(iv)**? (1 line, 1 mark)

5 Name the colourless gas evolved during experiment **(iv)**. (1 line, 1 mark)

6 What is present in yeast which produces the reaction in experiment **(iv)**? (1 line, 1 mark)

7 Draw a labelled diagram of apparatus that could be used in order to obtain a fairly pure sample of ethanol from the solution remaining from experiment **(iv)**. Give the name of this process. (diagram, 3 marks)

(East Anglian Examinations Board)

11 Three essential elements in plant foods are nitrogen, phosphorus and potassium. Compounds of nitrogen increase the size of leaves, rate of growth and final amounts of crop. Compounds of phosphorus stimulate root development and encourage early ripening. Compounds of potassium encourage healthy growth and make the crops more resistant to drought and extremes of temperature.

	N *% of nitrogen compounds*	*P* *% of phosphorus compounds*	*K* *% of potassium compounds*
Fertilizer 1	10	10	20
2	6	15	15
3	14	6	20
4	8	12	8

The table above shows the percentage of N, P and K in four synthetic fertilizers. Use this information to do the following.

(a) Construct a pie chart of the essential plant foods in Fertilizer 1. (1 mark)

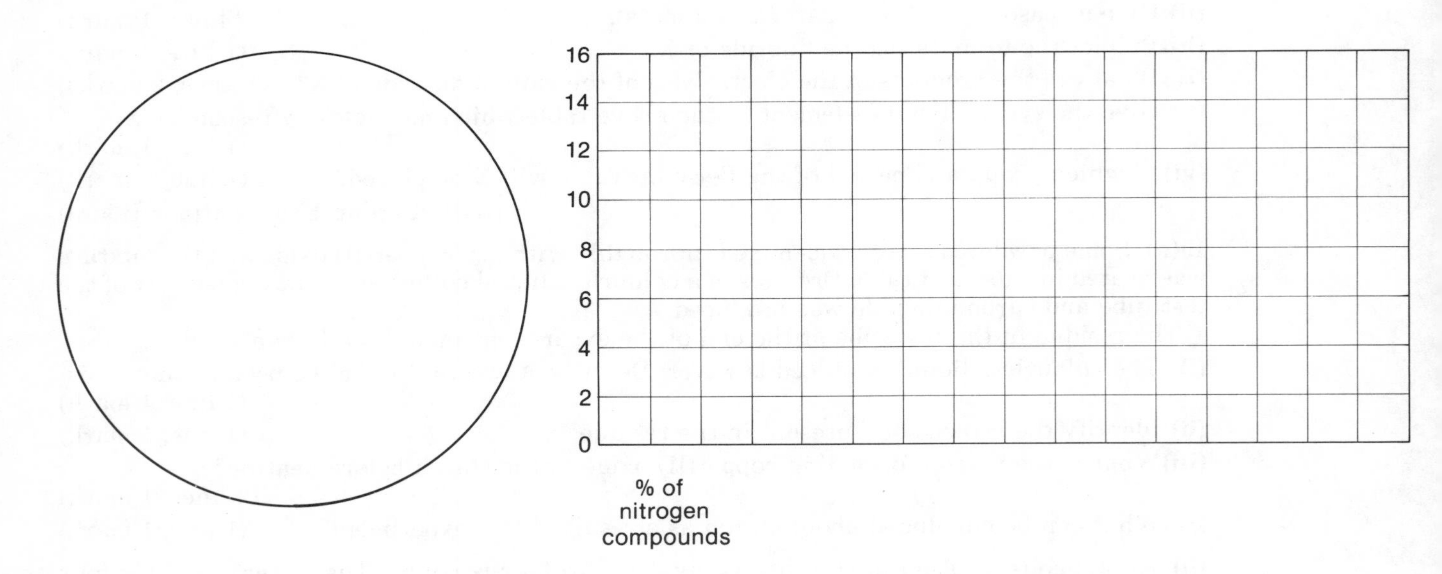

(b) Construct a bar graph showing the percentage of nitrogen in each of the four fertilizers. (4 marks)

(c) Which of the four fertilizers described has the largest percentage of nitrogen? (1 mark)

(d) For potatoes a fertilizer should be chosen which has an equal N and P content with a high K content. Which *one* of the four fertilizers best fits the description? (1 mark)

12 Various government publications list serial numbers for food additives with their specific names. Here is a part of the list.

E100 Curcumine	E200 Sorbic acid	E400 Alginic acid
E101 Riboflavine	E210 Benzoic acid	E405 Propane-1,2-diol alginate
E102 Tartrazine	E211 Sodium benzoate	E406 Agar
E104 Quinoline Yellow	E220 Sulphur dioxide	E410 Locust bean gum
E110 Sunset Yellow FCF	E221 Sodium sulphite	E412 Guar gum
E123 Amaranth	E222 Sodium hydrogensulphite	E413 Tragacanth
E140 Chlorophyll	E223 Sodium metasulphite	E414 Gum Arabic
E150 Caramel	E226 Calcium sulphite	E415 Zanthan gum

Here is a label from a fruit drink bottle.

1 litre e

ORANGE & PINEAPPLE DRINK

DILUTE TO TASTE

Ingredients
Water, sugar, oranges, pineapple juice, citric acid, stabilisers (E412 E415 E405), preservatives (E211 E223), artificial sweetener (saccharin), flavouring, colours (E102 E110)

(a) Use the list to name the stabilizers, preservatives and colours in the drink.

E412 ______________ E211 ______________
E415 ______________ E223 ______________
E405 ______________ E102 ______________
E110 ______________

(4 marks)

(b) Suggest a reason why the names do not appear on the label instead of serial numbers. (1 mark)

(c) Why are E102 and E110 added to the drink? (1 mark)

(d) If a food additive had a serial number of 416 what kind of substance would it be? (1 mark)

(e) Predict what food additive E234 would do. (1 mark)

(SEB Standard Grade Chemistry sample papers (Ass. 5 → 3) Q1&2)

***13** Toothpastes contain a polishing agent, a compound to lessen tooth decay and a flavouring agent.

(a) A compound added to lessen tooth decay is sodium fluoride. Suggest:

(i) the formula; and

(ii) the colour of sodium fluoride. (2 marks)

(b) One of the flavourings added is menthol. Menthol is a covalent compound which is insoluble in water.

Suggest a suitable solvent for menthol. (1 mark)

(c) Substances used for cleaning false teeth contain potassium percarbonate ($K_2C_2O_6$).

When potassium percarbonate is heated it decomposes into potassium carbonate, carbonate dioxide and oxygen.

Construct the symbol equation for this reaction. (1 mark)

(MEG Chemistry Paper III Q2)

***14** Ammonia is manufactured by making nitrogen and hydrogen react together at a moderate temperature and a high pressure in the presence of finely divided iron. The process is exothermic and the equation for the reaction is

$$N_2(g) + 3H_2(g) \rightleftharpoons 2NH_3(g)$$

The graph below shows the percentage yield of ammonia at different pressures and three different temperatures.

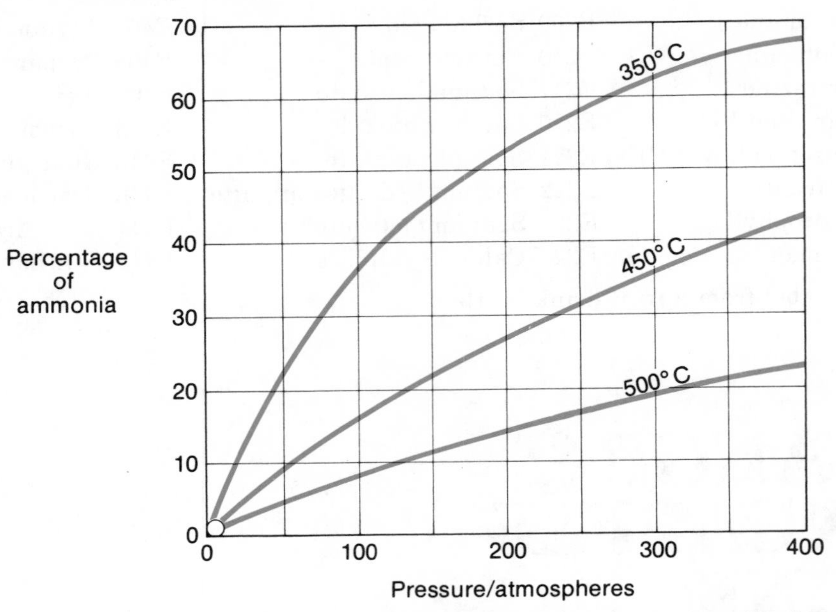

(a) Using the information contained in the graphs, describe how the amount of ammonia produced is affected by:

(i) changes of temperature. **(ii)** changes in pressure. (1 mark each)

(b) What yield of ammonia would you expect to obtain if the reaction were carried out at 400 °C and 300 atmospheres? (1 mark)

(c) What is the purpose of the finely divided iron? (1 mark)

(d) Ammonia solution can be used as a fertilizer. It has to be inserted in the soil via pipes placed well below the roots of the plants. Give *two* possible reasons why ammonia solution is not used on the surface of the soil. (2 marks)

(e) Copper metal sheet sometimes softens during heating in an atmosphere of ammonia gas. The metal emerges from the process with a bright, shiny surface without the usual tarnishing coat of oxide. Explain how ammonia could have this effect on the surface of copper. (2 marks)

(Total 8 marks)

(LEAG Syll. B Paper III Q3)

***15** Titanium is the seventh most abundant element in the earth's crust. One form in which it occurs is rutile, TiO_2. In extracting titanium from its ore, rutile is first converted to titanium(IV) chloride, $TiCl_4$, and this is then reduced to the metal by heating it with sodium or magnesium in an atmosphere of argon. Titanium(IV) chloride is a simple molecular covalent substance.

(a) Given that the titanium atom has four outer electrons used for bonding, draw a diagram to show the bonding in titanium(IV) chloride (only the outer electrons of the chlorine atoms should be shown). (2 marks)

(b) Write a balanced equation for the reaction of titanium(IV) chloride with sodium. (1 mark)

(c)(i) In which physical state would you expect to find titanium(IV) chloride at room temperature?

(ii) Explain why the physical state of titanium(IV) chloride differs from that of sodium chloride at room temperature. (3 marks)

(d) Suggest a reason why it is necessary to carry out the reaction of titanium(IV) chloride with sodium in an atmosphere of argon. (1 mark)

(e) Titanium is expensive in spite of the fact that it is relatively abundant in the earth's crust. Suggest a reason for this. (1 mark)

(f) Titanium is used in the structures of supersonic aircraft and space vehicles. Suggest *two* properties it might have that make it more suitable than other metals for this purpose. (2 marks)

(Total 10 marks)

(LEAG Syll. A Paper III Q5)

***16(a)** The first stage in the manufacture of sulphuric acid is the oxidation of sulphur dioxide to sulphur trioxide (SO_3) at 450 °C with a vanadium(V) oxide catalyst.

(i) What do you understand by the term *catalyst*? (2 marks)

(ii) Of which group of elements in the Periodic Table is vanadium a member? (1 mark)

Suggest one other property you would expect for vanadium compounds. (1 mark)

(iii) Construct an equation for the catalytic oxidation reaction described above. (2 marks)

(b) The exhaust system of American cars are fitted with special catalysts to convert carbon monoxide and nitrogen monoxide into safer products.

$$2CO + 2NO \rightarrow 2CO_2 + N_2$$

(i) Name an additional pollutant present in the exhaust fumes of British cars. How does it get there? (2 marks)

(ii) Explain briefly why carbon monoxide is dangerous. (2 marks)

(iii) Explain why carbon dioxide can also be considered an atmospheric pollutant.

(iv) How might nitrogen monoxide be formed in a car engine? (2 marks)

(v) From your knowledge of nitrogen monoxide, suggest one way in which its presence in the air might cause harmful effects. (2 marks)

(c) In hard water areas concentrated nitric acid is sometimes added to irrigation water in greenhouses to clear scale from the piping.

(i) What is the origin of the scale? (2 marks)

(ii) Write an equation to show how the scale is removed. (2 marks)

(iii) What is the other benefit of using nitric acid in the irrigation system? (1 mark)

(SEG Paper III Q4)

***17** Answer *either* A *or* B

A(a) Outline briefly how sulphur is produced commercially. (4 marks)

(b) Write equations for the reactions of oxygen with:

(i) sulphur to produce sulphur dioxide; (1 mark)

(ii) sulphur dioxide to produce sulphur trioxide. (2 marks)

(c) Describe with a reason how sulphur trioxide is converted into sulphuric acid commercially (equations are not required). (3 marks)

(d)(i) Describe *two* reactions which you have done in the laboratory which show how dilute sulphuric acid behaves as an acid. (6 marks)

(ii) Concentrated sulphuric acid can act as a dehydrating agent. Describe *two* reactions which illustrate this property. (4 marks)

B In the United Kingdom all the ethanol used for industrial and medical purposes is obtained from ethene whereas in the West Indies and elsewhere in the world fermentation processes are used.

(a) Describe and explain the main chemical reactions by which the ethanol is produced commercially:

(i) from ethene; (3 marks)

(ii) by fermentation. (4 marks)

(b) The fermentation processes being developed in the West Indies are creating much interest because the sugar raw material is a renewable resource. Explain how such a resource is better than the resource from which ethene is obtained in the UK. (3 marks)

(c) Describe, giving an equation and stating conditions and observations, how ethanol may be oxidized in the laboratory. (6 marks)

(d) What is meant by microbiological oxidation of ethanol? Why is this of importance to wine producers? (4 marks)

(NI Paper III Q2)

***18** Read this newspaper extract about Andi motor cars.

Andi join rust war

Andi is waging all-out war against corrosion in introducing galvanised bodies.

The new scheme, costing Andi £550 million, is to get the protection of zinc coating around the 100 and 200 models initially.

It is the biggest step yet taken by a single manufacturer to combat the 'red menace,' currently costing £1000 million a year in the UK through scrapping cars.

Adding just £50 to the price of the cheapest Andi 100 at £9944, the treatment involves the application of 14lb of zinc to internal and external surfaces.

It is expected to last longer than the worthwhile service life of the car, which goes up from around 12 years to 16 or 17.

(a) What is the 'red menace' referred to in the article? (1 mark)
(b) Explain what is meant by 'galvanized' bodies. (1 mark)
(c) The process described in the extract is to prevent the formation of iron(II) ions. Write the symbol for an iron(II) ion. (1 mark)
(d) In terms of electron flow explain how galvanizing helps to protect the steel car body. (2 marks)
(e) Some car bodies are made of aluminium instead of steel. Predict the effectiveness of galvanizing as a form of protection in this case. Justify your answer. (2 marks)
(SEG Standard Grade (1–3) Q6)

***19** A pupil was attempting to construct a reactivity series for metals.
The method used was to add some detergent to tiny pieces of the separate metals, and then to add dilute acid. The rates at which foam was produced were compared. The more active the metal, the quicker the foam was produced.
The pupil's results did not match the correct results quoted in the text book.
(a) From the brief description give *two* possible experimental errors which may have occurred. (2 marks)
(b) Describe in some detail how the experiment should be carried out to avoid errors. (4 marks)
(c) If aluminium and sulphuric acid had been used in the above experiment the products would have been aluminium sulphate and hydrogen. (3 marks)
Write a balanced equation for this reaction.
(SEG Standard Grade (1–3) Q7)

ANSWERS TO STRUCTURED QUESTIONS

1(a) Methane CH_4.
(b)(i) Carbon dioxide and water.
(ii) Limited amount of air (or oxygen). **(iii)** Carbon monoxide.
(c)(i) Polymerization.
(ii) A large number of the 'repeating unit' – the exact number being unknown and variable.
(iii)
```
H       H
 \     /
  C = C
 /     \
H       H
```
(iv) Double bond between two carbon atoms.

2(a) E.g. ammonium sulphate, ammonium nitrate, ammonia.
(b)(i) They were washed off fields which drain into rivers. **(ii)** From detergents.
(c) Fertilizers – nitrogen makes plants grow.
(d) Fish would die.

3(a) Propene.
(b) C_2H_3Cl.
(c)
```
 [ H   Cl ]
 [ |   |  ]
—[ C — C  ]—
 [ |   |  ]
 [ H   H  ]n
```

(d) There is a double bond between two carbon atoms.
(e) Red (or brown) to colourless (not white).
(f)(i) Petroleum (crude oil).
(ii) High density. Close packing produces a high density – more molecules in a given space.
(g)(i) It does not rust, is easy to mould, light, etc.
(ii) It is stronger.
(h)(i) They do not rot away (biodegrade).
(ii) Carbon dioxide, water, carbon monoxide (any two).
(ii) With a magnet.
(ii) Soak the refuse thoroughly with water. Scoop off polymers from the surface. The soaked paper sinks.
(j)(i) They are good heat insulators. The polymers melt easily and can be cheaply moulded into blocks.
(ii) Pure reclaimed polymers could be used for a wider range of uses. They would be more valuable and would save the earth's resources.

4(a) Put a lighted splint into the gas. The gas burns with a squeaky pop.
(b) Z, X, W, Y.
(c)(i) Copper.
(ii) Displacement reaction.

5(a) Equal masses of crushed ice.

(b)

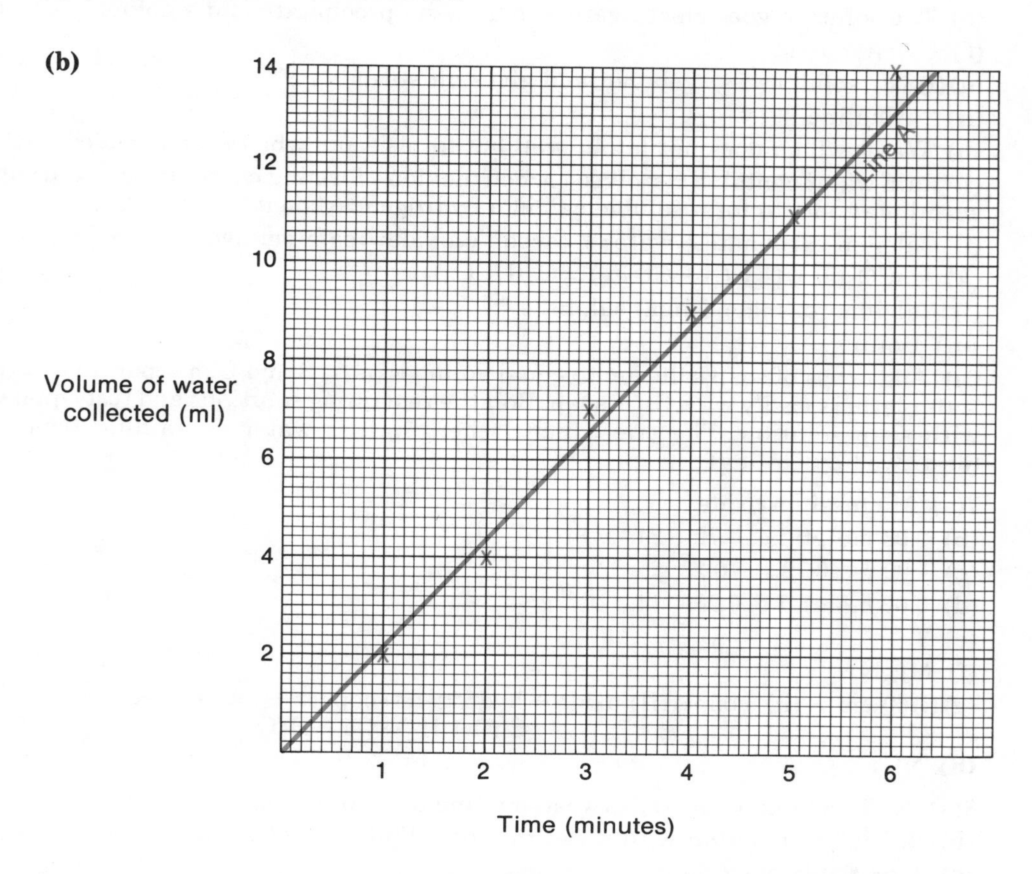

(c)

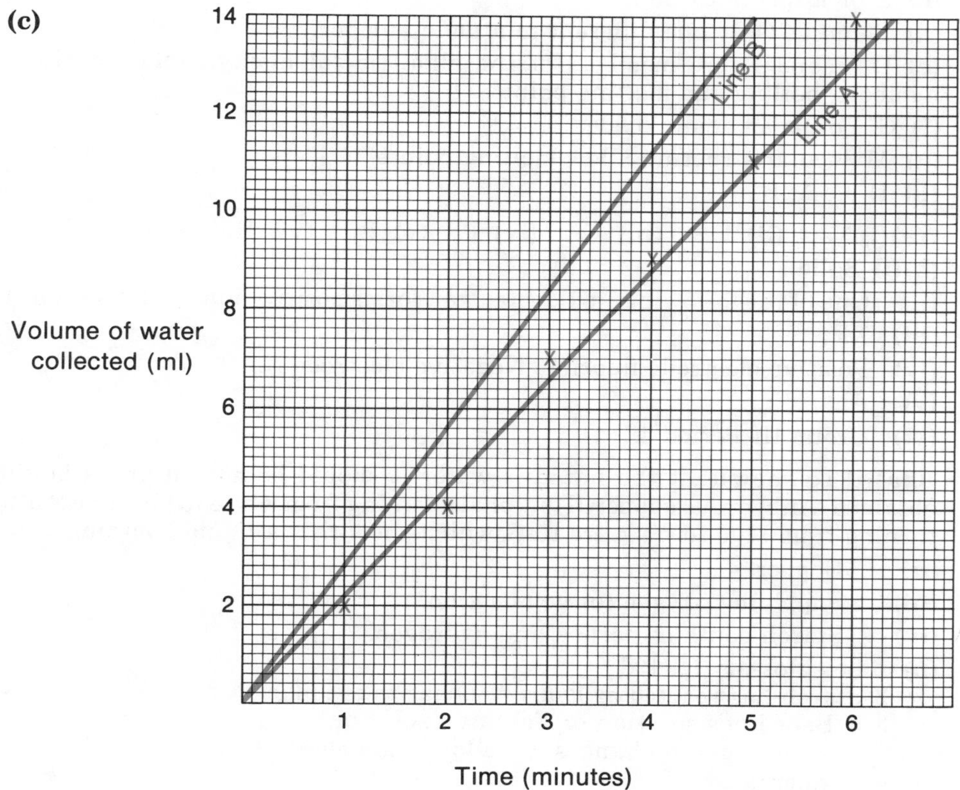

(d) Adding salt lowers the melting point of ice.
(e) It causes corrosion of motor cars and also damages concrete on motorway bridges, etc.

6(a) Chalk, limestone.
(b) $CaCO_3$.
(c)(i) Limewater.
(ii) Calcium hydroxide.
(d) It turns milky.
(e) The solution goes clear again – there is no precipitate and a colourless solution.
(f) Calcium oxide.
(g)(i) The solid 'fizzes', gets hot and gives off steam.
(ii) Alkaline.
(h) Water which does not lather well but forms scum when soap is added.
(i) Rain water contains dissolved carbon dioxide – forming carbonic acid. Carbonic acid reacts with limestone, to form soluble calcium hydrogencarbonate.
(j) Calcium carbonate + water + carbon dioxide $\rightleftharpoons$ calcium hydrogencarbonate.
(k) $CaCO_3(s) + H_2O(l) + CO_2(g) \rightleftharpoons Ca(HCO_3)_2$.
(l) Calcium (or magnesium) sulphate.
(m) Boiling.
(n) Add drops of soap solution to equal volumes of A, B and C in separate test tubes. Distilled water lathers easily with little soap. Boil the remaining solutions and test equal volumes again with soap solution. The one which lathers well is water containing temporary hardness (destroyed by boiling).

7(a) Magnesium oxide.
(b) 0.03 moles, i.e. $0.72 \div 24$.
(c) $1.00\ g - 0.72\ g = 0.28\ g$.
(d) 0.02 moles, i.e. $0.28 \div 14$.
(e) 3.
(f) Mg_3N_2.
(g)(i) 60 cm^3 **(ii)** copper(II) oxide + hydrogen → copper + water.
$CuO + H_2 \rightarrow Cu + H_2O$
(h) NH_3.

8(a) E. Room temperature between melting and boiling points.
(b) B. Violent reaction with water or most alkaline oxide.
(c) A or E. pH of oxide is 7.
(d) F and C. Only one different in atomic number.
(e) D, F and B are members of the alkali metal family. Relationship is between atomic numbers, boiling and melting points.
(f) G. Forms an acidic oxide.
(g) Molecular structure – low melting point.
(h) D, C, A, E.

9(a)(i) Na; **(ii)** He; **(iii)** Cr; **(iv)** F; **(v)** Br; **(vi)** He; **(vii)** Si.
(b)(i) Solid.
(ii) One of the following – oxide neutral, forms positive ions, conducts electricity when solid.
(iii) XF_3.
(iv) Positive electrode fluorine. Negative electrode X.
(v) Al.
(vi) Group III Period 4.

10(a)(i) To show that the liquid is water it should be heated and a boiling point of 100 °C obtained. Anhydrous copper(II) sulphate turning from white to blue or cobalt(II) chloride paper turning from blue to pink are tests which show that a liquid contains water.
(ii) Copper.
(iii) Supplying oxygen for combustion.
(iv) Starch contains carbon and hydrogen.
(b) 1 See glossary (p. 198.)
2 Starch has reacted or hydrolysis has taken place.
3 a Benedict's solution or Fehling's solution.
b Blue or green changes to yellow/orange/red.
4 Fermentation.
5 Carbon dioxide.
6 Enzymes.
7 Refer to Fig. 2.4.
Fractional distillation.

11(a)

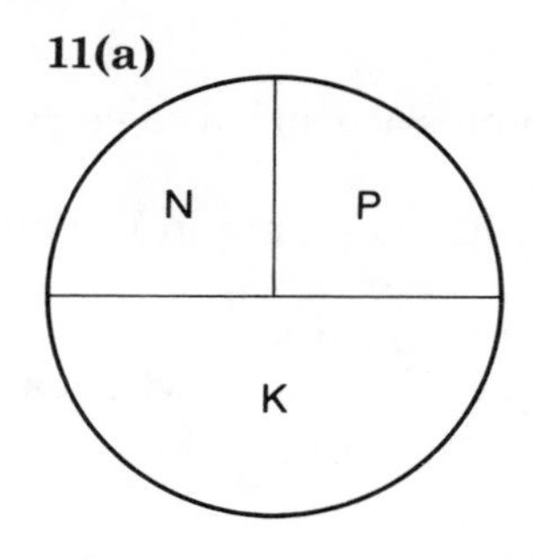

(b)

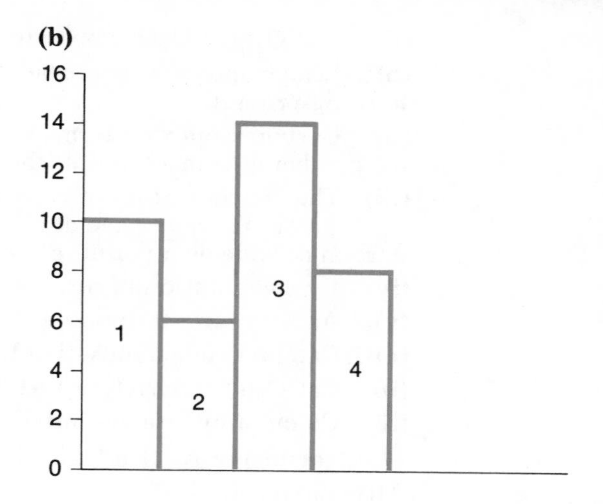

(c) Fertilizer 3.
(d) Fertilizer 1.

12(a)

E412	Guar gum	E211	Sodium benzoate
E415	Zanthan gum	E223	Sodium metasulphite
E405	Propane-1,2-diol alginate	E102	Tartrazine
		E110	Sunset yellow FCF

(b) Numbers take up less space on the label.
(c) E102 and E110 are colourings.
(d) A stabilizer.
(e) It would be a preservative.

13(a) Like sodium chloride **(i)** NaF; **(ii)** white.
(b) Ethanol, hexane.
(c) $2K_2C_2O_6 \rightarrow 2K_2CO_3 + 2CO_2 + O_2$.

14(a)(i) Increasing temperature decreases the percentage of ammonia.
(ii) Increasing pressure increases the percentage of ammonia.
(b) *About* 50 per cent.
(c) As a catalyst.
(d) Ammonia escapes into the air. Ammonia solution runs off the soil.
(e) Ammonia reacts with any copper(II) oxide reducing it to copper.

$$3CuO(s) + 2NH_3(g) \rightarrow 3Cu(s) + 3H_2O(l) + N_2(g)$$

15(a)

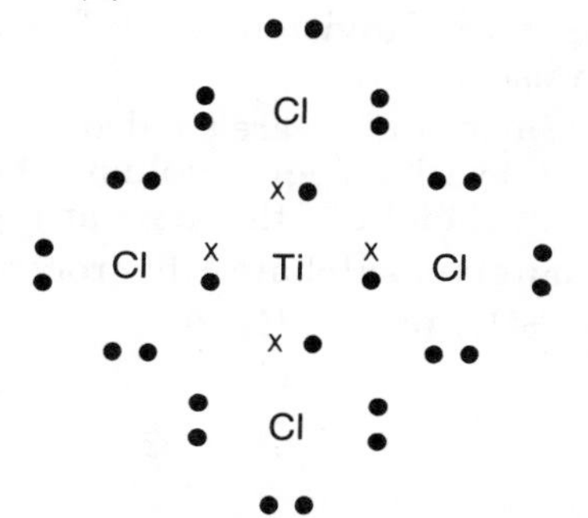

(b) $TiCl_4 + 4Na \rightarrow Ti + 4NaCl$.
(c)(i) Liquid.
(ii) Sodium chloride consists of a giant structure of sodium and chloride ions. Titanium(IV) chloride consists of molecules with only weak forces between molecules.
(d) Sodium would react with oxygen in the air forming sodium oxide – argon is inert and does not react with sodium.
(e) It is very expensive to extract – sodium is expensive and a great deal of energy is required.
(f) It has a higher melting point and is less reactive than other metals.

16(a)(i) A substance that alters the rate of a chemical reaction without being used up.
(ii) Transition metal.
They have coloured compounds.

(iii) $2SO_2(g) + O_2(g) \rightleftharpoons 2SO_3(g)$.

(b)(i) Lead compounds (e.g. lead bromide). Lead compounds are added to petrol to improve the burning properties.

(ii) Carbon monoxide forms a stable compound with haemoglobin in the blood, preventing oxygen being transported in the body.

(iii) The concentration of carbon dioxide is increased by burning, etc.

(N.B. There is some doubt here. Some books and examination syllabuses do not regard carbon dioxide as a pollutant because it is normally present in air.)

(iv) By combination of nitrogen and oxygen.

(v) As acid rain. It dissolves in water after oxidation producing nitric acid.

(c)(i) Calcium compounds dissolve in water. Calcium compounds deposit in pipes, etc.

(ii) $CaCO_3(s) + 2HNO_3(aq) \rightarrow Ca(NO_3)_2(aq) + CO_2(g) + H_2O(l)$.

(iii) Calcium nitrate and nitric acid act as nitrogen fertilizers.

17 Alternative A. See Unit 27.

Alternative B

(a)(i) Ethene is dissolved in concentrated sulphuric acid and the resulting mixture is diluted with water.

$C_2H_4(g) + H_2SO_4(l) \rightarrow C_2H_5HSO_4(l)$

$C_2H_5HSO_4(l) + H_2O(l) \rightarrow C_2H_5OH(l) + H_2SO_4(l)$

(ii) By the action of yeast on glucose solution in a warm place.

$C_6H_{12}O_6(aq) \rightarrow 2C_2H_5OH(l) + 2CO_2(g)$

(b) The fermentation of sugar to produce ethanol was first done in Brazil. The sugar cane can be grown fresh every year. The supply should not run out. Fossil fuels are finite. Also this process was developed at a time when sugar prices were very low and selling sugar was difficult.

(c) Ethanol can be oxidized to ethanoic acid by warming ethanol with excess potassium dichromate(VI). The orange solution turns to green.

$C_2H_5OH + 2[O] \rightarrow CH_3COOH + H_2O$

(d) Bacteria present in the air will oxidize ethanol to ethanoic acid. During fermentation the fermenting mixture is kept out of contact with air. The carbon dioxide is allowed to escape but the air is not allowed to come into contact with the wine. An open bottle of wine will sour, i.e. oxidize to ethanoic acid. This was the original way of making vinegar.

18(a) Rust.

(b) Steel is coated with a layer of zinc.

(c) Fe^{2+}

(d) Electrons flow from the zinc to the iron, preventing ion formation.

(e) Aluminium is more reactive than zinc; the flow of electrons this time would be from aluminium to zinc and the corrosion of aluminium would be speeded up.

19(a) The metals may not have been in the same state of division, i.e. different surface areas. Different quantities of detergent may have been used.

(b) Equal masses of the different metals with the same surface area added to equal volumes of dilute hydrochloric acid (of the same concentration) with an equal volume of detergent added to each, kept at the same temperature. Measure the height of the foam at regular intervals.

(c) Aluminium + sulphuric acid → aluminium sulphate + hydrogen.

$2Al(s) + 3H_2SO_4(aq) \rightarrow Al_2(SO_4)_3(aq) + 3H_2(g)$

Comprehension Questions

Unstarred questions are for all candidates. Starred questions are for A, B, C candidates.

1 Read the following passage and answer the questions which follow:

Aircraft manufacturers are preparing for one of the biggest technological revolutions in the way they build planes since aluminium replaced canvas-and-wood frames more than 50 years ago. It involves the use of aluminium lithium, a new alloy that is up to 20 per cent lighter than traditional materials.

So enthusiastic are the manufacturers that British Alcan Aluminium, one of the two companies leading the worldwide charge of the light brigade, believes 'al-li' alloys could replace up to 75 per cent of conventional metals over the next decade.

Lithium, the lightest metallic element, is mined in America, Australia and Africa. Research into its properties when mixed with aluminium has been going on since the 1920s.

The industry really started to get interested in the 1970s after the oil crisis, when soaring prices made fuel saving important. Lighter planes fly further.

Several technological problems have to be overcome. Though lithium is already widely used in batteries and in atomic power generation, it can be tricky stuff to handle. It oxidizes rapidly in air and in its molten state is highly reactive.

In recent years these problems have largely been overcome. Only the financial problem – the high price of lithium alloys – remained. However, oil prices rose so fast in the 1970s that high strength al-li alloys are now cost effective at three times the price of conventional aluminium.

These alloys are not only lighter, they are also stiffer, stronger and more corrosion resistant.

Sunday Times

(a) What is meant by the word *'alloy'*? (1 mark)

(b) Why is it an advantage to use lighter alloys in aircraft construction? (1 mark)

(c) What information in the passage suggests that lithium is a reactive metal? (1 mark)

(d) Give *three* advantages, apart from lightness, of using aluminium–lithium alloys rather than aluminium. (3 marks)

(e) Give one disadvantage of aluminium–lithium alloys. (1 mark)

(EAEB Syll. B/14/–CN 1985)

*2 Read through the following account and study the map carefully. Then answer the questions.

THE INVERGROG RESERVOIR PROJECT

A. L. McHol and Co are investigating the possibility of building a new whisky distillery on the Invergrog Industrial Estate. The proposed site is to the north-east of the town (see map). Their big problem is where to get pure water from to make the whisky.

There are many streams flowing from the hills to the west of Invergrog. At first the company thought there were five possible places to build a small reservoir. These are shown and numbered on the map (1, 2, 3, 4 and 5). When samples of water taken from two of these locations were examined, they were found to contain too high a level of pollution. An analysis of water from a third possible site showed that it contained too many dissolved mineral salts.

Analysis of the mineral water from the third site (mg/litre)

hydrogencarbonate	180.60	calcium	44.80
sulphate	10.00	magnesium	20.00
chloride	15.60	fluoride	0.06
nitrate	0.20	sodium	12.50

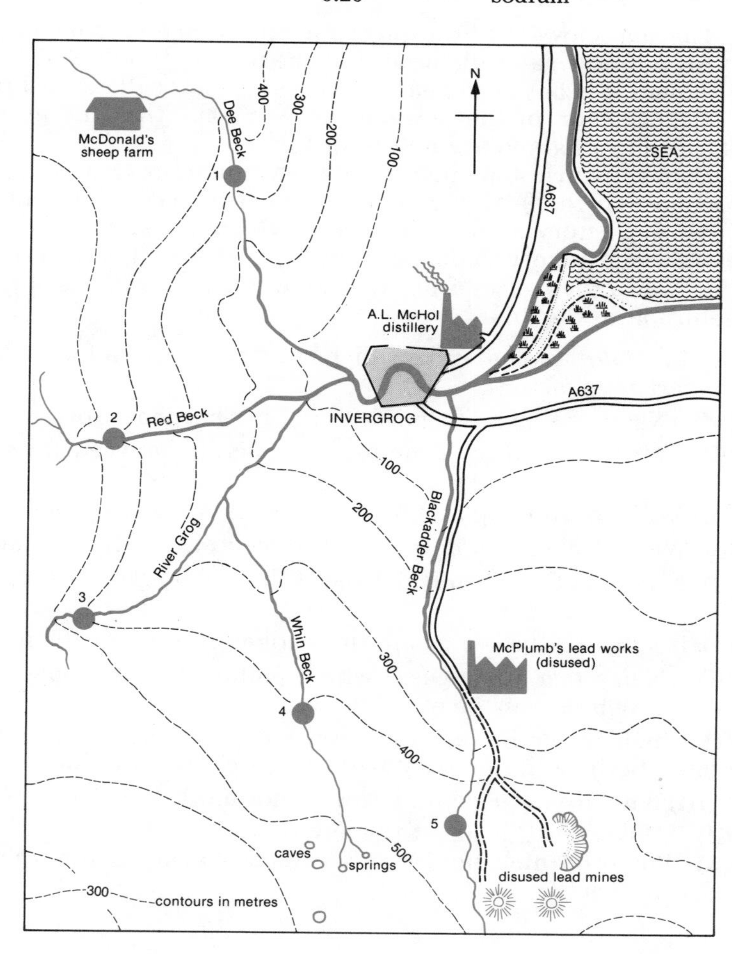

To help them choose between the two remaining sites the company carried out chemical tests to find the hardness of the water.

(a) Explain as fully as you can why the water at two of the sites was polluted. (2 marks)

(b) What is the molarity of the mineral water with respect to sodium ions? (2 marks)

(c) Which of the ions present in this spring water are associated with hardness in water? (2 marks)

(d) Describe a simple method you could use to compare the hardness of the water from the two remaining sites. Your account should include the list of apparatus and chemicals you would use and details of how it is possible to make a fair comparison. (4 marks)

(e) If the distillery needed to use very soft water, it could be treated by using an ion exchange column. Explain briefly how this process works. (2 marks)

(NEA Syll. A Paper III QB4)

***3** Carefully read the following passage and then answer the questions which follow it.

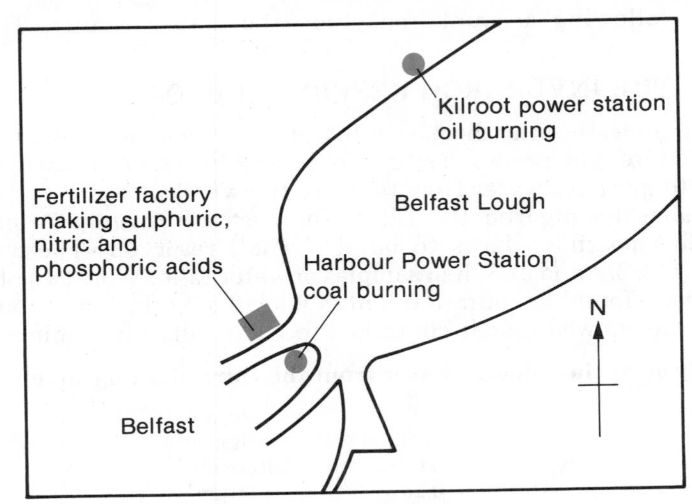

The map shows the location of a fertilizer factory and two power stations in the Belfast area.

All the raw materials needed at the power stations and fertilizer factory must be imported although it is hoped to generate power from locally mined lignite at some time in the future.

The location of the power stations means that the giant cooling towers seen at power stations in England are not needed.

Both sulphuric and nitric acids are manufactured in the fertilizer factory. The sulphuric acid produced at the fertilizer factory is used to make phosphoric acid from phosphate rock. The fertilizer factory markets fertilizers which are mixtures of the ammonium salts of the acids made at the factory. A plume of brown smoke is often seen rising from the nitric acid plant, and this, and other gases from the factory and power stations, contribute to the atmospheric pollution in the area.

(a) List *four* raw materials which must be imported for use either in the power stations, or in the fertilizer factory. (4 marks)

(b)(i) Name the main chemical found in phosphate rock. (1 mark)

(ii) Write a word equation for the reaction between sulphuric acid and phosphate rock. (1 mark)

(iii) Write a word equation for the reaction between ammonia and phosphoric acid. (1 mark)

(c) Write a symbol equation for the reaction between ammonia and sulphuric acid(2 marks)

(d) Name another chemical made at the factory which is an ingredient of the mixed fertilizers. (1 mark)

(e)(i) Name the brown gas in the smoke given off at the fertilizer factory. (1 mark)

(ii) Name *two* other gases which pollute the atmosphere over these industries and state which process gives off the gas. (4 marks)

(f) Winds in the Belfast area blow mostly from the south west. Does this have any effect on atmospheric pollution over Belfast caused by these industries? (1 mark)

(g)(i) Why are large cooling towers not needed at the power stations? (1 mark)

(ii) How does this affect the environment? (1 mark)

(h) Would burning lignite make much change to atmospheric pollution? Give a reason for your answer. (2 marks)

(NI Paper III Q4 Comprehension)

ANSWERS TO COMPREHENSION QUESTIONS

1(a) An alloy is a mixture of metals.
(b) Lighter aircraft fly further on the same amount of fuel.
(c) It oxidizes rapidly in air.
(d) It is stiffer, stronger and more corrosion resistant.
(e) High price.

2(a) Site 1: water polluted by waste from sheep farm – ammonia.
Site 5: water polluted by disused lead mines – lead compounds.
(b) Concentration of sodium ions = 0.0125 g/l
= 0.0125/23 mol/l
= 0.00054 mol/l
(c) Calcium, magnesium.
(d) Measuring cylinder, burette, soap solution, burette stand, flasks, clock.
Measure out equal volumes of the two water samples into separate flasks. Add soap solution in small portions to each flask until a lasting lather is formed. (Lather lasts for at least half a minute.) Compare the two volumes of soap solution used.
(e) See Unit 19.

3(a) Sulphur, phosphate rock, coal (or coke), oil.
(b)(i) Calcium phosphate.
(ii) Calcium phosphate + sulphuric acid → phosphoric acid + calcium sulphate.
(iii) Ammonia + phosphoric acid → ammonium phosphate.
(c) $2NH_3(g) + H_2SO_4(l) \rightarrow (NH_4)_2SO_4(aq)$.
(d) Nitric acid.
(e)(i) Nitrogen dioxide.
(ii) Carbon dioxide and sulphur dioxide – burning coal.
(f) Gases are blown away from Belfast over the Lough.
(g)(i) Cooling water is discharged into the sea.
(ii) It may raise the temperature of water with some effect on the natural life of Belfast Lough.
(h) The answer should argue possible pollutant effects of burning lignite.

Essay Questions

These are used only for assessing A, B or C grade GCSE candidates.

1 Discuss the manufacture and importance of sulphuric acid. Your answer should include reference to the chemistry of the process, the factors affecting the siting of a plant, problems associated with storage and transport, and the economic importance and uses of sulphuric acid.
(10 marks)
(SEG Syll. 14 Paper II Q7)

2(a) Sodium hydroxide is manufactured from brine on a large scale, using electrolytic methods. Describe *one* such method. In your account, you should draw a simple labelled diagram of the cell, state the materials of which the electrodes are made, write ionic equations for the reactions taking place at the electrodes and list any byproducts which are obtained.
(b) For *two* of the byproducts which are obtained, give two uses of each and say how the materials so obtained have benefited society.
(25 marks)

3(a) Describe briefly how aluminium can be extracted from purified bauxite. Include in your answer a diagram of the plant and the equations. (10 lines, diagram, 8 marks)
(b) The map shows an imaginary island with no bauxite deposits. It is intended to set up a works to produce aluminium from imported bauxite to supply aluminium for this and neighbouring islands. Three sites labelled A, B and C have been discussed. List the advantages and disadvantages of each site. Which one of the three sites would you choose?
(20 lines, 12 marks)

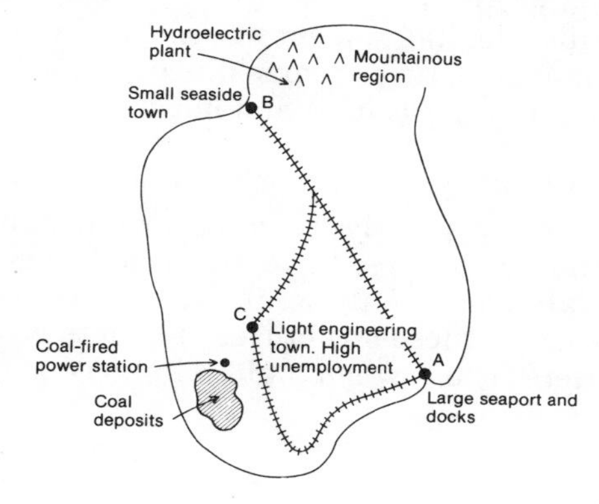

4(a) Describe the bonding that is present in the three different types of solids, making clear the nature of the particles in the solid lattice and of the forces holding them together. You may use sodium chloride, iodine and carbon (graphite), or examples of your own choosing, in answering this question.
(b) Compare the physical properties of the three chosen substances. Explain the differences or similarities in terms of the bonding.

(Total 15 marks)
(LEAG Syll. B Sample paper III Q6)

SKELETON ANSWERS TO ESSAY QUESTIONS

1 There are four parts to this question:

chemistry of the process,
factors affecting the siting,
problems of storage and transport,
importance of sulphuric acid and uses.

Your answer should cover each of these and full marks (or good marks) would only be given if all aspects were fully answered.

2(a) A longer question which emphasizes the trends towards a social angle to chemistry. The two possible methods are given in Unit 28. The Diaphragm cell method is becoming far more popular.
(b) The byproducts are hydrogen and chlorine in each method. Hydrogen is used for hardening oils for margarine and soapless detergent manufacture. It could also be developed as a pollution free fuel to replace petrol. Chlorine is used in the production of a wide range of chemicals (e.g. PVC), as a bleach and in the chlorination of water supplies.

3(a) See Unit 23.
(b) *The following answer is the skeleton on which to hang your essay. Make sure that you write in complete sentences and that you express your arguments as logically as possible.*
Site A. Good port facilities for import of bauxite and export of finished aluminium. Electricity could be transported from hydroelectric plant or coal-fired power station.
Site B. Bauxite and finished aluminium would have to be transported by rail. Electricity from hydroelectric plant readily available. Possible difficulties with availability of labour and environmental complaints.
Site C. Bauxite and finished aluminium transported by rail. Possible to use some of the aluminium in local industry. Electricity available from coal-fired power station and labour readily available.

On balance, site A is probably best. It is easier to transport electricity than bauxite or aluminium. If there were problems with labour or exact siting, site C would be possible.

4(a) Molecules – atoms (1) jointed together by covalent (1) bonds in which electrons are shared (1). (3 marks)

E.g. I_2:

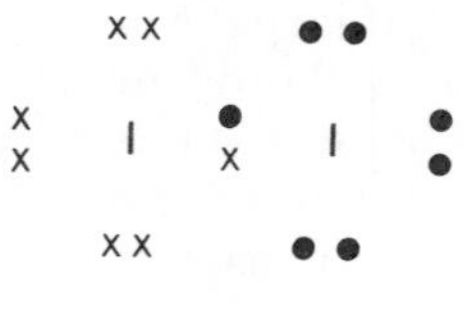

```
   xx    ••
x  I  •  I  •
x     x     •
   xx    ••
```

Giant structure of atoms joined by covalent bonds (1) to form an extended structure of undefined size (1).
E.g. graphite. (2 marks)
(Suitable diagram showing layers and arrangement within layers (1)). (1 mark)
Giant structure of ions joined by nondirectional electrostatic forces (1) to form an extended structure (1). Positive and negative ions formed by transfer of electrons (1). (3 marks)
(Suitable diagram of Na^+Cl^- or other example (1).) (1 mark)
(b) Main properties to consider are melting/boiling points and electrical conduction.
Giant structures have much higher melting points than molecular compounds (1) because the latter have only weak forces between the molecules (1). (2 marks)
None of these structures conducts electricity *as solid* (1) because none has mobile charge carriers (1) but when molten ionic compounds do conduct ($\frac{1}{2}$) because the ions become mobile ($\frac{1}{2}$). (3 marks)

(Total 15 marks)

COURSEWORK ASSESSMENT

It was explained on pages xix–xx that all GCSE Chemistry schemes contain coursework assessment. This can be of various types:

1 Assessment of practical work carried out during the course. This should account for at least 20 per cent of the total marks.
2 Project work or writing extended essays.
3 Oral tests.

In this section you will find help with these aspects of your GCSE Chemistry. Remember that these parts of your work are very important and a good performance in this internal assessment is 'marks in the bag' when the final examination takes place.

Assessment of Practical Work

This will probably be done by your teacher during normal chemistry lessons. This should be much fairer than relying upon the performance in a single practical examination. Your teacher should not give you the actual marks that you are awarded but I am sure that he or she will all the time be giving you advice on how you are performing and how you could do better.

WHAT IS BEING ASSESSED?

Various skills will be assessed throughout the course, not necessarily in every experiment that you do. Your teacher should tell you when assessments are being made and what skill is being assessed.

The skills being assessed vary from one syllabus to another, although much of the difference is in the wording. Skills that are usually assessed include your ability to:

follow instructions;
use apparatus and chemicals safely and be able to carry out basic techniques;
make accurate observations and measurements, recording them accurately;
plan an experiment or part of an experiment;
evaluate methods and suggest possible improvements;
interpret experimental data and draw conclusions.

Some of these skills could be assessed in a written examination but others must be done during practical sessions.

WHAT ARE YOU EXPECTED TO BE ABLE TO DO?

It is most important for you to know what is expected of you in each of these skills. For your help, three levels of achievement have been identified.

Level I is the standard which nearly all candidates should achieve by the end of the course.

Level II is a good acceptable standard of practical work.

Level III is practical work of the highest standard from candidates with a special ability in practical work who have developed these skills thoroughly throughout the course.

You will realize, of course, that the best practical work does not necessarily come from the students who are best at theory.

ABILITY TO FOLLOW INSTRUCTIONS

Level I candidates should be able to follow simple instructions, e.g. to heat some water in a beaker to boiling or filter some solid from a suspension. These instructions may be written or given orally.

Level II candidates would be able to follow more complex instructions, e.g. crystallize or distil.

Level III candidates would be able to follow complex series of instructions. For example, 'add magnesium carbonate to warm, dilute hydrochloric acid until no further reaction takes place, filter and evaporate to produce some crystals of hydrated magnesium chloride'.

Ability to use apparatus and chemicals safely

Level I candidates would be able to carry out simple operations such as filtering and heating a solid chemical safely. He or she would use a spatula to transfer chemical to the test tube and heat the test tube in the flame so that it points away from other people.

Level II candidates would be able to use a wider range of apparatus safely and would see possible dangers and correct them, e.g. heating flammable liquids safely.

Level III candidates would be able to carry out operations safely with more complex apparatus, e.g. distillation of a flammable liquid.

It is important when doing practical work that you read any safety warnings on bottles of chemicals and that you use chemical apparatus carefully and for the purpose it is designed.

Figure 1 shows a school laboratory with various unsafe practices going on. List all the errors in the drawing that you can find.

Fig 1. Possible dangers in a school laboratory

Ability to make accurate observations and measurements and record them accurately

Unless you record your observations and measurements accurately there is nothing for your teacher, or anybody else, to judge. It is essential to record all observations or measurements immediately. Do not try to remember them to write down later.

Level I candidates can observe a single change which takes place, e.g. the formation of a precipitate or a colour change when copper(II) sulphate crystals are heated. They can also read the measuring instruments in Fig. 2 accurately.

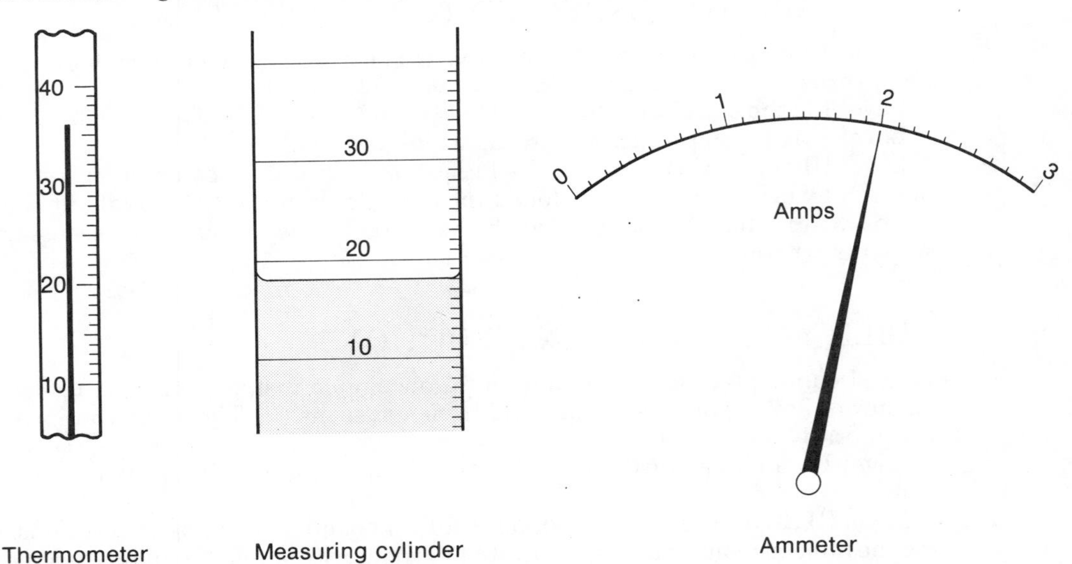

Fig 2 Reading simple laboratory instruments

(Answers: 36°C; 18 cm³; 2A)

Level II candidates can make more detailed observations. For example, they would note a colour change from brown to yellow when lead(IV) oxide is heated and also identify the colourless gas evolved as oxygen. They would be able to read the measuring instruments shown in Fig. 3.

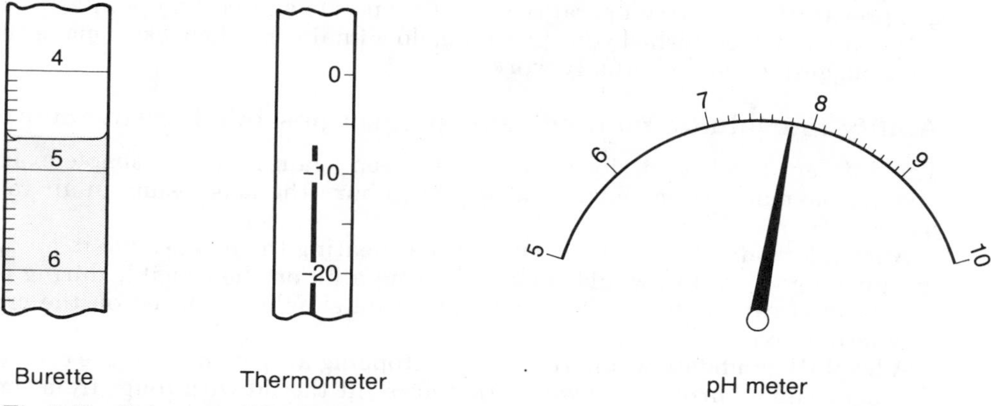

Fig 3 Reading more difficult scales

(Answers: 4·7 cm³; −7°C; 7.8)

It is important to have your eyes level with the instrument you are reading.

Level III candidates would record all of the observations in an experiment including colour changes, temperature changes, changes of state or form, evolution of a gas, etc. They would also be able to read all measuring instruments accurately.

Candidates usually lack detail when making observations. With practice you can become so much better when making observations. For example, when describing the colour of a precipitate formed pale yellow is so much better than just yellow. Look for colour changes, changes of state or form, evolution of a gas, formation of a precipitate, temperature changes associated with exothermic and endothermic reactions and crackling noises when substances are heated. Do not overlook any change taking place and do not forget to record it.

The detail required is best seen with an example. A student is given the following task:

Heat some cobalt(II) chloride crystals in a test tube. When the residue has cooled, add water drop by drop.

The observations should be as follows:

The **pink crystals** turn to a **dark blue liquid** which **boils** and a **colourless liquid condenses** on the **top of the test tube**. A **pale blue powder** remains.

When the water is added, a **hissing sound** is heard, the **temperature rises** and **steam is seen escaping from the test tube**. The colour of the residue becomes **dark blue** and then **pink** and finally the residue **dissolves** forming a **pink solution**.

This should give some idea of the detail required when making observations.

Ability to plan an experiment or part of an experiment

Level I candidates would be able to plan a very simple experiment. For example, which of two metals is more reactive? A level I candidate might add the metals to dilute hydrochloric acid in separate test tubes and see which fizzes the most.

A level II candidate might carry out a similar investigation but attempt to ensure that similar quantities of the different chemicals are used.

A level III candidate would ensure that the different metals had the same surface area and try to make some allowance for the temperature rise which might occur and speed up a reaction. So even in one very simple experiment, you can see different levels of performance.

If you are asked to plan an experiment or part of an experiment, think clearly about similar things you have done during your course. Often these exercises include filtering, crystallizing, gas preparations, drying operations, melting points and boiling points.

When you have finished your planning, do a final mental check to make sure that what you have suggested would actually work.

Ability to evaluate methods and suggest possible improvements

The different levels of performance can be seen again in a very simple example: a student is given some magnesium ribbon and is told to burn the magnesium in air to form magnesium oxide.

A level I candidate, having been told that heating the magnesium in an open crucible does not give a good result, would suggest putting a lid on the crucible during heating.

A level II candidate would realize that a lid should be placed on the crucible to prevent magnesium oxide escaping.

A level III candidate would realize that stopping air getting into the crucible would prevent the magnesium burning and would therefore lift the lid with tongs from time to time.

In another example, a candidate is given three possible methods for preparing and collecting a dry sample of a gas. The gas is less dense than air and can be dried by passing it over lumps of calcium oxide. The three methods are shown in Fig. 4. The candidate should realize that none of these methods would be suitable. In (a) the gas is collected over water and cannot therefore be dry. In (b) the apparatus is sealed and so the displaced air cannot escape and the apparatus could explode. In (c) the fact that the gas is less dense than air has been ignored. The apparatus in (b) would be suitable if the cork labelled X was not there.

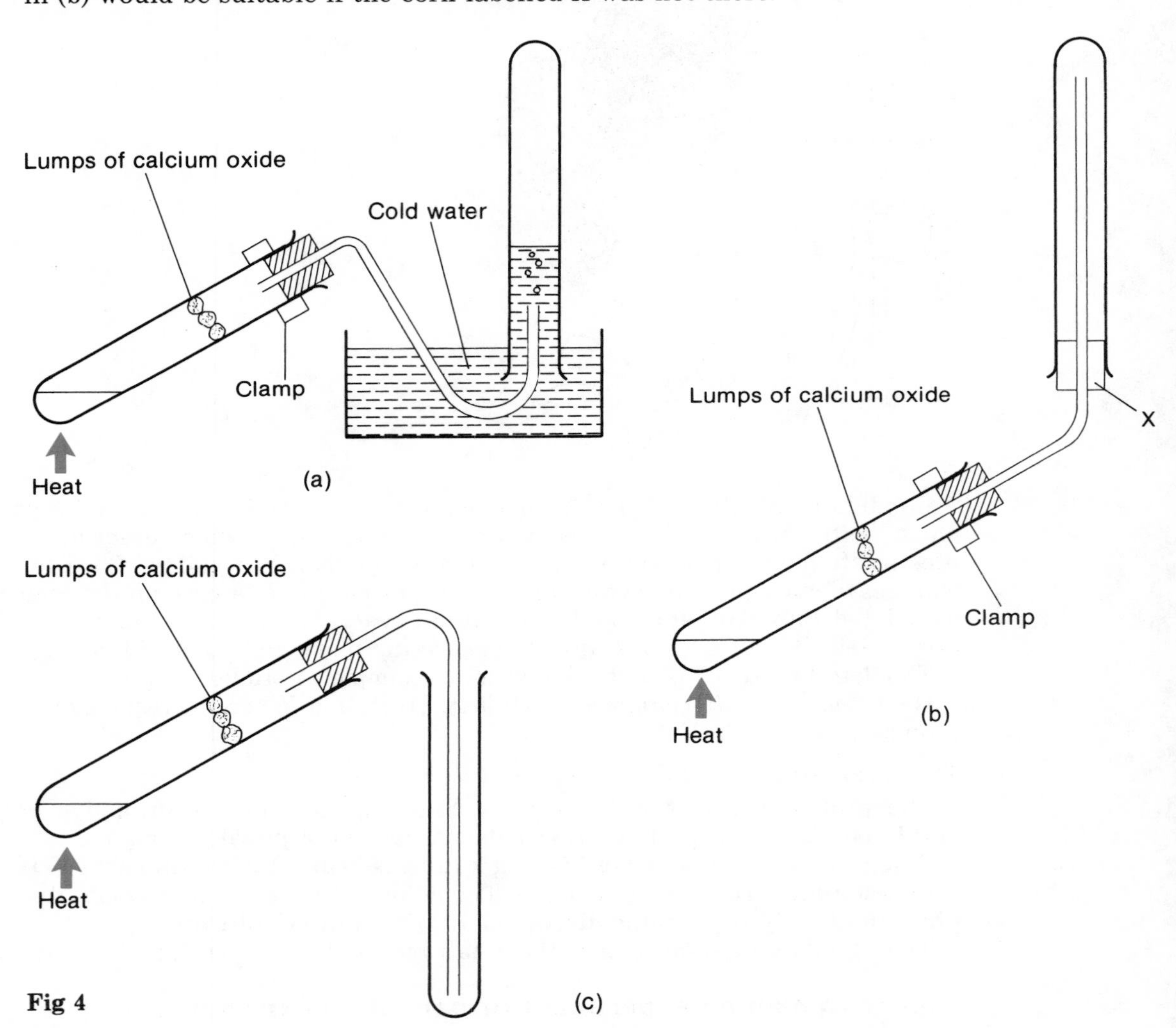

Fig 4

Ability to interpret data and draw conclusions

Interpreting data and drawing conclusions can cause problems for many GCSE candidates. A level I candidate would be able to identify a gas from an appropriate test if he/she is in possession of a data sheet giving common gas tests. Alternatively, a level I candidate, given the results in Table 1, could conclude that all metal oxides are basic.

Table 1

Metal oxide	*pH of oxide solution*
Calcium oxide	10
Magnesium oxide	9
Sodium oxide	11

You will notice that none of the results in Table 1 give conflicting information. They all confirm the conclusion made.

A level II candidate would be able to make the conclusion even if some of the information given was conflicting.

A level III candidate would be able to make more detailed conclusions from this type of information, possibly linking the pH of metal oxides with the position of the metal in the reactivity series.

Before drawing conclusions using quantitative results it is often advisable to get it into an ordered form. For example, five different masses of magnesium were heated (0.4 g, 0.1 g, 0.5 g, 0.3 g and 0.2 g) until they burned in air. The five residues formed weighed 0.67 g, 0.16 g, 0.84 g, 0.46 g and 0.36 g respectively. If they are arranged in order of increasing mass of magnesium, a table (Table 2) can be drawn and it is easy to conclude that the mass of residue increases as the mass of magnesium increases. It would now be easy to estimate the mass of residue which would be formed if 0.24 g of magnesium were burned.

Table 2

Mass of magnesium (g)	*Mass of residue (g)*
0.1	0.16
0.2	0.34
0.3	0.46
0.4	0.67
0.5	0.84

Do not overlook conclusions which seem obvious to you. Many candidates can actually see the conclusion but think it is too obvious and do not write it down.

DRAWING GRAPHS

Chemistry practical coursework should involve quantitative work and probably graphical work. If you are given a plain piece of graph paper choose your scales carefully to fill at least half of the graph paper. Always label your axes clearly.

Mark each point with a small cross or dot. Then draw a line with a sharp pencil through or as close to as many points as possible. Do not just join them up. You should not expect all of your points to lie directly on the line because of errors in experiments.

Figure 5 shows a graph of the results in Table 2 above. Most of the points lie on a straight line but the point (0.3 g, 0.46 g) is slightly off the line due to a slight error.

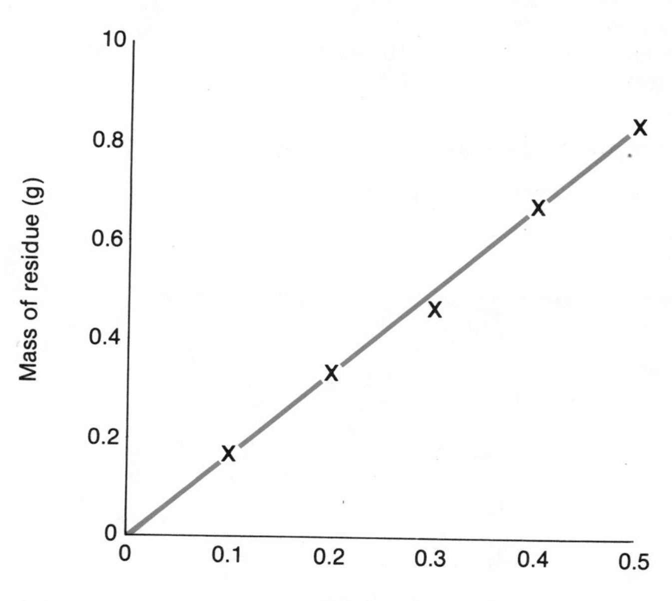

Fig 5

AREAS SUITABLE FOR ASSESSMENT

Your teacher will choose which experimental situations to use for assessment. Certain parts of the course are more suitable than others. The following list suggests topics of the course which might be suitable for assessment:

1 Rate of reaction experiments (Unit 13).
2 Mass changes when substances are heated (Unit 29).
3 Precipitation of hydroxides using aqueous ammonia and sodium hydroxide solution (Unit 32).
4 Separation of mixtures (Unit 2).
5 Preparation of crystals of a soluble salt (Unit 2).
6 Investigation of electrolysis of solutions (Unit 2).
7 Simple cation/anion analysis (Unit 32).
8 Tests for common gases (Unit 32).
9 Investigation of hardness or pollution of water samples (Unit 19).
10 Investigation of redox reactions (Unit 10).
11 Investigation of the order of reactions (Unit 13).
12 Investigation of the decomposition of metal nitrates and metal carbonates (Unit 8).
13 Simple experiments involving volumes of solutions reacting (Unit 30).
14 Investigation of energy changes (Unit 15).
15 Hydrolysis of starch (Unit 21).
16 Finding the formula of a hydrated salt (Unit 29).
17 Patterns of solubility (Unit 11).

QUESTIONS TESTING PRACTICAL SKILLS

The following questions are designed to give you practice at the same skills as you will require for practical coursework. They are designed for all candidates and you should expect to achieve high marks – 60 per cent or more – if you are to achieve a GCSE grade C.

1 Experiments were carried out with **equal** masses of zinc and 50 cm^3 samples of 2M sulphuric acid. The equation for the reaction is:

$$Zn(s) + H_2SO_4(aq) \rightarrow ZnSO_4(aq) + H_2(g)$$

zinc + sulphuric acid → zinc sulphate + hydrogen

In each experiment hydrogen was collected and the volume recorded at intervals until the reaction stopped.
Table 3 summarizes the conditions for each experiment.

Table 3

Experiment number	*Form of zinc used*	*Temperature (°C)*
1	powder	30
2	sheet	20
3	sheet	30

The graphs obtained are shown in Fig. 6.

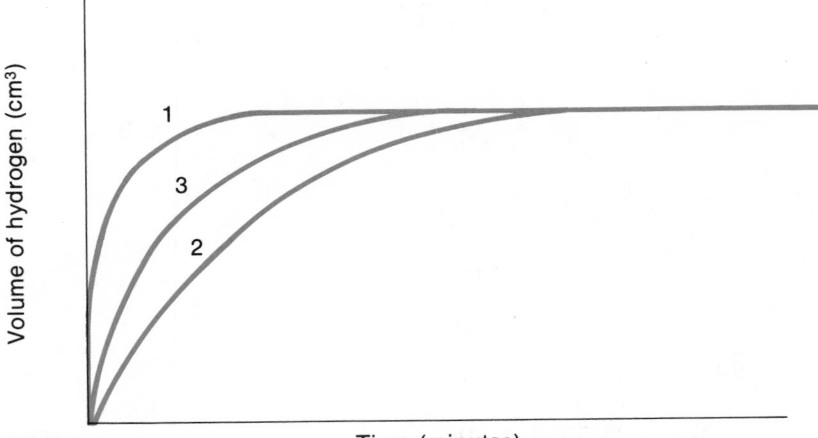

Fig 6

(a) Using the apparatus in Fig. 7 draw a diagram of the apparatus set up for these experiments. (3 marks)

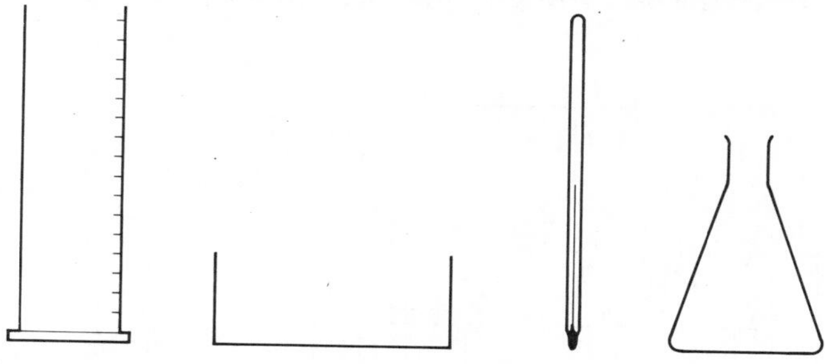

Fig 7

(b) What piece of apparatus could be used for measuring out 50 cm^3 of sulphuric acid? (1 mark)

(c) How is it possible, without using a chemical balance, to obtain pieces of zinc sheet of equal mass? (1 mark)

(d) What can be concluded by comparing the results of:

(i) experiments 1 and 3?

(ii) experiments 2 and 3?

(2 marks)

(Total 7 marks)

2 Copper(I) oxide can be reduced to copper by heating in a stream of hydrogen gas in the apparatus in Fig. 8.

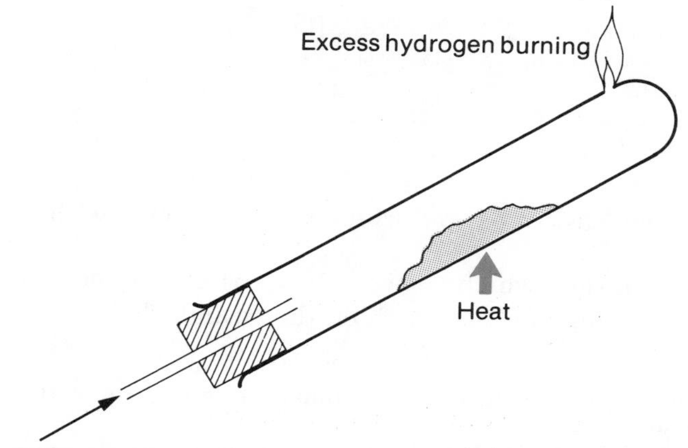

Fig 8

Dry copper(I) oxide was placed inside the dry test tube. The hydrogen supply was turned on and the hydrogen was allowed to pass through the test tube before lighting. The test tube was heated until all of the copper(I) oxide was converted to copper. The apparatus was allowed to cool with hydrogen passing before dismantling.

(a) Mark on the diagram the best place for clamping the test tube. (1 mark)

(b) Why was the hydrogen passed through the apparatus for some time before lighting? (1 mark)

(c) Why was the apparatus allowed to cool before dismantling? (1 mark)

A class carried out a series of reduction experiments with dry copper(I) oxide. Each group used a different mass of oxide and the mass of copper produced was found in each case. The results are shown in Table 4.

Table 4

Group	*Mass of copper oxide (g)*	*Mass of copper produced (g)*	*Mass of oxygen lost (g)*
1	0.62	0.55	0.07
2	0.90	0.80	0.10
3	1.12	1.00	0.12
4	1.69	1.50	—
5	1.80	1.60	—

(d) Complete the table of results. (1 mark)

(e) Plot the mass of copper (on the x axis) against the mass of oxygen lost (on the y axis) on graph paper as in Fig. 9. Draw the best straight line through the points and through the origin. (3 marks)

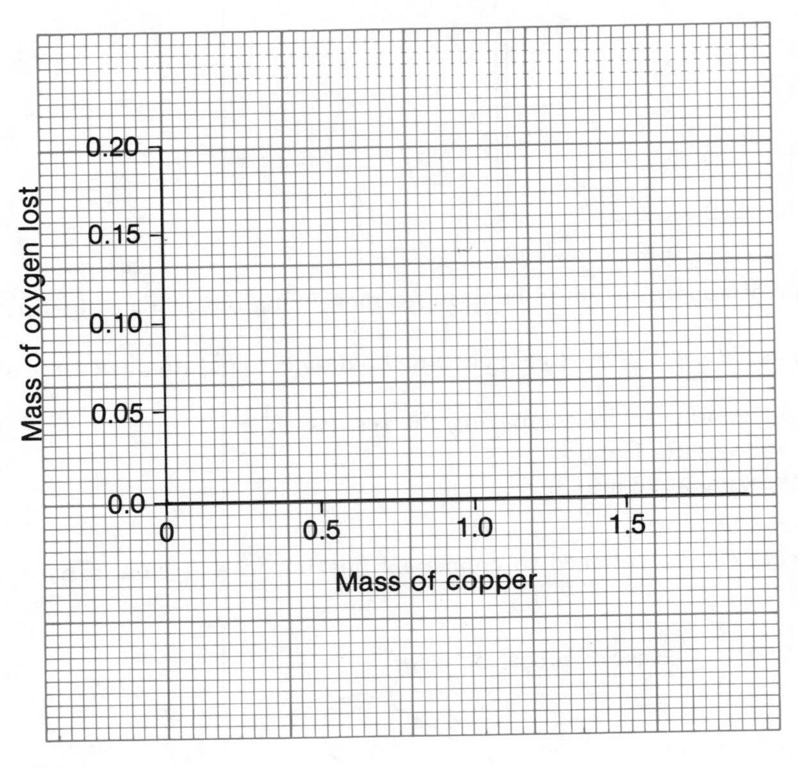

Fig 9

(f) From the graph find the mass of oxygen which would combine with 1.28 g of copper. (1 mark)

(g) Calculate the mass of oxygen which combines with 64 g of copper. (1 mark)

(h) Why are not all of the points on the graph on the straight line? (1 mark)

(Total 10 marks)

3 25 cm^3 of distilled water were measured into a plastic cup and the water left for some time. The temperature of the water was taken.

Lumps of white solid were added and the temperature taken as soon as the lumps had dissolved.

(a) Why was the distilled water left for some time before the temperature was taken? (1 mark)

(b) Why is a plastic cup better than a glass or metal beaker for this experiment? (1 mark)

(c) What simple improvement can you suggest to prevent heat losses during the experiment? (1 mark)

(d) Figure 10 shows the thermometer before and after dissolving the white solid. Read the temperatures before and after. (2 marks)

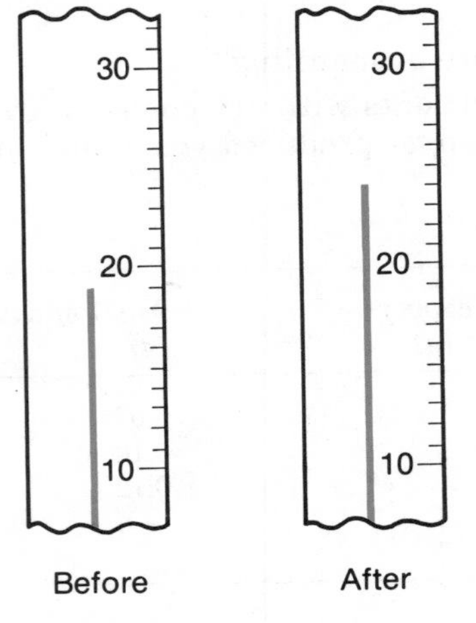

Fig 10

(e) What temperature change was obtained during this experiment? (1 mark)

(f) What can be concluded about the dissolving of the white solid from this experiment? (1 mark)

(g) Without heating the water, give two ways of speeding up the dissolving of the white solid. (2 marks)

(h) What temperature change would be expected if an equal mass of the white solid were dissolved in 50 cm^3 of distilled water? (1 mark)

(i) Mixing water and concentrated sulphuric acid can be a dangerous process. Give three precautions which should be taken when doing this. (3 marks)

(Total 13 marks)

4 Using the apparatus in Fig. 11, together with glass tubes, corks and a source of heat, draw diagrams of apparatus set up to produce and collect:

(a) a sample of pure water from ink; (2 marks)

(b) a test tube filled with a colourless gas produced when a solid reacts with a liquid. The colourless gas is less dense than air. (2 marks)

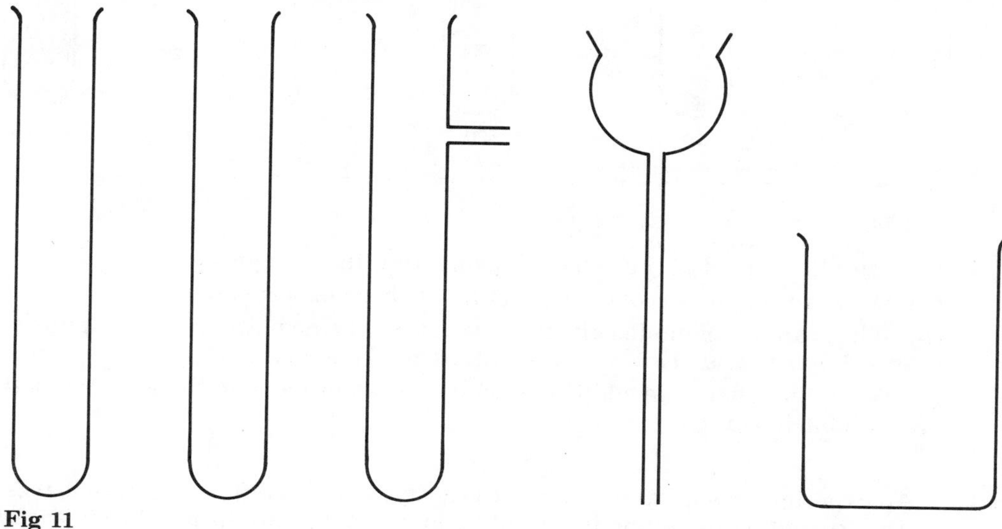

Fig 11

Suggest any problems which the apparatus would have and any ways in which it could be improved. (3 marks)

(Total 7 marks)

5 Ethene is made in the laboratory from ethanol. Ethanol vapour is passed over a heated porous pot in the apparatus in Fig. 12. Ethene is almost insoluble in water.

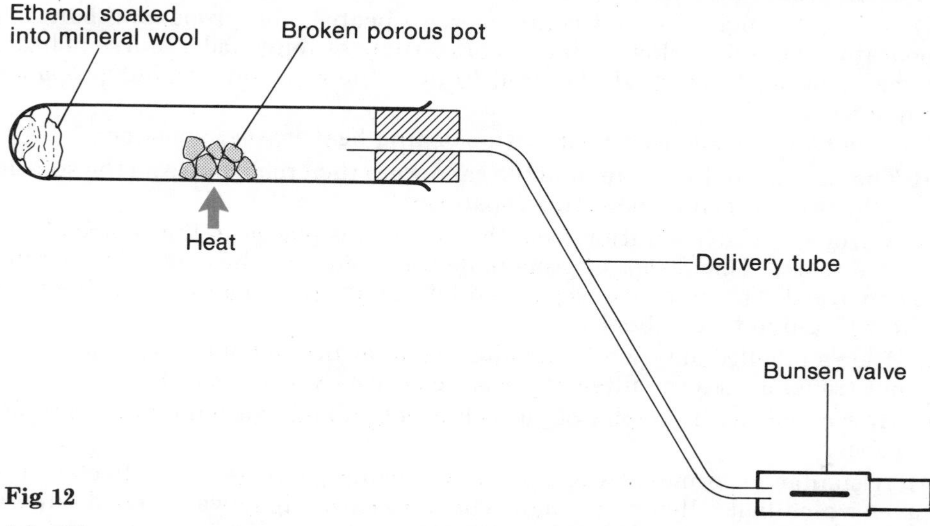

Fig 12

(a) What is the reason for using the mineral wool in the test tube? (1 mark)

(b) Complete the diagram to show how the ethene gas may be collected. (2 marks)

(c) Why is a Bunsen valve used? (1 mark)

(d) Label the diagram to show the best place for clamping the apparatus. (1 mark)

(e) Why does the first test tube of gas collected contain very little ethene? (1 mark)

(Total 6 marks)

6 A series of reactions was carried out to compare the reactions of zinc foil, zinc powder, copper powder and a mixture of zinc and copper powders with dilute sulphuric acid. If a reaction occurs, hydrogen gas is produced. Figure 13 shows the results of these experiments.

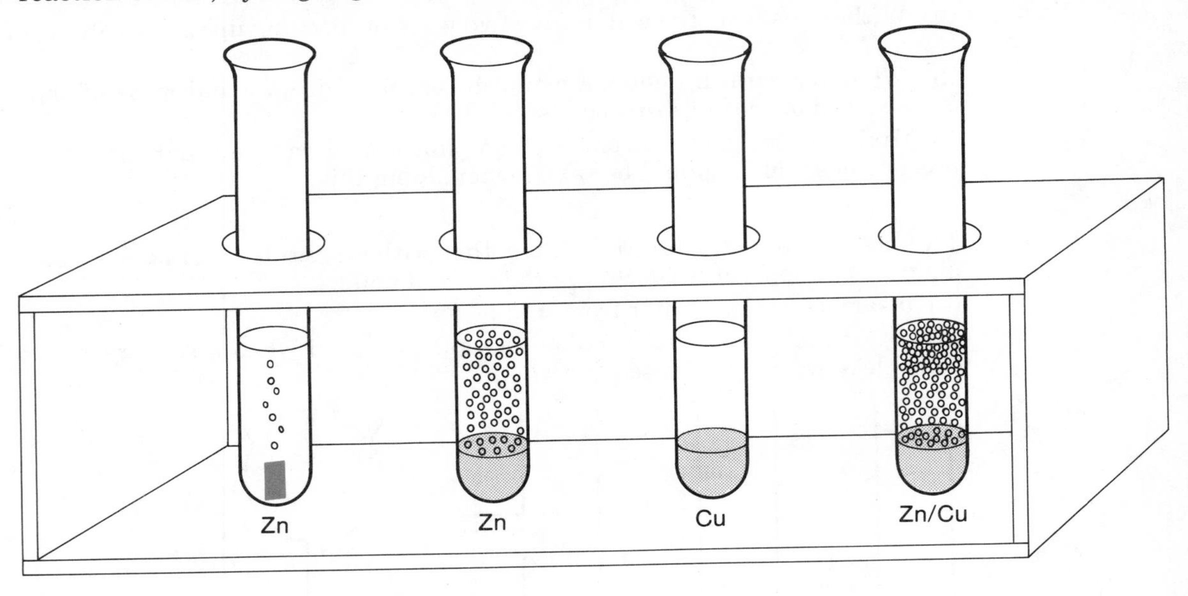

Fig 13

(The sulphuric acid is the same concentration in all experiments.)

(a) Draw a table and record the results of these experiments. (5 marks)

(b) What can be concluded about the reaction of zinc powder with sulphuric acid and zinc foil with sulphuric acid? How can this difference be explained? (2 marks)

(c) What effect does the addition of copper powder have on the reaction between zinc powder and sulphuric acid? (1 mark)

(Total 8 marks)

7 An experiment was carried out with the dye extracted from elm tree leaves using propanone as the solvent. (Propanone is a liquid which catches alight and burns easily.)

The leaves were put into a mortar with a little sand and some propanone. The mixture was then crushed with a pestle. The mixture was then added to a boiling tube and a further 10 cm^3 of propanone added. The boiling tube and contents were heated to about 50 °C.

(a) What apparatus was used to crush the elm leaves to extract the dye? (1 mark)

(b) Why was the sand added? (1 mark)

(c) How could the boiling tube and its contents be heated safely? (1 mark)

(d) As the boiling tube and contents were heated, the propanone boiled away. Using the apparatus in the list below, plus any corks, rubber tubing and a source of heat, draw a diagram of the apparatus that could be used to heat the solution without propanone being lost by evaporation.

Flask, condenser, flask with sidearm, beaker, thermometer (1 mark)

(e) The remains of the leaves and the sand were then removed from the solution. What process could be used to bring about this separation? (1 mark)

(f) A drop of the dye solution from the leaves was placed in the centre of the filter paper and allowed to dry. Then drops of propanone were added to the centre of the filter paper; the spot was enlarged. The green dye separated into two rings – a green ring due to chlorophyll and a yellow ring due to xanthophyll.

(i) Draw a labelled diagram of the filter paper at the end of the experiment. (Chlorophyll does not travel across the filter paper as readily as xanthophyll.) (2 marks)

(ii) How could small samples of pure chlorophyll and xanthophyll be obtained from this filter paper? (1 mark)

(g) A similar experiment was carried out using green leaves collected in spring and some leaves which had fallen in autumn. The chromatography was carried out with strips of filter paper and the results are shown in Fig. 14.

What can be concluded about the dyes present in the leaves in spring and autumn? (1 mark)

(h) At the end of the experiment the teacher wanted to recover the propanone from the remaining dye solution. Using the apparatus listed in (d) draw a diagram of the apparatus set up to recover the propanone. (1 mark)

(Total 10 marks)

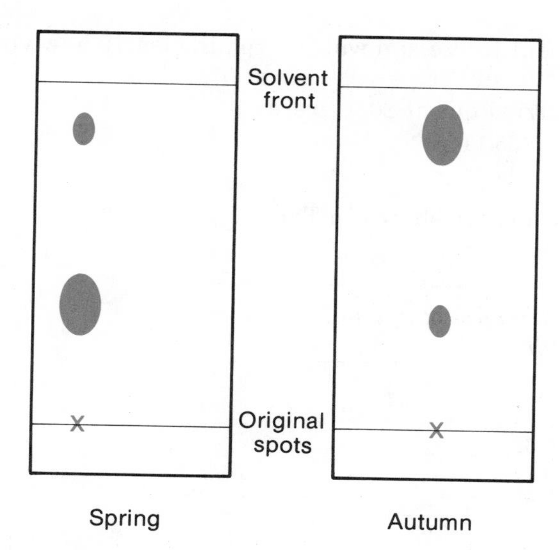

Fig 14

8 Use the information in Unit 32 when attempting this question. Find it before you start.

(a) Four solutions A, B, C and D are sodium hydroxide, hydrochloric acid, sulphuric acid and distilled water, but not in that order.

The results of experiments to find out which is which are given in Table 5.

Table 5

	Add Universal Indicator	*Add barium chloride solution*
A	Turns red	White precipitate formed
B	Turns purple	No precipitate
C	Turns yellow	No precipitate
D	Turns red	No precipitate

Identify A, B, C and D, giving reasons for your answers. (4 marks)

(b) Four white solids E, F, G and H are potassium sulphate, sodium sulphate, magnesium chloride and zinc chloride, but not in that order.

The results of experiments to find out which is which are given in Table 6

Table 6

	Add barium chloride solution to a solution of each substance	*Add sodium hydroxide solution to a solution of each substance*
E	White precipitate formed	No precipitate formed
F	No precipitate formed	White precipitate formed; dissolves in excess sodium hydroxide
G	No precipitate formed	White precipitate formed; does not dissolve in excess sodium hydroxide
H	White precipitate formed	No precipitate formed

Identify, as far as is possible, these substances giving reasons for your answers. What further tests would you carry out to complete the identification? (6 marks)

(Total 10 marks)

9 Antifreeze used in the cooling system of a motor car engine can cause the dissolving of metal components.

An investigation was made into the speed of dissolving of iron in antifreeze at different temperatures. The following procedure was used on five pieces of iron foil.

Step 1. Five pieces of iron foil of similar size were rubbed with wire wool, dipped into detergent solution, rinsed in distilled water and dried.

Step 2. The mass of each piece of foil was found.

Step 3. The pieces of foil were placed in 100 cm^3 of 25 per cent antifreeze solution kept at different temperatures.

Step 4. After 1 week the pieces of corroded iron were removed from the antifreeze solutions, washed in detergent, rinsed in distilled water and dried. The masses of each sample were found.

(a) Given a supply of 100 per cent antifreeze and water, explain clearly how you would make 100 cm^3 of 25 per cent antifreeze solution. (3 marks)

The results of the experiment carried out at 20 °C were:

mass of foil at the start 25.11 g
mass of foil after 1 week 24.95 g

The results of the other investigations are shown in Table 7.

Table 7

Temperature of antifreeze (°C)	*Mass loss of iron foil (g)*
20	—
30	0.24
40	0.32
50	0.40
60	0.48

(b) Work out the mass loss of the foil at 20 °C. Put this answer in the space in the table. (1 mark)

(c) What is the effect of temperature on the speed of dissolving of iron in antifreeze? (1 mark)

(d) Predict the mass loss when a similar piece of iron is left in 25 per cent antifreeze at 80 °C for 1 week. (1 mark)

A further investigation was made into the effects of various treatments on the speed of corrosion of iron in antifreeze. Three similar samples of iron were treated in the ways shown in Table 8 and left in 25 per cent antifreeze for 1 week at 40 °C.

Table 8

Treatment	*Mass loss of foil (g)*
Copper coating	0.48
Zinc coating	0.00
Painting	0.11

(e) What was the effect on the speed of corrosion of:

(i) copper coating;
(ii) zinc coating;
(iii) painting? (3 marks)

(Total 9 marks)

ANSWERS TO QUESTIONS TESTING PRACTICAL SKILLS

1(a) See Fig. 15: 1 mark for the measuring cylinder correctly collecting the gas; 1 mark for air-tight apparatus; 1 mark for thermometer dipping in the liquid. (3 marks)

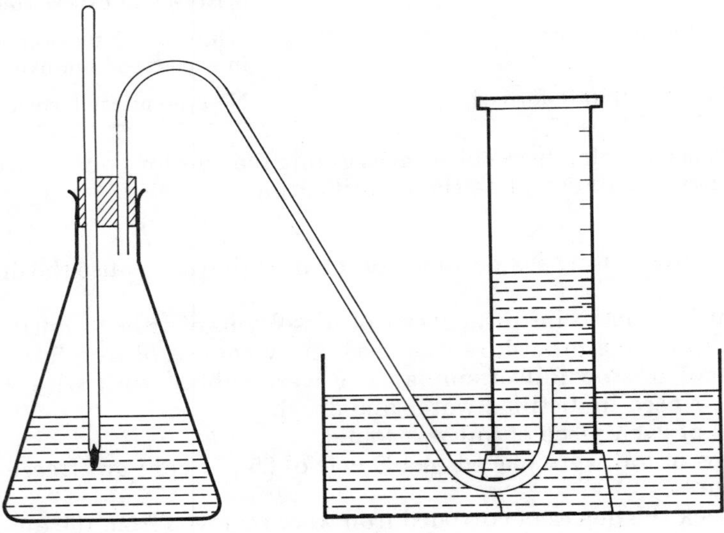

Fig 15

(b) Measuring cylinder (100 cm^3) (1 mark)
(c) Cut pieces of equal area. This assumes that the sheet is the same thickness all over. (1 mark)
(d)(i) Rate of reaction increased with powder compared with sheet. (1 mark)
(ii) Rate of reaction increased with increasing temperature. (1 mark)
(Total 7 marks)

2(a) Figure 16 (1 mark)

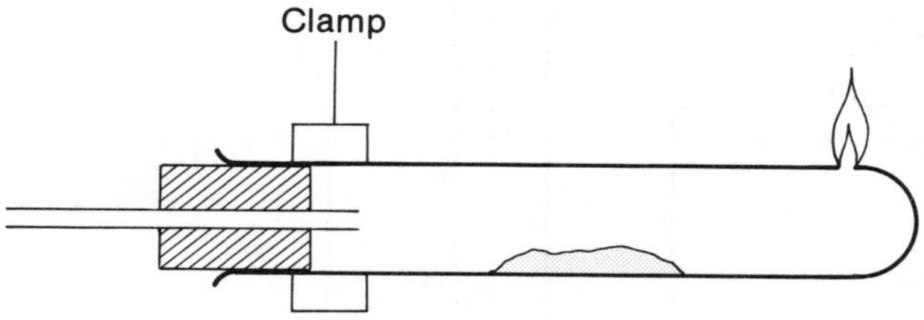

Fig 16

(b) To push out the air. Mixtures of hydrogen and air (or oxygen) are explosive. (1 mark)
(c) If copper is hot when the apparatus is dismantled, it reacts with oxygen in the air forming copper(II) oxide (1 mark)
(d) 0.19 g and 0.20 g. (both are needed for 1 mark)
(e) Graph – Fig. 17. 1 mark for graph of the largest possible size with axes labelled. 1 mark for a straight line through the origin. 1 mark for correct plotting of the points. (3 marks)

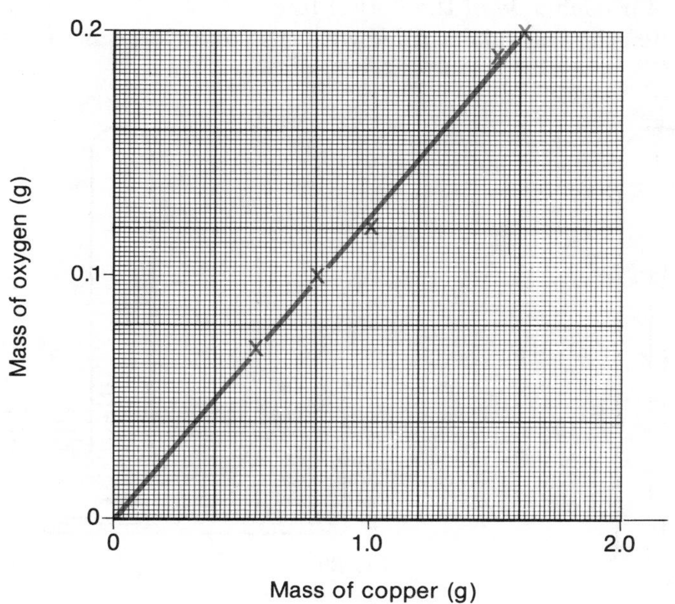

Fig 17

(f) 0.16 g. (1 mark)
(g) 8.0 g. (1 mark)
(h) Experimental error. (1 mark)
(Total 10 marks)

3(a) To ensure that the water is at room temperature before the experiment starts. (1 mark)
(b) Plastic is a good heat insulator – reduces heat losses. (1 mark)
(c) A lid on the cup. (1 mark)
(d) Before 19 °C (1 mark); after 23 °C (1 mark). (2 marks)
(e) 4 °C. (1 mark)
(f) Heat is given out when the solid dissolves or the process is exothermic. (1 mark)
(g) Crushing the lumps into a powder. (1 mark)
Stirring. (1 mark)
(h) 2 °C. (1 mark)
(i) Add acid to water; small portions at a time; stirring on each addition; wear goggles. (any 3 – 1 mark each)
(Total 13 marks)

4(i) Figure 18 – 1 mark for each side of the dotted line. (2 marks)

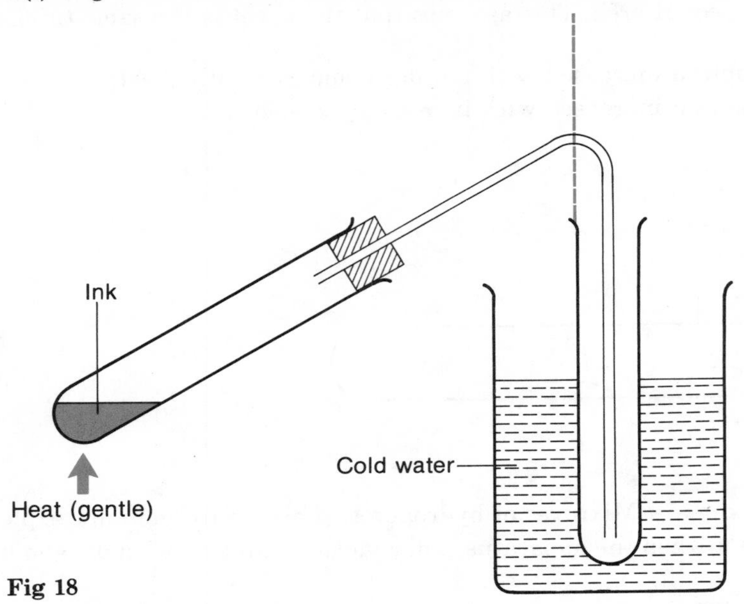

Fig 18

It is difficult to prevent ink boiling over and condense all of the steam (either answer 1 mark). Improvements: soak up ink with mineral wool or use a condenser (either answer 1 mark).

(ii) Figure 19 – 1 mark for each side of the dotted line. (2 marks)

It is difficult to know when the test tube is filled with the colourless gas. No easy improvement is possible. (1 mark)

(Total 7 marks)

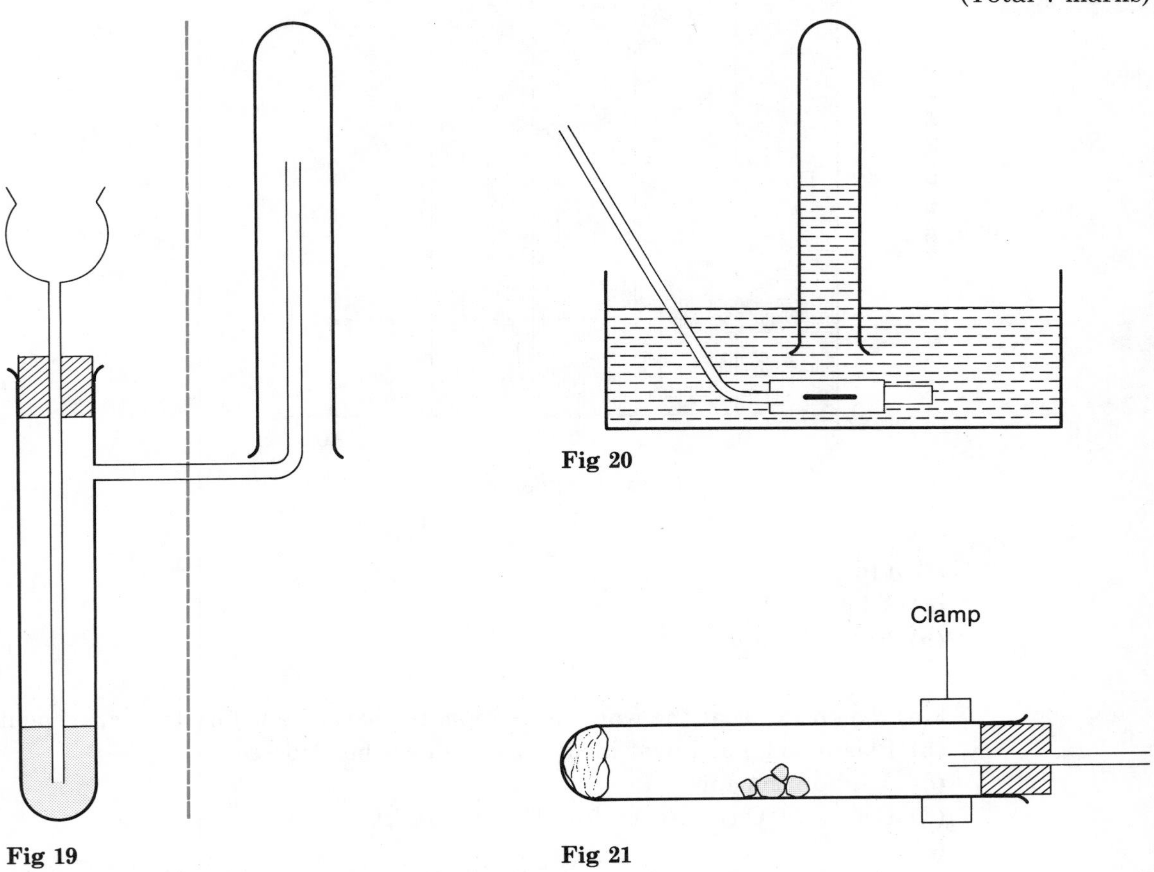

Fig 19

Fig 20

Fig 21

5(a) To prevent ethanol running down the test tube. (1 mark)

(b) Figure 20: Bunsen valve below water level – 1 mark. Test tube correctly placed – 1 mark. (2 marks)

(c) To prevent 'sucking back'. When the apparatus is heated the air inside the apparatus expands and bubbles out of the delivery tube. Without the Bunsen valve on cooling water would travel back up the delivery tube, spoiling the experiment. The Bunsen valve prevents this backward movement of water. (1 mark)

(d) Figure 21. (1 mark)
(e) Air inside the apparatus is pushed out first and collected. (1 mark)
(Total 6 marks)

6(a) A correctly drawn table: (1 mark)

Table 9

Metal	*Observations on adding sulphuric acid*	
Zinc foil	Few bubbles of gas	(1 mark)
Zinc powder	Steady stream of bubbles	(1 mark)
Copper powder	No bubbles	(1 mark)
Zinc and copper powders	Rapid stream of bubbles	(1 mark)

(b) Zinc powder reacts faster than zinc foil (1 mark). Zinc powder has a larger surface area (1 mark). (2 marks)

(c) Copper powder speeds up the reaction of zinc and sulphuric acid. (Copper is, in fact, a catalyst for this reaction.) (1 mark)
(Total 8 marks)

7(a) Mortar and pestle. (1 mark)
(b) To break down the plant cells. (1 mark)
(c) Figure 22. (1 mark)
(d) Figure 23. (1 mark)
(e) Filtering (or decanting or centrifuging). (1 mark)
(f) (i) Figure 24: Two rings one inside the other – 1 mark. Xanthophyll on the outside, chlorophyll on the inside – 1 mark. (2 marks)

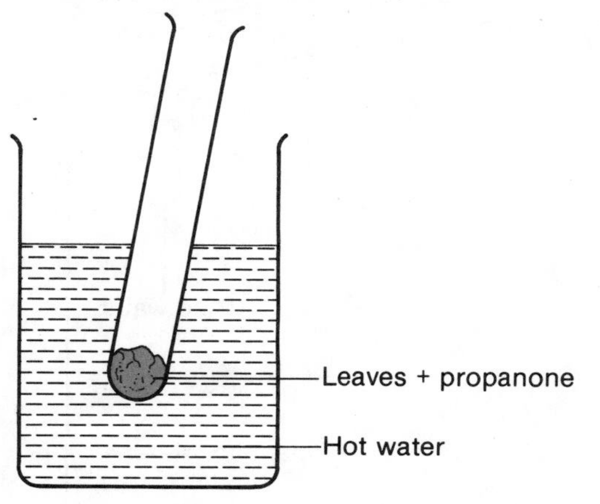

Fig 22

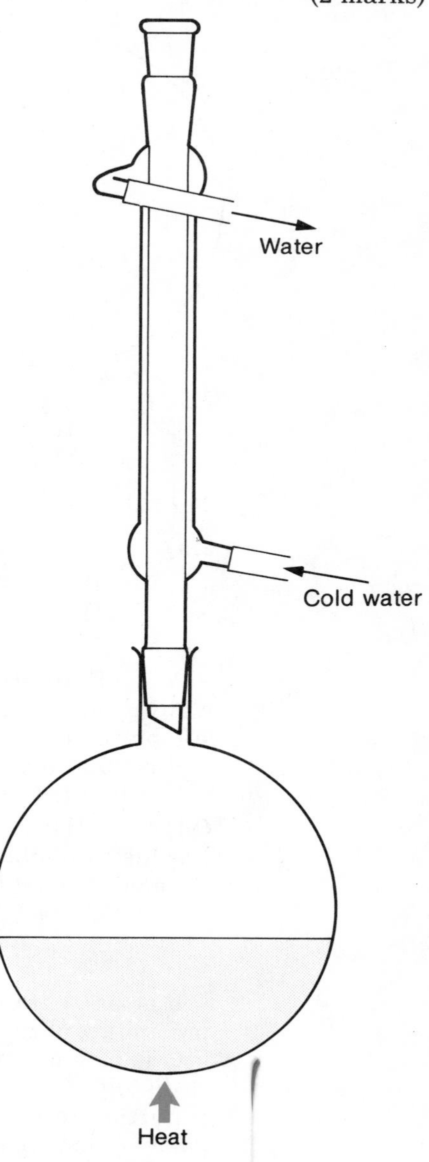

Fig 23

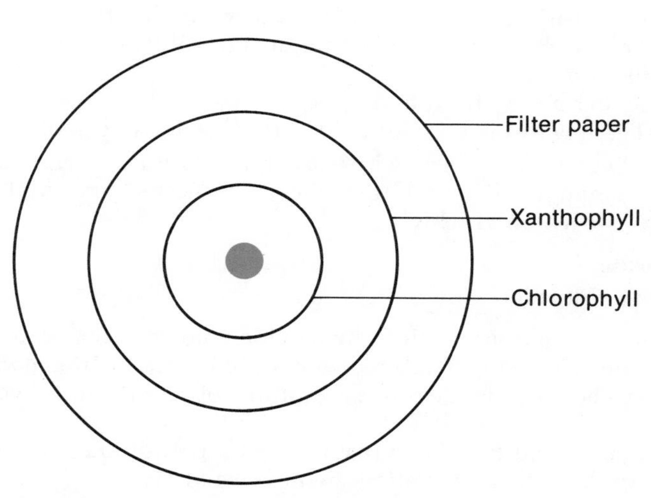

Fig 24

(ii) Cut a small piece of green-coloured paper and dissolve the chlorophyll in the minimum volume of propanone. You have a solution of pure chlorophyll. (1 mark)

(g) Leaves in spring contain more chlorophyll and less xanthophyll than the leaves in autumn. (1 mark)

(h) Figure 25. (1 mark)

(Total 10 marks)

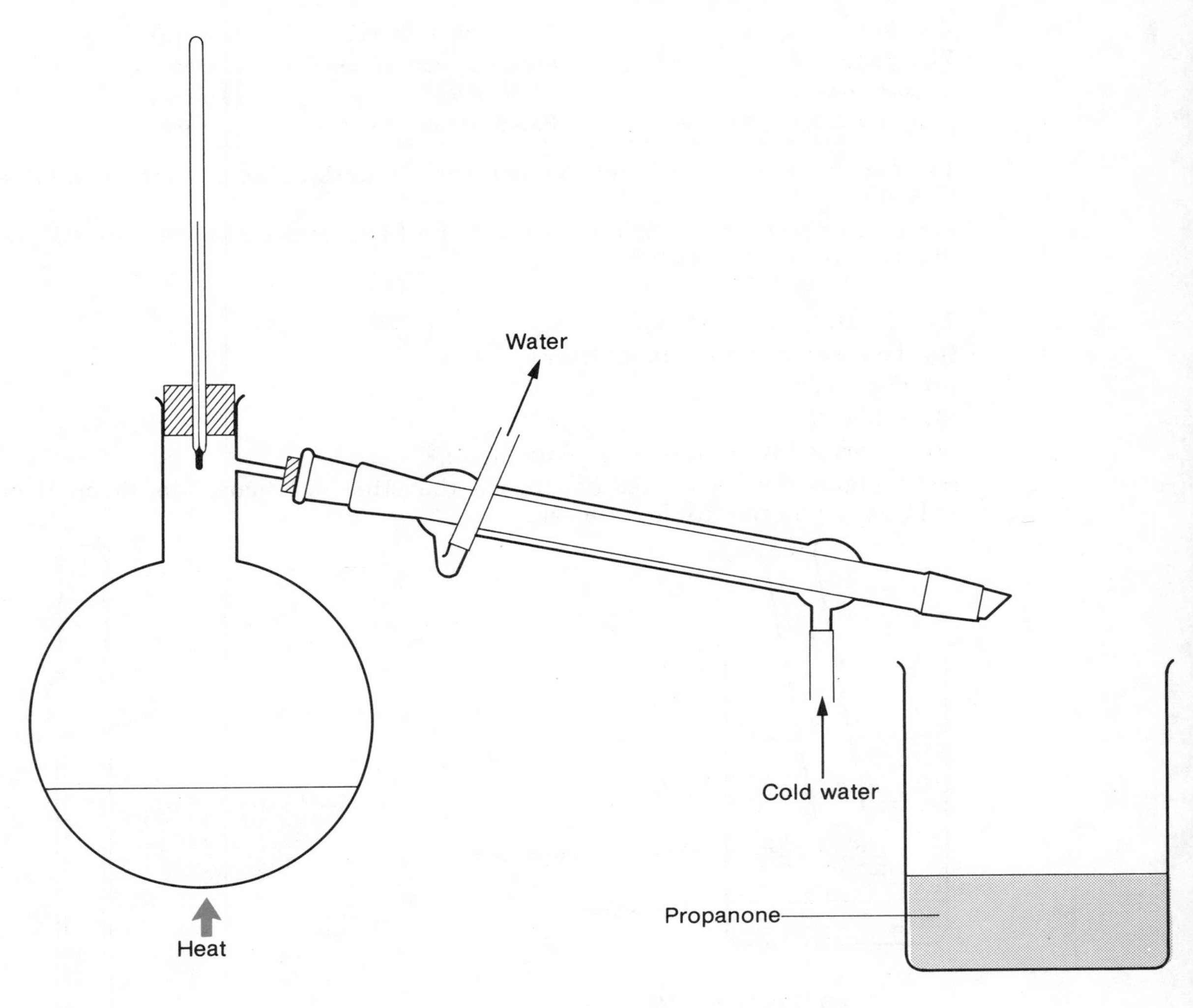

Fig 25

8(a)

A	sulphuric acid	(1 mark)
B	sodium hydroxide	(1 mark)
C	distilled water	(1 mark)
D	hydrochloric acid	(1 mark)

A and D are strongly acidic (pH 1). A and D are sulphuric and hydrochloric acids. A gives a white precipitate of barium sulphate when barium chloride solution is added. A is therefore sulphuric acid; D is hydrochloric acid.

B is alkaline. B must be sodium hydroxide.

C is distilled water. Distilled water is not neutral in practice due to dissolved carbon dioxide.

(b) E and H are sulphates because they form white precipitates when barium chloride solution is added to solutions of E and H. E and H are therefore potassium sulphate and sodium sulphate but we do not know which is which. (1 mark)

F	zinc chloride	(1 mark)
G	magnesium chloride	(1 mark)

F is zinc chloride because on adding sodium hydroxide solution to a solution of F a white precipitate of zinc hydroxide is formed, which redissolves in excess sodium hydroxide solution.

G is magnesium chloride because the white precipitate of magnesium hydroxide does not re-dissolve.

Further test to distinguish E and H – flame test (1 mark); potassium sulphate (pinkish-lilac flame) (1 mark); sodium sulphate (orange-yellow flame) (1 mark). (3 marks)

(Total 10 marks)

9(a) Measure out 25 cm^3 of antifreeze (1) and 75 cm^3 of distilled water (1 mark); using measuring cylinders (1 mark); mix the two liquids thoroughly (1 mark). (any 3 – 3 marks)
(b) 0.16 g. (1 mark)
(c) Increasing temperature causes more iron to dissolve. (1 mark)
(d) 0.64 g; this is only a prediction. (1 mark)
(e)(i) Corrosion increased, more iron dissolved. (1 mark)
(ii) No corrosion. (1 mark)
(iii) Some corrosion but less than untreated iron. (1 mark)
(Total 9 marks)

The total mark for the test is 80. If you achieved 70+ you are in line for a grade A at GCSE. 60–9 is a grade B. 48–59 is a grade C. 40–7 is a grade D. 34–9 is a grade E. 28–33 is a grade F.

Project Work or Writing Extended Essays

Coursework assessment often requires the completion of a project or an extended essay. This may include some practical work on the part of the student or may be entirely based upon research in books, etc. The first thing to remember is that other subjects that you are studying will also require extensive coursework and all of this coursework is going to be due in at about the same time. Do not leave it but get started on your coursework as soon as possible.

GCSE syllabuses have to reflect social, economic, environmental, industrial and technological aspects of chemistry and project work is an ideal place to show this.

MAKING A START

Before you start it is most important to choose your topic carefully. Often titles which seem attractive at first cause problems because they do not contain enough chemistry. For example, a girl chose a topic on 'fingernails'. From biology books she found out about the chemical nature of fingernails and the structure of fingernails. She then did something on the making of nail varnish. The resulting project contained little chemistry and could not be marked highly. It is a good idea to discuss your project at an early stage with your teacher and get his or her suggestions as to its feasibility.

Every project should have a clear title. This title could be in the form of a question.

E.g. Corrosion and its prevention in a motor car *or* How can the corrosion in a motor car be reduced?

Having chosen a title it is important to collect together material from different sources before you start. Starting your project by just copying out the section on Corrosion from the *Encyclopaedia Britannica* is not acceptable. The assessor of your project wants to see evidence of your work and your ability to use different souces. You might find information in newspapers, magazines, a visit to a local library or museum, in booklets from firms and organizations, etc. Later in this section there is a list of useful magazines and addresses that could be used for source material. Try and find local source material. In addition to these sources you might use questionnaires, surveys, interviews, industrial visits, field work or computers.

Most of your project will consist of written work. There is no need to try and type it or have it typed. In fact assessors prefer to have it handwritten. However, try and present it well, with clearly defined sections and headings. Try and reach some sort of conclusion or satisfactory ending – don't just finish. If you could see possible limitations in your project or you could see extensions of your project, they should be included. At the end of your project include a section headed 'Bibliography'. Here you list all of the sources you have used.

You should consider the use of drawings, graphs, charts, pictures and photographs. In addition you could use audiotapes, videotapes and computer programs.

Remember that you will receive credit only for what you can actually present. Make sure that you use your source material to the full.

SUGGESTIONS ON POSSIBLE PROJECTS

The ideas here are only suggestions and I am sure you will come up with many more.

1 Projects involving the whole class for the collection of data

Individual pupils are able to select and use data plus other information.

(a) A visit to the local sewage works and discussions with staff of local water authority could be a prelude to projects on water treatment and water pollution.

(b) A visit to a local factory producing chemicals or using large quantities of chemicals could lead to projects on chemical manufacture, use of chemicals and disposal of chemical wastes.

(c) Experiments on water samples from a local river or sample of air could lead to projects on pollution.

2 Projects involving surveys of materials

Individual students could study different kinds of plastics found around the home – their manufacture, use, identification and problems in their disposal. Alternatively, a study could be made of different building materials or, perhaps in an agricultural area, different types of commercial fertilizers.

3 Projects involving practical work

There are many examples suitable. These would include:

(a) a study of different catalysts on the rate of reaction;
(b) a study of household chemicals, e.g. bleaches, polishes or detergents;
(c) a study of corrosion and methods of preventing corrosion.

4 Projects involving computers

(a) Location of industry study.
(b) Factors affecting the amount of ammonia produced in the Haber process when conditions are altered.

5 Projects from books, magazines, etc.

(a) Histories of soap making, paint manufacture, detergents, etc.
(b) Recycling of materials.
(c) Uses of electrolysis.
(d) Acid rain.
(e) Renewable energy sources and depletion of energy sources.

MARKING OF PROJECTS

Projects can be marked in different ways. However, to give you some guidelines an assessor is often asked to mark:

1 The quality of the content of the project.
2 The quality of the investigation carried out.
3 The quality of the conclusions drawn.
4 The standard of presentation of the project.

SUGGESTED ADDRESSES FOR INFORMATION

When contacting any organization listed or another organization, it is advisable to write a polite letter requesting specific information rather than make vague enquiries.

The Department of the Environment, Warren Springs Laboratory, Stevenage, Hertfordshire (information on air pollution).

Your local council's Environmental Health Officer (information on air pollution).

The National Society for Clean Air, 136 North Street, Brighton BN1 1RG (information on air pollution).

Friends of the Earth, 377 City Road, London EC1V 1NA (information on environmental matters).

Health and Safety Executive, PO Box 109, Maclaren House, 19 Scarbrook Road, Croydon, Surrey CR9 1QH (industrial safety).

Britoil PLC, 150 St Vincent Street, Glasgow G25 5LJ (oil and natural gas).

Esso UK PLC, Victoria Street, London SE1 5JW (oil and natural gas).

Cement Makers' Federation, Terminal House, 52 Grosvenor Gardens, London SW1W 0AH (cement manufacture and use).

The Royal Mint, 7 Grosvenor Gardens, London SW1 0BH (use of alloys and minting coins).

Imperial Chemical Industries PLC, School Liaison Section, PO Box 6, Bessemer Road, Welwyn Garden City, Herts AL7 1HD (wide range of industrial chemical manufacture).

Van den Berghs and Jurgens Ltd, Sussex House, Civic Way, Burgess Hill, West Sussex RH15 9AW (margarine manufacture).

Vegetable Protein Association, 43 Great Marlborough Street, London W1 (synthetic proteins).

British Sugar Bureau, 140 Park Lane, London W1Y 3AA (production and use of sugar).

British Gas Corporation, 326 High Holborn, London WC1V 7PT (natural gas and its use).

Department of Energy, Thames House South, Millbank, London SW1P 2RD (matters concerned with energy sources and use of energy).

The Electricity Council, 30 Millbank, London SW1R 2RD (generation and use of electricity).

Solid Fuel Advisory Service, Hobart House, Grosvenor Place, London SW1X 7AE (types of solid fuel and their use).

British Ceramic Manufacturers Association, Federation House, Stoke-on-Trent ST4 2SA (china and ceramic materials).

Rayner and Co. Ltd, Edmonton, London N18 1TQ (flavouring essences and colourings).

Your council's County Analyst (information about food additives and consumer protection).

Crown Decorative Products Ltd, Product Training Centre, Crown House, Darwen, Lancashire BB3 0BG (paints and decorating materials).

Reckitt Household and Toiletry Products, Reckitt House, Stoneferry Road, Hull HU8 8DD (household chemicals).

Lever Brothers Ltd, Lever House, 3 St James Road, Kingston-on-Thames, Surrey KT1 2BA (detergents).

Procter and Gamble Educational Service, PO Box 1EE, Gosforth, Newcastle upon Tyne NE99 1EE (detergents).

Shell UK Ltd, Shell Education Service, Shell-Mex House, Strand, London WC2R 0DX (oil and natural gas).

Monsanto Ltd, Monsanto House, 10/18 Victoria Street, London SW1H 0NQ (man-made fibres).

Domestos Hygiene Advisory Service, 114 New Bond Street, London W1Y 9AB (household chemicals including bleaches).

The Royal Society for Prevention of Accidents, Cannon House, The Priory, Queensway, Birmingham B4 6BS (accident prevention).

British Oxygen Co. Ltd, North Circular Road, London NW2 (distillation of liquid air and uses of oxygen and nitrogen).

Tin Research Institute, Fraser Road, Perivale, Greenford, Middx (uses of tin).

Lead Development Association, 34 Berkeley Square, London W1 (production and uses of lead).

Copper Development Association, 35 Audley Street, London W1 (production and uses of copper).

Alcan Aluminium (UK) Ltd, Southam Road, Banbury OX16 7SN (manufacture and uses of aluminium and aluminium alloys).

Institute of Petroleum Information Service, 61 New Cavendish Street, London W1M 8AR (petroleum and natural gas).

National Sulphuric Acid Association Ltd, Piccadilly House, 16 Jermyn Street, London SW1Y 4NF (sulphuric acid).

British Steel Corporation, Information Officer, 151 Gower Street, London WC1E 6BB (iron and steel manufacture and use).

Understanding British Industry, The Resource Centre Director, Sun Alliance House, New Inn Hall Street, Oxford OX1 2QE.

Schools Information Centre on the Chemical Industry, Polytechnic of North London, Holloway Road, London N7 6DB.

Central Electricity Generating Board, Department of Industry and Public Affairs, Sudbury House, 15 Newgate Street, London EC1A 7AU (air pollution and acid rain).

Unilever Education Section, Unilever House, Blackfriars, London EC4 (detergents).

Tioxide Group PLC, 10 Stratton Street, London W1A 4XP (paint pigments and disposal of wastes).

Tyne and Wear County Council, 4th Floor, Hadrian House, Higham Place, Newcastle upon Tyne NE1 8BX (reclamation of materials from rubbish – Byker Reclamation Plant).

Glass Manufacturers Federation, 19 Portland Place, London W1N 4BH (recycling glass and manufacture of glass).

Brazilian Embassy (Science and Technology Section), 32 Green Lane, London W1 (making ethanol by fermentation for use as a motor fuel).

Albright and Wilson, Hagley Road, West Oldbury, Birmingham (manufacturer of phosphorus compounds/pollution).

Severn Trent Water Authority, Abelson House, 2297 Coventry Road, Sheldon, Birmingham B26 3PU (information on water supply, water treatment and water pollution).

Anglian Water Authority, Ambury Road, Huntingdon, Cambs. PE18 6NZ.

Northumbrian Water Authority, Northumbria House, Regent Centre, Gosforth, Newcastle upon Tyne NE3 3PX.

North West Water Authority, Dawson House, Great Sankey, Warrington WA5 3LW.

Southern Water Authority, Guildbourne House, Chatsworth Road, Worthing, West Sussex BN11 1LD.

South West Water Authority, 3–5 Barnfield Road, Exeter, Devon EX1 1RE.

Thames Water Authority, New River Head, Rosebery Avenue, London EC1R 4TP.
Welsh Water Authority, Cambrian Way, Brecon, Powys LD3 7HP.
Wessex Water Authority, Wessex House, Passage Street, Bristol BS2 0JQ.
Yorkshire Water Authority, West Riding House, 67 Albion Street, Leeds LS1 5AA.

This list is not intended to be exhaustive but merely to suggest possibilities. I hope you will have some ideas of your own.

USEFUL BOOKS AND MAGAZINES

Studies in Industrial Chemistry by Harrison and Wright published by Edward Arnold. A very useful background to industrial chemistry.

New Scientist. A weekly magazine of current scientific interest. Your chemistry teacher could perhaps arrange for you to subscribe to this magazine at a reduced cost.

Chemistry in Britain. Monthly magazine of the Royal Society of Chemistry.

WHICH, the magazine of the Consumers Association, sometimes has useful articles. For example, in August 1984 there was an interesting article on lead in petrol and its effects on health.

STEAM. A most useful magazine published by ICI twice a year for science teachers. Your teacher might have copies you could borrow. Interesting articles include:

Steam 1 – published January 1984. Articles on paint technology, electrolysis of brine producing sodium hydroxide and chlorine by three different processes, drugs to control heart disease, sulphuric acid manufacture and history of alkali works.

Steam 2 – published June 1984. Articles on ammonia production by the Haber process, transporting chemicals, protecting the environment, biotechnology, single cell proteins and a new type of concrete.

Steam 3 – published January 1985. Articles on fertilizers, agrochemicals and the environment, a biodegradable thermoplastic and chemical engineering.

Steam 4 – published July 1985. Articles on sodium carbonate, development of a drug, new agrochemicals, speciality chemicals and Winsford salt mine.

Steam 5 – published in January 1986. Whole magazine was devoted to a young person's guide to science and engineering.

Steam 6 – published in September 1986. Articles on biotechnology, chemistry of fluorine and its compounds.

PLACES TO VISIT

The following places might be suitable for visits when collecting material for your project.

Science Museum, Exhibition Road, London SW7. Open 10.00–18.00 Monday to Saturday, 14.30–18.00 Sunday. The museum has a wide variety of exhibits including fine galleries devoted to chemical exhibits. There are free lectures and films.

Geological Museum, Exhibition Road, London SW7. Open 10.00–18.00 Monday to Saturday, 14.30–16.00 Sunday. Exhibits include the story of the Earth exhibition, rocks and minerals and treasures of the Earth, which describes several industrial processes.

Lion Salt Works Ltd, Marston, Northwich, Cheshire (telephone (0606) 2066). A working salt works. Open May–September 14.00–17.00 every day.

Oral Work

If you are to be assessed in oral work it is almost certain that the oral will be conducted by your teacher. He or she may record the discussion for marking and/or assessing later. The oral will probably last only a few minutes. You will talk either about project work you have carried out, some aspect of economic, environmental or social chemistry or some practical situation or situations in the laboratory. Prepare yourself thoroughly beforehand. If you are asked a question try to be positive in your answer and try to give reasons to support your answer. Don't worry if there is any question that you cannot answer. The teachers, assessors and examiners understand the difficulties of giving oral answers to questions.

LAST MINUTE HELP AND REASSURANCE

Unlike the GCE O level and CSE examination which it replaced, GCSE is intended to be an examination in which you can show positively what you can do. The questions are intended to give you a chance of scoring good marks.

Read the questions carefully. Too many candidates do not answer the question asked but answer their own question. Write as neatly as possible and do not worry if there are things you cannot do. Every examination throws up unexpected questions. If you find something difficult, probably everybody else does.

Remember if you have prepared thoroughly and you do the examination carefully your chances of success are high.

GOOD LUCK!

GLOSSARY

The following words may be met during your Chemistry lessons.

Absolute temperature There is a minimum temperature below which it will never be possible to cool anything. This is called **absolute zero** and is −273°C. This is the starting point for the **Kelvin** or absolute temperature scale: e.g. 0°C is the same as 273K on the absolute temperature scale.

Acid A substance that dissolves in water to form a solution with a pH below 7. An acid contains hydrogen which can be replaced by a metal to form a salt. The three mineral acids are sulphuric acid H_2SO_4, hydrochloric acid HCl and nitric acid HNO_3.

Alcohol An alcohol is an organic compound containing an OH group. A common alcohol is ethanol C_2H_5OH.

Alkali A base that dissolves in water to form a solution with a pH above 7. Alkalis are neutralized by acids to form salts. Common alkalis include sodium hydroxide NaOH, potassium hydroxide KOH and calcium hydroxide $Ca(OH)_2$.

Alkali metal A metal in group I of the Periodic Table. Common alkali metals include lithium, sodium and potassium.

Alkaline earth metal A metal in group II of the Periodic Table. Common alkaline earth metals include calcium and magnesium.

Alkane A family of hydrocarbons with a general formula C_nH_{2n+2}. The simplest alkane is methane CH_4. This is the main ingredient of natural gas.

Alkene A family of hydrocarbons with a general formula C_nH_{2n}. The simplest alkene is ethene C_2H_4. This is a most important chemical in industry.

Allotropy When an element can exist in two or more forms in the same physical state, it is said to show allotropy. The different forms are called **allotropes**. Diamond and graphite are two solid allotropes of carbon. Different allotropes exist because of different arrangements of atoms.

Alloy A metal made by mixing two or more metals together e.g. brass is an alloy of copper and zinc.

Amalgam Many metals form alloys when mixed with mercury. These alloys are called amalgams. The mixture used to fill teeth is an amalgam.

Amorphous Without definite or regular shape.

Analysis Finding out the elements present in a substance is called **qualitative analysis**. **Quantitative analysis** is finding out how much of each element is present.

Anhydride An anhydride (sometimes called an acid anhydride) is an oxide of a non-metal which dissolves in water to form an acid. Carbon dioxide is an anhydride, dissolving in water to form carbonic acid.

Anhydrous A substance without water. Often used to describe salts which have lost water of crystallization.

Anion A negatively charged ion which moves towards the anode during eloctrolysis e.g. Cl^-.

Anode A positively charged electrode in electrolysis.

Aqueous solution A solution made by dissolving a substance in water. The solvent in an aqueous solution is always water.

Atom The smallest part of an element that can exist.

Atomic number The atomic number is the number of protons in the nucleus of an atom. It is equal to the number of electrons in the atom. The elements in the Periodic Table are arranged in order of atomic number.

Base A substance which reacts with an acid to form a salt and water only. Metal oxides are bases. A base which is soluble in water forms an alkaline solution.

Battery A battery is a source of electricity. A carbon-zinc battery is the type of battery used in a torch. The battery in a car is a lead-acid battery which stores electricity. It can be re-charged.

Boiling When a liquid turns rapidly to its vapour at a fixed temperature called the **boiling point**. The boiling point of a liquid varies with pressure. The lower the pressure the lower the boiling point.

Calorimeter Apparatus used for measuring heat.

Carbohydrates Compounds of carbon, hydrogen and oxygen. The number of hydrogen atoms in each molecule is twice the number of oxygen atoms. These compounds are energy foods e.g. glucose $C_6H_{12}O_6$.

Catalyst A substance which alters the rate of a chemical reaction but is not used up in the reaction.

Cathode The negatively charged electrode in electrolysis.

Cation Positively charged ion which moves towards the cathode in electrolysis e.g. H^+.

Chemical change A change which results in the formation of new substances. A chemical reaction is not easily reversed.

Chromatography A way of separating mixtures, especially of coloured substances, by letting them spread across filter paper or through a powder.

Combination The joining together of atoms of different elements to form a compound (see **synthesis**).

Combustion Burning is a combination of a substance with oxygen. Combustion is another word for **burning**.

Compound A substance formed by joining atoms of different elements together. The properties of a compound are different from the elements that make it up. The proportions of the different elements in a particular compound are fixed.

Condensation When a vapour turns to a liquid on cooling. Heat is given out during this change. Condensation is the opposite of evaporation.

Conductor A conductor will allow electricity to pass through it (electrical conductor) or heat to pass through it (heat conductor). Metals are good conductors of heat and electricity. Carbon, in the form of graphite, is a good electrical conductor but a poor heat conductor.

Corrosion Corrosion is the wearing away of the surface of a metal by chemical attack. The rusting of iron is an example of corrosion. Rusting requires the presence of oxygen and water.

Cracking Cracking is the breaking down of long hydrocarbon molecules with heat and/or a catalyst to produce short hydrocarbon molecules. The short molecules are much easier to sell, especially for making plastics.

Crystal A piece of a substance that has a definite regular shape. Crystals of the same substance have the same shape. Slow **crystallization** will produce larger crystals.

Decomposition A chemical reaction that results in the breaking down of substances into simpler ones. This is often brought about by heating, when it is called **thermal decomposition**.

Dehydration A reaction where water (or the elements of water – hydrogen and oxygen) are removed. Dehydration of ethanol produces ethene. The substance which brings about this reaction e.g. concentrated sulphuric acid, is called the **dehydrating agent**.

Density The mass of a particular volume of a substance. It is expressed as kg/m^3 or g/cm^3.

Detergent A detergent is a cleansing agent. There are two main types of detergent – soaps and soapless detergents.

Diatomic An element whose molecules are composed of two atoms is said to be diatomic. The common gases oxygen, hydrogen, nitrogen and chlorine are all diatomic and are written as O_2, H_2, N_2 and Cl_2.

Diffusion This is the spreading out of a substance to fill all of the available space. Diffusion takes place quickly with gases and liquids.

Discharge An ion may be converted to a neutral atom at an electrode during electrolysis by loss or gain of electrons. This is called discharging of ions.

Dissolving When a substance is added to water it can disappear from view when stirred. This disappearance is called dissolving. The substance is still there and can be recovered by evaporation.

Distillation A way of purifying a liquid or obtaining the solvent from a solution. The liquid is vaporized and the vapour condensed to reform the liquid. The condensed liquid is called the **distillate**.

Ductile A metal is said to be ductile as it can be drawn into thin wires.

Effervescence If a gas is produced during a chemical reaction bubbles of the gas can be seen to escape from the solution. This 'fizzing' is called effervescence and this word is frequently confused with **efflorescence**.

Electrode The conducting rod or plate which carries electricity in and out of an electrolyte during electrolysis. Graphite and platinum are good unreactive or inert electrodes.

Electrolysis The passing of a direct electric current (d.c.) through an electrolyte, dissolved in water or in the molten state, resulting in the splitting up of the electrolyte at the electrodes. Lead bromide, for example, is split up into lead (formed at the negative electrode) and bromine (formed at the positive electrode).

Electrolyte A chemical compound which, in aqueous solution or when molten, conducts electricity and is split up by it. Acids, bases, alkalis and salts are all electrolytes.

Element A single pure substance that cannot be split up into anything simpler.

Endothermic reaction A reaction which takes in heat.

Environment The surroundings in which we and other animals and plants live. A person who studies the environment may be called an **environmentalist**.

Enzyme An enzyme is a protein which acts as a biological catalyst. Certain enzymes only work with certain reactions. The action of an enzyme is best under certain temperature and pH conditions. Enzymes are part of the growing subject of **biotechnology**.

Evaporation The process by which a liquid changes to its vapour. This happens at a temperature below its boiling point but is fastest when the liquid is boiling.

Excess In a chemical reaction there is a connection between the quantities of substances reacting. In practice one of the reactants is present in larger quantities than is required for the reaction and is said to be in excess.

Exothermic reaction A reaction that gives out heat e.g. the burning of coal. A reaction which takes in heat is called an **endothermic reaction**.

Extraction The removal of one thing from a group of other things e.g. separating iron from iron ore.

Fermentation Enzymes in yeast convert glucose into ethanol and carbon dioxide. This process can be used to dispose of waste sugars in industry.

Filtrate The liquid that comes through the filter paper during filtration.

Filtration (or filtering) A method of separating a solid from a liquid. The solid is 'trapped' on the filter paper and the liquid runs through.

Flammable Describes a substance, e.g. petrol, that catches fire easily.

Fractional distillation A method of separating a mixture of different liquids that mix together. The process depends upon the different boiling points of the liquids. The liquid with the lowest boiling point boils off first and is condensed. As the temperature is raised, liquids with higher boiling points distil over.

Freezing When a liquid changes to a solid. It will do this at the **freezing point**. A pure substance will have a definite freezing point.

Fuel A substance that burns easily to produce heat and light. A **fossil fuel** is present in the earth in only limited amounts and cannot be readily replaced, e.g. coal, petroleum.

Funnel A piece of glass or plastic apparatus used for filtering. A **Buchner funnel** is a particular type of funnel usually made of china. It produces quicker filtration because the filtrate is sucked through the filter paper.

Giant structure This is a crystal structure in which all of the particles are strongly linked together by a network of bonds extending through the crystal, e.g. diamond.

Group Vertical column in the Periodic Table. Elements in the same group have similar chemical properties.

Halogen An element in group VII of the Periodic Table. The word halogen means 'salt producer'. Common halogens are chlorine, bromine and iodine.

Homologous series The name given to a family of organic compounds, e.g. alkanes.

Hydrated Contains water.

Hydrocarbon Compounds made up from the elements hydrogen and carbon only.

Hydrolysis The splitting up of a compound with water.

Igneous Rocks have cooled and solidified from molten rock material produced deep in the earth. Granite is an example of an igneous rock. In an igneous rock there are interlocking crystals.

Immiscible Two liquids that do not mix are said to be immiscible, e.g. oil and water.

Indicator A chemical that can distinguish between an alkali and an acid by changing colour e.g. litmus is red in acids and blue in alkalis.

In situ It is sometimes necessary to prepare a chemical, as it is required, by mixing other chemicals. Ammonium nitrate, which is inclined to be explosive because of impurities, can be prepared by mixing ammonium chloride and sodium nitrate. The ammonium chloride is said to be prepared *in situ*.

Insoluble Describes a substance that will not dissolve in a particular solvent.

Insulator A substance which does not conduct electricity e.g. rubber or plastic. Insulators may be called **non-conductors**.

Ion A positively or negatively charged particle formed when an atom or group of atoms lose or gain electrons.

Ion exchange A process in which ions are taken from water and replaced by others. In an ion exchange column used to soften hard water, calcium and magnesium ions are removed from the water and replaced by sodium ions.

Malleable Metals are very malleable as they can be beaten into thin sheets or different shapes.

Melt A solid changes to a liquid at the **melting point**.

Metal An element that is shiny, conducts heat and electricity, can be beaten into thin sheets (malleable) or drawn into wires (**ductile**) is probably a metal. Metals usually have high melting points and boiling points and high densities. Metals burn in oxygen to form neutral or alkaline oxides.

Metamorphic Rocks were originally either igneous or sedimentary rocks which were thoroughly altered by heat or pressure within the crust of the Earth, without melting. Marble is an example of a metamorphic rock.

Mineral A naturally occurring substance of which rocks are made.

Mixture A substance made by just mixing other substances together. The substances can easily be separated again.

Molecule The smallest part of an element or compound that can exist on its own. A molecule usually consists of a small number of atoms joined together.

Neutralization A reaction where an acid is cancelled out by a base or alkali.

Non-aqueous solution Solution where the solvent is not water e.g. iodine dissolved in hexane.

Oxidation This is a reaction where a substance gains oxygen or loses hydrogen.

Oxides are compounds of an element with oxygen. A **basic** oxide is an oxide of a metal. It reacts with an acid to give a salt and water only. Some basic oxides dissolve in water to form **alkalis**. A **neutral** oxide such as carbon monoxide CO does not react with acids or alkalis and has a pH of 7. An **acidic** oxide dissolves in alkalis to form a salt and water only. It has a pH of less than 7. An **amphoteric** oxide can act as either an acidic or a basic oxide depending upon conditions. Examples of amphoteric oxides include zinc oxide ZnO and aluminium oxide Al_2O_3.

Oxidizing agent An oxidizing agent, e.g. concentrated sulphuric acid, oxidizes another substance. It is itself reduced.

Period A horizontal row in the Periodic Table.

pH A measure of the acidity or alkalinity of a solution. The scale is from 0 to 14. Numbers less than 7 represent acids; the smaller the number the stronger the acid. Numbers greater than 7 represent alkalis; the larger the number the stronger the alkali. pH 7 is neutral.

Polar solvent The molecules of some solvents, such as water, contain slight positive and negative charges. A stream of a polar solvent is deflected by a charged rod. A polar solvent dissolves substances containing ionic bonds. Solvents without these charges are called **nonpolar solvents**.

Pollution The presence in the environment of substances which are harmful to living things.

Polymer A long chain molecule built up of a number of smaller units, called **monomers**, joined together by a process called **polymerization**. Polymers are often called plastics, e.g. poly(ethene) is a polymer made up from ethene molecules linked together.

Precipitate An insoluble substance formed in a chemical reaction. This usually causes a cloudiness to appear in the liquid and eventually the solid sinks to the bottom. The precipitate can be removed by filtering or centrifuging.
OR

Precipitate If a solid which is insoluble in water is produced during a chemical reaction in solution, the particles of solid formed are called a precipitate and the process is called **precipitation**.

Product A substance formed in a chemical reaction.

Properties A description of a substance and how it behaves. **Physical properties** include density and melting point. **Chemical properties** describe chemical changes.

Proportional If a car is driving along at a constant speed the amount of petrol used increases regularly as the distance increases. The amount of petrol used is said to be proportional to the distance travelled.

Pure substance A single substance that contains nothing apart from the substance itself. Pure substances have definite melting and boiling points.

Qualitative A qualitative study is one which depends upon changes in appearance only. A **quantitative** study requires a study of quantities, e.g. mass, volume, etc.

Reactant A chemical substance which takes part in a chemical reaction.

Redox reaction A reaction where both oxidation and reduction take place.

Reduction Reduction is the opposite of oxidation. This is a reaction where oxygen is lost or hydrogen is gained. A **reducing agent**, e.g. carbon monoxide, reduces another substance left on a filter paper during filtration.

Reversible reaction A reversible reaction is a reaction which can go either forwards or backwards depending upon the conditions. A reversible reaction will include the sign $\rightleftharpoons$ in the equation.

Salt A substance which is formed as a product of a neutralization reaction. A salt is the product obtained when hydrogen in an acid is replaced by a metal.

Saturated compound A saturated compound is a compound which contains only single bonds e.g. methane CH_4.

Saturated solution A solution in which no more of the solute will dissolve providing the temperature remains unchanged.

Sedimentary Rocks, e.g. sandstone, are composed of compacted fragments of older rocks and other minerals which have accumulated on the floor of an ancient sea or lake, etc.

Semi-conductor Some substances, e.g. silicon, have a very slight ability to conduct electricity. They are called semi-conductors and are used to make microchips.

Solubility The number of grams of a solute that will dissolve in 100 g of solvent at a particular temperature.

Solute The substance that dissolves in a solvent to form a solution.

Solvent The liquid in which a solute dissolves.

Spectroscopy The study of the light coming from a substance. Helium was first discovered by examining light from the sun. Helium must be present on the sun.

Sublimation When a solid changes straight from a gas to a solid **or** solid to a gas, missing out the liquid. The solid collected is called the **sublimate**.

Surface tension This is a measure of the attraction between molecules at the surface of a liquid. Water has a high surface tension.

Suspension A mixture of a liquid and an insoluble substance where the insoluble substance does not sink to the bottom but stays evenly divided throughout the liquid.

Synthesis The formation of a compound from the elements that make it up. This is usually accompanied by a loss of energy.

Titration A method of investigating the volumes of solution that react together.

Transition metal A block of metals between the two parts of the main block in the Periodic Table. Transition metals are usually dense metals much less reactive than alkali metals.

Vapour A vapour is a gas that will condense to a liquid on cooling to room temperature.

Viscous A viscous liquid is thick and 'treacle-like'. It is difficult to pour.

Volatile Describes a liquid which is easily turned to a vapour, e.g. petrol.

Water of crystallization A definite amount of water bound up in the crystal, e.g. $CuSO_4.5H_2O$.

INDEX